Pleist
olstice Y
assium
Almana
Power Spe
ocermat
yramiteAm
Iodine A
Glu
arsec
ute Zero Amalgam
Trigonometry Rea
Volcano Irrat
Universe Zodiac In
mium Uranium Spi
us Argon
Rhino
esterol Ungulate Cal
Hydrophobia
llistics Colloid

Words
of Science

Words
of Science

and the History behind Them

ISAAC ASIMOV

HARRAP   LONDON

*First published in the United States as two
volumes in 1959 and 1972.*

*First published in Great Britain, fully revised
and updated in one volume, in 1974
by* GEORGE G. HARRAP & CO. LTD
182-184 High Holborn, London WC1V 7AX

ISBN 0 245 52228 X

Composed in VIP Times by
Amos Typesetters, Hockley, Essex
Printed and bound in
Great Britain by
REDWOOD BURN LIMITED
Trowbridge & Esher

Preface

ALMOST any specialized field of human knowledge, whether it be carpentry, mountain-climbing, or horse-racing, has its special vocabulary. Without a knowledge of that vocabulary, it is difficult to discuss the subject with those who are acquainted with the field. Any introduction to the field must, in fact, include as part of the education an explanation of the vocabulary, and a definition of its terms.

No specialized field has so large and intricate a vocabulary as has science. It is, indeed, the vocabulary, more than any other aspect of the field, that may give the student the initial concept that science is difficult to understand.

Entering the world of science turns into a meeting with a whole realm of new words: words that look and sound odd; words that are long and hard to pronounce; words that the ordinary person never meets with in ordinary life. There is coacervate for instance, and lepidoptera, and diatom, and porphyrin.

Worse yet, there are ordinary words that one thinks one knows, which somehow take on a new and strange life when used in science. An example is the word 'work'. Scientists use it to mean motion against a resisting force, so that holding a heavy piece of luggage, motionless, ten centimetres above the ground for two hours is *not* work as far as a physicist is concerned, even though you may think it is.

Worst of all, though the vocabulary of science is huge and intricate, it is never sufficiently huge and intricate to satisfy the scientists of all the various specialities. From year to year more words are invented to convey new concepts and new discoveries. In 1960 no one had ever heard the words quasar, pulsar, mascon, laser or transfer-RNA. Those words did not exist because they represent findings made since 1960.

Yet it is a shame to allow the scientific vocabulary, enormous, formidable, and continually expanding though it is, to become too difficult a hurdle to overcome. It cannot be made less enormous or forced into smaller tendency to expand, but it can certainly be made less formidable. One need only look closely at the words in order to see the meaning hidden in them. They will then be found more interesting, more dramatic, more fascinating than anyone would suspect from the mere dictionary definitions.

Within that pungent gas ammonia is an Egyptian god. Behind magnet and magnesium and manganese is an ancient Greek city. Mascon is an abbreviation of mass concentration, radar of radio directing and ranging, quasar of quasi-stellar object. Holography means whole writing, isotope means same place, ecology means study of the house, and so on, and so on.

There's not a word in science but has some interesting story behind the way it is built up, or the manner in which it was first used. Once you know the story you can more easily remember the word. Having made friends with it, it is no longer as formidable as it was before.

And in this book, you have the chance to make friends with many hundreds of them.

Isaac Asimov

Ablation

IN SOME WAYS a spacecraft is in greatest danger when it is within a hundred miles of the surface of earth. It is then not in space but in earth's atmosphere, which heats it by friction.

Moving away from earth is not so bad. The spacecraft moves upward very slowly at first, so that there is little friction with the air. As the rocket engines drive it faster and faster the air it penetrates is thinner and thinner, and offers less and less friction. By the time the spacecraft is moving really fast the air about it is so thin that there is no danger.

Returning to earth and re-entering the atmosphere is different. The space capsule, with its human passengers, is moving faster and faster as it approaches earth, thanks to the steady pull of earth's gravity. Eventually the air will be dense enough to allow the use of parachutes to slow the fall, but by that time frictional heat will have been deadly.

To avoid this the capsule is shaped like a cone. The blunt bottom end hits the atmosphere, pushing the air ahead of it and forming a shock wave that cushions the capsule and acts like a spring, slowing it down. This takes care of from 90 to 95 per cent of the capsule's motion.

What's more, the blunt end of the capsule is covered with a layer of resin, strengthened with glass fibres. The resin chars, melts and flakes away, carrying heat with it. The charred resin left behind is black and radiates additional heat.

This system of removing heat by flaking away a layer is an effect called ablative cooling. The resin undergoes ablation, which is from Latin words meaning to carry away. The word was once used to mean the removal of part of the body by surgery, but has now joined the world of space flight.

Absolute Zero

IN SOME CASES, a zero from which to start counting can be set anywhere. Zero longitude (see MERIDIAN) is set arbitrarily. The same is true of zero in temperature. In the Celsius scale of temperature (see CELSIUS), zero degrees is set at the point where ice melts; while in the Fahrenheit scale, it is set at a point some way below the melting point of ice. In either case, though, temperatures below zero may be experienced.

Toward the end of the 1700s it began to seem that there might be a limit to cold. The French physicist Jacques A. C. Charles discovered in about 1787 that gases contracted $1/273$ of their volume at

0°C for each Celsius degree they were cooled. (This is called Charles's Law.) If this were to continue, then, at about –273°C the gas should disappear entirely. Of course, this does not happen. The gas invariably turns first liquid, then solid, as it is cooled.

The British physicist William Thomson extended the idea in the 1860s. He treated temperature as an expression of the velocity of movement of molecules in a substance. The colder the substance, the slower the motion, until at a certain temperature (–273·18 °C) there was no motion at all. There could not be less than no motion, hence no temperature less than that. The temperature –273·18 ° was a real zero.

Now, the Latin *solvere* means to loosen or to free and the prefix *ab-* means from. Therefore, something that is absolute is something that is freed from all restraint. We speak of an absolute monarch, for instance. Anything that is an extreme can then be absolute, since it is freed from all qualifications or maybes. You can have absolute opinions or be an absolute fool (or both). And a zero like –273·18 °C that is really zero is absolute zero.

Thomson's temperature scale, starting at absolute zero, is called the absolute scale, or the Kelvin scale (since Thomson was made Baron Kelvin of Largs in 1892). A temperature on that scale can be abbreviated either A or K.

Academy

THE ANCIENT Athenian hero Theseus once carried off Helen of Sparta (who was later carried off by Paris, thus starting the Trojan War), and Helen's brothers, Castor and Polydeukes, went searching for her. Another Athenian, Akademos, revealed her hiding-place. For this reason, during the wars between Athens and Sparta, the Spartans (who held Castor and Polydeukes in special honour) always spared the site of the estate of Akademos (about 1½ kilometres north-west of Athens) whenever they invaded Athenian territory. This estate, the Akademeia, became a symbol of peace in a war-torn time.

Plato lived near the Akademeia and resorted to that pleasant place with his students. There he taught for fifty years, and his successors for another eight hundred. The Academy (as we call it) was the most famous school of antiquity, and schools still occasionally call themselves academies today as a result.

The term academic is applied to anything pertaining to these academies, particularly to the type of learning they encouraged. Because Plato's philosophy was highly theoretical and abstract and did not concern itself with practical everyday matters, an academic question has now come to mean one that has no practical meaning—that is, of theoretical interest only.

Other ancient philosophers also left the names of their places of teaching in the language. Aristotle used to teach in a gymnasium called the Lykeion, or in Latin, Lyceum. It was named in honour of a near-by temple to Apollo in his capacity as wolf-killer. As such he was called Lykeios, probably from the Greek *lykos* (wolf). In France the derived word for what we call a high school is *lycée*.

Again, the philosopher Zeno taught in Athens at a place called Stoa Poikile (painted porch in Greek). His school of philosophy was termed stoicism because of it. Since Zeno taught that the way to happiness was to avoid undue emotion, a person who does not show his emotions is called a stoic to this day.

Acetylcholine

IN 1861 the German chemist Adolph Strecker isolated a hitherto unknown nitrogen-containing compound from liver bile. He named it choline, from a Greek word meaning bile.

In 1914 the English biologist Henry Dale isolated a substance with a molecule composed of two parts linked in a fashion called an ester linkage. (The word ester was manufactured from the first and last syllables of the German name for the best-known compound of this type, back about 1880.) Dale separated the two parts and identified one as choline and the other as a well-known compound, acetic acid. He therefore named the combination acetylcholine, and observed that it had a strong effect on various organs.

In 1921 a German physiologist, Otto Loewi, discovered that nerve-endings liberated small traces of a chemical that activated near-by cells and allowed nerve impulses to leap the intervals between one nerve cell and the next. He called the chemical substance *Vagusstoff*, which is German for vagus material, because he obtained evidence for its existence by stimulating the vagus nerve.

Since *Vagusstoff*, by stimulating nerves, produced effects on organs similar to that produced by acetylcholine, Dale suggested that *Vagusstoff* was acetylcholine, and by 1933 had proved his case.

Acetylcholine remains in existence at nerve-endings only long enough for the impulse to cross from one cell to another and then is broken up to acetic acid and choline. This break-up is hastened by the presence of an enzyme.

Enzymes are usually named for the type of reaction they hasten, with an -ase suffix added. Thus the enzyme which brings about the break-up of the ester linkage in acetylcholine is named cholinesterase.

Acid

SOURNESS IS one of the four fundamental tastes (the others being sweetness, saltness and bitterness), and it occurs naturally in such things as unripe fruits and some ripe ones. It was there that primitive man first became acquainted with the taste. In addition, it was already known in prehistoric times that certain liquids, such as milk, would develop sourness if allowed to stand. Fruit juices on standing fermented and became wine, which on further standing also soured. The expression sour wine in Old French is *vin egre,* and from this comes our own word vinegar.

The Latin word for 'to be sour' is *acere*. (The Old French *aigre* is a form of that.) From *acere,* two other words are derived: *acidus* as the adjective sour, and *acetum,* meaning vinegar.

The medieval chemists were particularly interested in the sour substances. Strong vinegar could attack or corrode a number of metals, and in its presence certain chemical changes took place that did not take place otherwise. About 1300, new and stronger chemicals of this sort were discovered. By their greater activity, metals and other substances could be brought into solution more quickly than by use of even the strongest vinegar. Virtually a chemical revolution took place.

All such compounds were named acids after their most prominent characteristic, sourness. Vinegar and fruit juices contained organic acids (see ORGANISM) while the new, stronger substances which were obtained from non-living sources were the mineral acids.

The particular acid in vinegar (*acetum*, remember) was named acetic acid, so that, as it happened, the two words of that name both came from the same Latin word *acere*, and the substance bears a double dose of testimony to the fact that it is sour.

As in so many other cases, modern science has given the lie to the old name. To the modern chemist, an acid is any compound that has a tendency to lose a proton. If the tendency is great enough the acid will be sour to the taste; if not great enough, it will not be sour, but it will still be an acid.

Actinide

SINCE THE 1870s it has been known that the chemical elements exist in families. Tables of elements (periodic tables) have been prepared which present those families in rows down, or lines across. Thus the 57th element, lanthanum, is followed by fourteen elements (58 to 71 inclusive) that are similar to lanthanum and make up the rare earth family. This is usually presented as a horizontal line in the table. Lanthanum is also part of a family of elements arranged in a vertical row. Above it are scandium (21) and yttrium (39).

In 1899 element 89 was discovered and named actinium because it gave off radioactive rays and the Greek word for ray is *aktinos*. Actinium fits under lanthanum in the periodic table and has properties like scandium, yttrium and lanthanum.

Was actinium followed by a series of similar elements as lanthanum was? At first chemists thought not. They felt the rare earths to be a special case, and considered the element after actinium (which was thorium, element 90) to be like hafnium (element 72), which appeared in the table after all the rare earths were over and done with.

Until 1940, however, only three elements beyond actinium were known, so the evidence was inconclusive. In that year chemists began to construct atoms of higher elements in the laboratory —neptunium (93), plutonium (94), and so on. With more elements to study, new evidence appeared, and in 1944 the American chemist Glenn T. Seaborg was able to show that the elements after actinium made up a second series very like the rare earths.

A name was needed to distinguish one series from the other and it was decided to name each series after its first member. The first series of rare earths were the lanthanides, after lanthanum, and the newly discovered series were, naturally, the actinides.

Adamantane

THE ANCIENT GREEKS imagined a material so hard that it could not be scratched or dented in any way. They called it *adamas,* which in their language meant untamable.

There is nothing that is infinitely hard, but in the form of adamant the expression came to be applied to any metal that was particularly hard. Finally it came to be applied to a glossy form of carbon, rarely found, which was harder than any other substance found in nature. If you drop the initial a and distort the rest of the word slightly you end up with diamond, which is what this hardest substance is now called in English.

The carbon atoms that make up diamond are very small and can approach each other quite closely. What's more, they do it in so symmetrical a fashion that each carbon atom is closely approached by four others at equal distances from each other. It is the equal, tight grip that each carbon atom has on its neighbours that makes diamond so hard.

Organic compounds are made up of chains and rings of carbon atoms, but generally the carbon atoms are not arranged as in diamond. Instead they are arranged in lines or rings that are less symmetrical than the carbon distribution in diamond.

Tiny quantities of a compound with a molecule containing ten carbon atoms were first prepared in large amounts in 1957. The ten carbon atoms were found to be arranged in three interconnected rings, with a distribution exactly as one would find in ten neighbouring carbon atoms in diamond. This molecule (which carried sixteen hydrogen atoms attached to its carbons) was particularly stable, and was named adamantane. The -ane suffix is used by chemists for compounds made up of carbon atoms attached to all the hydrogen atoms they can hold. It turns out that modified adamantane compounds may be useful in medicine, since they block virus action in some cases.

Adsorption

A SPONGE or towel or a piece of blotting-paper will take up water in defiance of gravity. The reason for it lies in the force of capillarity (see CAPILLARITY).

Before capillarity was understood, however, it looked as though the sponge, etc., were simply sucking up the water. (Sucking was one well-known way of lifting a fluid against gravity.) The phenomenon was therefore called absorption, from the Latin *ab-* (from) and *sorbere* (to suck up). A dry sponge, placed in a water-filled pan, sucked water up from the pan.

Chemists came across a similar phenomenon involving finely powdered substances. If a mixture of gases, for instance, were forced through a layer of finely powdered charcoal some of the gas molecules would stick firmly to the surface of the charcoal particles and remain trapped there. The larger molecules would stick more tightly and the smaller ones would be shouldered aside so that they would get through the charcoal without sticking at all.

Poison-gas vapours have larger molecules generally than do the oxygen and nitrogen of the atmosphere. Air containing poison gas, if forced through a canister containing powdered charcoal, will therefore lose its poison. The poison gas molecules will remain attached to the charcoal particles while the ordinary breathable air will get through. Such a charcoal canister, fitted into an air-tight mask, is called a gas-mask.

The poison gas has been 'sucked out' of the air by charcoal as water is 'sucked up' by a sponge. However, the reason is different. In the case of charcoal it is not capillarity that does the work, but all the many surfaces of the tiny particles to which the gas molecules adhere. Taking the ad- prefix (meaning to) from adhere, you have instead of absorption, adsorption.

However, adsorption plays a part in capillarity too, and sometimes it is difficult to tell whether a phenomenon should be truly spoken of as absorption or adsorption. Some scientists have suggested the word sorption to cover both types of phenomena.

Aflatoxin

A CERTAIN FUNGUS attacks rye, causing infected grains to darken and curve into the appearance of a rooster's spur, which in medieval French was called an *argot*. In English this has been corrupted to ergot, which is the common name given to the fungus. This fungus produces certain compounds that have powerful effects on the body even in small quantities. It causes contractions of the uterus, for instance, so that medieval midwives used it sometimes to ease difficult childbirths. It was a dangerous remedy, however, for too much of it produces a serious and even fatal condition known as ergotism. Every once in a while there are reports of an epidemic of such a disease among humans or animals that have been eating contaminated rye.

Nor is ergot the only fungus capable of producing compounds which, in small quantities, have profound and usually unpleasant effects on men. These compounds are called mycotoxins. The prefix myco- is from a Greek word meaning mushroom, which is the most familiar fungus because it is so large. (Most fungi are microscopic organisms.) Toxin is from the Greek word for bow, since arrows were so frequently poisoned in warfare. A mycotoxin is therefore a fungus poison.

In 1960 an epidemic of serious liver disease was traced to the eating of mouldy peanuts. The mould concerned was a very common one of the Aspergillus group. This name was used because such moulds were made up of strands swollen at one end so that they looked like tiny versions of a device (aspergillum) used in churches. The aspergillum had small holes in its swollen end and was used to sprinkle holy water, its name coming from the Latin word for sprinkle. The particular variety of mould was *Aspergillus flavus* (the latter word from a Latin word meaning yellow, because of its colour). From the initials *A. fla.*, the poisons it produced were called aflatoxins. There is also a lining disease caused by

Aspergillus, with symptoms resembling tuberculosis; this is called aspergillosis.

Airglow

ON A CLEAR MOONLESS NIGHT there can sometimes be detected a very faint general luminosity in the sky, which seems to originate in the upper atmosphere. It is called airglow. It is there both day and night, but it is so much more intense at night, and so much more easily detected then, that it is sometimes called nightglow.

The nature of the airglow is revealed by a careful analysis of its light as photographed from the ground and by rockets sent into the upper atmosphere.

The upper atmosphere, it turns out, is rich in atomic oxygen —that is, single oxygen atoms rather than the oxygen molecules of the lower atmosphere, which are made up of two joined oxygen atoms each.

Apparently the ultra-violet light of the sun is absorbed by oxygen molecules in the upper atmosphere, and as a result the oxygen atoms in the molecule gain too much energy to remain together. They split apart. Every once in a while two oxygen atoms collide, give off their excess energy as a tiny flash of visible light, and combine again. During the day the reunion is less frequent than the split-up, and by the time night falls there is a large supply of atomic oxygen in the upper atmosphere. Through the night this combines, producing the airglow.

Other upper-atmosphere phenomena are the very thin clouds one can see at heights of eighty kilometres (fifty miles) or more. They are most easily seen just after sunset in high latitudes in the summer, when they seem to shine very faintly against the darkening sky. They are noctilucent clouds (from Latin words meaning night shining). Rockets have brought back particles of matter from those regions, and chemical analysis has shown the presence of nickel. It may be, then, that noctilucent clouds are made up of very fine dust particles produced by tiny meteors that flash into the upper atmosphere, heat, and disintegrate into powder.

Alcohol

WOMEN, APPARENTLY, have for many centuries been darkening their eyelids to make their eyes seem large and lustrous. Arabic women used a very finely divided powder, antimony sulphide, for the purpose, and the Arabic expression for that powder was *al koh'l,* meaning the finely divided.

The medieval chemist took up the expression, converted it to alcohol, and used it for any fine powder, particularly a powder so fine it could not be felt—one that was impalpable, in other words, from the Latin *in-* (not) and *palpare* (to feel).

Some time in the early 1500s, chemists began applying the term to vapours that could be forced out of certain liquids.These vapours, you see, were impalpable, too. Wine, when heated, gave off a vapour which was first called alcohol of wine and then simply alcohol.

When wine is heated the alcohol it contains is more easily boiled than the water. The vapour is richer in alcohol than the wine originally was. If the vapour is cooled the resulting liquid is a stronger drink than the original. This process is called distillation, from the Latin *de-* (down) and *stilla* (a small drop). And, indeed, the vapours, when cooled, are collected as small cold drops falling into a container. Alcoholic drinks strengthened in this way are distilled liquors, and the device in which the process is carried out is popularly called a still. Actually, distillation is an important chemical process that is used to separate individual compounds from many types of liquid mixtures.

The -ol suffix of the word alcohol is now used by chemists to name any compound which, like alcohol, contains a hydroxyl group in its molecule (so called because it consists of a hydrogen and oxygen atom in combination). The alcohol of wine (C_2H_5OH) contains also a two-carbon group in its molecule similar to that in ether (see ETHER), so it is called ethyl alcohol or ethanol.

Oddly enough, the Arabs call alcohol *spir't* from the English word spirits, so if we use their word for the purpose, they use ours.

Aleph-One

WHEN WE COUNT we match one set of objects with another. We point to a set of objects and say 'one' when we point to the first, 'two' when we point to the second, and so on, matching the set we are counting to the set of whole numbers.

If two sets can be made to match exactly we know the number representing one set is equal to the number representing the other. For instance, the set of even numbers 2, 4, 6, 8 . . . can be matched evenly with the set of whole numbers 1, 2, 3, 4 . . . . . Each even number is matched with the whole number equal to half itself. There would be a different whole number for each even number. We can therefore say that the set of even numbers is equal to the set of whole numbers. This seems paradoxical, since the set of whole numbers includes odd numbers as well as even numbers and we would think there would be twice as many whole numbers as even numbers.

In 1895 the German mathematician Georg Cantor worked out the mathematics of infinite sets (whole numbers and even numbers continue endlessly, and are therefore infinite sets) and showed that they do not follow the ordinary rules of arithmetic. He called the numbers representing endless sets transfinite numbers (from Latin words meaning beyond the end).

Transfinite numbers are symbolized by aleph, the first letter of the Hebrew alphabet. The number of whole numbers is the lowest transfinite number, aleph-null. Any infinite set which can't be matched with the set of whole numbers has a higher transfinite number. The set of points in a line is higher, and may be equal to aleph-one, the second lowest transfinite number. In the late 1960s, however, the American mathematician Paul J. Cohen showed that it wasn't possible either to prove or disprove the statement that the set of points in a line is equal to aleph-one.

Algol

BEFORE THE DAYS of modern astronomy, any change in the changeless, perfect heavens was cause for concern (see COMET). Even stars that merely varied in brightness from day to day could be regarded with alarm.

Actually there are many such variable stars, but only a few are bright enough and variable enough to be noted with the naked eye, and for some reason the ancients never commented on those few.

The most famous example is a star in the constellation Perseus, called Beta Persei, because at its brightest it is the second brightest star in the constellation (beta is the second letter of the Greek alphabet, so the star's astronomical name means 'the second in Perseus'. Other stars are named in similar fashion).

The variation isn't really much and it wasn't till 1669 that its fluctuations in light were noted by a European. However, the star already had a significant Arabic name, Algol. (During the night of the western Dark Ages, the Arabs carried the main weight of progress in astronomy.) Algol comes from the Arabic *al* (the) and *ghul* (demon), so that the star is often called the Demon Star today, a sharp reminder of the fears its behaviour gave rise to. As a matter of fact, *ghul* has come down to us as ghoul. Algol is 'the ghoul'.

But Algol is not really variable. It is an eclipsing binary (two stars revolving with an orbit edge-on to us). Every few days the dimmer star eclipses the brighter and the light reaching us is cut down.

A more remarkable star, in the constellation Cetus, is Omicron Ceti. (Omicron is the fifteenth letter of the Greek alphabet.) It actually pulsates and varies in brightness over irregular periods sometimes almost two years long. It can be as bright as the Pole Star or far too dim to be seen. The German astronomer David Fabricius noted it finally in 1596. By then astronomers were growing more sophisticated, and were less troubled by odd events in the heavens. So Fabricius named it Mira, from the Latin *mirus* (wonderful). Though Algol is the Demon Star, Mira, which is much more extreme, is the Wonderful One. There seems no justice.

Almanac

ACTUALLY, MAN is a very long-lived creature. The only other creatures that may live a hundred years or more are certain trees and large tortoises. At the other end of the scale, at least among animals visible to the naked eye, are certain insects which live their entire adult life in a day or less. (In their immature form, to be sure, such insects may live a total of one to three years.)

The common name for these short-lived insects is mayfly, but the scientific name for the group to which they belong is Ephemeridae, from the Greek *epi-* (over) and *hemera* (day). Their lives are, indeed, over in a day.

But an ephemeris is something astronomical as well. It is a table or series of tables which gives the exact position of various heavenly bodies at certain times. This is valuable in navigation, since position at sea can be plotted by observing the position of heavenly bodies. Since the table giving the position for a certain time is no longer useful once the time is past, it is good for the day only, so to speak, and hence its name.

Tables of astronomical data are also included in almanacs, a word sometimes used as a synonym for ephemeris. The word almanac comes from the Arabic *al manakh,* meaning the calendar or the weather. The calendar and the weather are much the same, of course, since the seasons of the year come and go with the months. People have always believed in a more intimate connection, too, and have tied the weather to the phases of the moon, for instance.

The very ephemeral nature of the almanac—that is, the fact that its tables grow out of date quickly—means that a new almanac must be put out periodically, usually every year. It is natural, then, to include information concerning the past year, such as news events and late statistics. As a result, our best-known almanacs these days have grown into a kind of annual one-volume encyclopedia.

Alpha Rays

THE DISCOVERY of radioactivity, which revolutionized science, came about this way: the French physicist Henri Antoine Becquerel was studying the way in which uranium salts glowed, or fluoresced, when exposed to sunlight. He wondered if the fluorescence contained X-rays, which had been discovered the year before (see X-RAY), so he exposed the uranium salt to the sunlight and placed a carefully wrapped photographic plate near by. Sure enough, the plate was fogged despite its protection.

However, on sheer impulse, he developed some wrapped plates which had been near some of the uranium compound in a dark drawer, and that plate was also fogged. It seemed that invisible radiation was being emitted by the uranium whether sunlight was present or not. What's more, this radiation, like the X-rays, was much more penetrating than ordinary light.

In 1899 Becquerel (and others) noticed that some of the uranium radiation could be bent to one side by a magnet, so there were at least two different kinds of rays. Since their nature was unknown, it would be easiest just to call them rays A and B. The New Zealand-born physicist Ernest Rutherford did just that, using the first two letters of the Greek alphabet, alpha and beta. He named them alpha rays and beta rays.

In 1900 the French physicist P. Villard discovered a new and particularly penetrating radiation emanating from uranium. These were automatically named gamma rays (probably by Rutherford again) since gamma is the third letter of the Greek alphabet.

In that same year Pierre and Marie Curie showed that beta rays consist of a stream of electrons moving at terrific speed. By 1909 Rutherford proved alpha rays to consist of streams of comparatively heavy particles, each consisting of two neutrons and two protons. As a result, speeding electrons are now called beta particles, while speeding two-neutron-two-proton combinations are alpha particles. These alpha particles, moreover, are now known to be the nuclei (that is, without the electrons) of helium, the second lightest element. (Gamma rays are not made up of particles, but are similar in nature to X-rays, being, however, more energetic and therefore more penetrating.)

18

Amalgam

IN THE ORDINARY SENSE, an alloy is a foreign substance added to something that is otherwise desirable, and added in such a way that it is not easily detectable. For instance, lead might be added to silver, water to milk, or sand to sugar. The word comes through the French from the Latin *ad-* (to) and *ligare* (bind). The impurity is bound to the original.

But an impurity can improve the original, and metal-workers particularly realized this in very early days. A little zinc added to copper produced a material called brass, which was yellower than copper and more decorative. If tin were added to copper the result was bronze, a much harder material than either metal separately. In fact, before the introduction of iron, bronze was the hardest metal known. Armour was made of it (as is well described in Homer's *Iliad*), and the period was known as the Bronze Age. Even today pure metals are hardly ever used. Instead metals are deliberately mixed to produce hundreds of new substances with desirable qualities not otherwise available.

Alloy has thus come to mean a mixture of metals. So important is iron in our day that all alloys are divided into two groups, ferrous alloys which contain iron, and non-ferrous alloys which do not. The stem ferr-, used by chemists for naming compounds containing iron, comes from the Latin *ferrum* (iron), and the chemical symbol for iron is Fe for that reason.

The one other type of alloy that has a special group name are those involving mercury. Mercury is a liquid metal, and its alloys are either liquids or soft solids. A soft solid metal is an odd phenomenon in a world which values metals for their hardness and toughness, and this softness gave mercury alloys the name of amalgams. This is a corruption of the Greek word *malagma,* meaning any soft, dough-like material. It is a silver amalgam (mercury plus several metals, mainly silver) and not silver itself that is used for 'silver fillings' in teeth. The amalgam is soft enough to knead into the cavity, where chemical reactions harden it quickly.

Amethyst

VARIOUS naturally occurring, but rare, stones are admired for a number of excellent reasons. They are beautiful to look at, and durable, so that they don't become less lovely with time. The pleasure men get from such jewels is found in the very word itself, which comes from the Old French *jouel,* meaning a little joy.

The romantic ancients, however, could not resist adding to the real virtues of jewels with occasional tales of magical properties. For instance, one purple jewel was considered to be a remedy against drunkenness (perhaps because its colour was like that of wine). Wine drunk out of a cup made of this jewel could never, it was said, intoxicate anyone. The Greek word for intoxicated is *methystos* while the prefix *a-* is used as a negative. The jewel that ensures no intoxication is therefore an amethyst.

The Greek word *methystos* (intoxicated) in turn comes ultimately from *methy* (wine). Ordinary wine contains ethyl alcohol (see ALCOHOL), but a compound similar to ethyl alcohol can be obtained by heating wood in the absence of air. This second compound is quite poisonous, and contains only one carbon atom in its molecule, whereas ethyl alcohol contains two.

This compound from wood is called wood alcohol, for obvious reasons, or the classical equivalent, methyl alcohol; the word methyl coming from *methy* (wine) and *hyle,* which means matter generally, but can be used to mean wood in particular. Methyl is the wine from wood.

Chemists use the meth- stem for other atom groupings containing a single carbon atom. A gas known as marsh gas (it is found over marshes, where it is formed from decaying vegetable matter) possesses a molecule made up of one carbon atom and four hydrogen atoms. Its proper chemical name is therefore methane, the -ane suffix being reserved for certain types of hydrocarbons (compounds with molecules made up of hydrogen and carbon atoms only).

Ammonia

ONE OF THE MORE important of the gods of ancient Egypt was the deity Amen or Amun, who was the patron god of the Egyptian city of Thebes on the upper Nile. When Greek culture spread throughout the Near and Middle East after the conquests of Alexander the Great the Greeks took to identifying their own gods with those of the other people with whom they mixed. For instance, they had already identified their own chief god, Zeus, with Amen (or Ammon, as they spelled it) and a temple to Zeus-Ammon was built in an oasis in the North African desert.

Any desert area has a problem when it comes to finding fuel. One available fuel in North Africa is camel dung. The soot that settled out of burning camel dung on the walls and ceiling of the temple contained white salt-like crystals, which were then called sal ammoniac. The word sal is Latin for salt, so that the phrase means salt of Ammon.

At various times in the centuries following a pungent gas was obtained from sal ammoniac, but Priestley (see OXYGEN), in 1774 was the first to collect the gas separately and study it. He called it alkaline air because it would dissolve in water and then exhibit alkaline properties (see POTASSIUM). However, the name ammonia, from sal ammoniac, won out, and is the name of the gas to this day.

The ammonia molecule is made up of three hydrogen atoms and one nitrogen atom. If a fourth hydrogen is added the ammonium ion is formed, and it is this that forms salt-like compounds. (Sal ammoniac turned out to be ammonium chloride.)

The ammonia molecule minus a hydrogen atom is an amine group. If instead of the missing hydrogen a carbon-containing group of atoms is attached the resulting compound is an amine. Proteins are made up of long strings of relatively simple compounds containing both amine groups and acid groups. These compounds are therefore called amino acids, so it turns out that in the most important substances in our bodies we carry a reference to the great god Amen of Egypt.

Amphibious

THE FIRST backboned animals to emerge from water and take to breathing air and living on land were ancestors of today's frogs and toads. (They were preceded by insects and snails, probably, but those were animals without backbones.)

These frog-ancestors did not make a complete change-over but emerged on land only in adult life. Their young had to pass through an immature stage in water, living a fish-like existence, before they too exchanged gills for lungs and became air-breathing. Because these creatures lived part of their lives in water and part on land, they were amphibious, from the Greek *amphi* (on both sides of) and *bios* (life). They led a double life.

(During the Second World War this word was also used for attacks made by land and sea, which is reasonable enough. However, when attacks by air, land and sea were made, the corresponding word should have been tribious (three lives). However, newspapers insisted on calling such a triple attack triphibious, a horrible word to anyone who knows his derivations.)

The most common amphibians today are frogs and toads, both of which words come from the Anglo-Saxon. The immature stage of either creature is a tadpole. The prefix tad- is just a corruption of toad. The pole is a corruption of poll, which is an old-fashioned word for head. (This word still lingers as a synonym for taking a vote, since that amounts, so to speak, to counting heads on both sides of a question. We also had poll tax, which was a tax laid on the head of each household.) In any case, a tadpole seems to be little more than a head swimming about, and that is what it means: toad-head.

A tadpole is also called, even more dramatically, a polliwog. Again, poll means head, while the wog portion of the word is a corruption of wiggle. A polliwog therefore is a wiggling head, and next time you see one, see if that isn't an exact description.

Analogue Computer

MOST SIMPLE MATHEMATICAL PROBLEMS deal with actual numbers: 2+3; 18×6; and so on. The answers are exact. If you manipulate separate objects, like pebbles or electricity pulses, you can get answers with what is called a digital computer.

Some mathematical problems, however, don't deal with exact numbers. The diagonal of a square is equal to 1·4142 . . . (an endless decimal) times a side. Or you might have to determine the direction in which to point a gun, firing a shell travelling at such and such a velocity, in order to hit an object moving in a given direction at a given velocity, which may be changing, making allowance for the wind, for the curvature of earth, and so on—where none of the values are absolutely exact.

In that case it might be handy to make use of something that varies just the way as do the conditions in your problem, but in a way that is easier to handle. A slide rule, for instance, gives approximate answers to numerical problems by having each number represented by a length. These lengths can be manipulated by sliding a piece of wood back and forth, which in this way imitates the arithmetical behaviour of the numbers and produces the answer. Because lengths are analogous to numbers here, the slide rule is a simple analogue computer.

In electronic devices it is current strength that can be treated as analogous to numbers. This is made to vary very quickly in accordance with conditions designed to resemble the variations of the numbers in the problem. The final current strength gives the answer desired, if the data fed into the machine is accurate in the first place.

Both analogue computers and digital computers do what the human brain can do, but many millions of times faster. So far, though, they only work as manipulated by human beings and are therefore still only super-complex slide rules.

Anechoic Chamber

SOUND CONSISTS of waves of compression in the air, moving outward from their starting-point at about 1190 kilometres per hour. If a sound wave hits a hard obstacle it is reflected. If the obstacle is far enough away it takes enough time for the sound to go there and bounce back to produce a second sound you can hear separately. This is an echo, from a Greek word meaning sound.

At shorter reflection distances, the echo is not heard separately, but makes the original sound last longer, and may cause it to get louder and softer in rapid succession as echo after echo returns. The result is reverberation, from a Latin word meaning to strike back. Too much reverberation can be disturbing, and in a poorly designed hall it can make speech unintelligible and music discordant.

Sound can be absorbed. If the waves strike a soft material with small openings, such as clothing, curtains or carpeting, the sound waves lose themselves in the openings and are not reflected. The study of room design to cut down undesirable reverberation is called acoustics, from a Greek word meaning to hear. Sometimes this word is applied to the study of sound generally.

We are used to a certain amount of reverberation, for sound always echoes from walls, floors, rocks, trees and ground. That is why sound seems strange in rooms that have been designed with surfaces so broken up and so lined with absorbing material as to allow virtually no sound reflections at all. Such a room is an anechoic chamber, from Greek words meaning no echoes. Sounds made in the room die out so quickly in the absence of echoes that such chambers are also called dead rooms. Anechoic chambers are used to adjust microphones and other acoustic devices, when one wishes to avoid the confusing existence of even the slightest echo or reverberation.

Aneroid

THE MOST FAMILIAR type of barometer involves a long glass tube full of mercury (see BAROMETER). While this may be very useful, it is not easily portable.

Another way of measuring air pressure is to make use of a thin-topped, disc-shaped metal box, hollow and evacuated. The outer air presses against the thin top and forces it in somewhat. The greater the pressure, the more the top is pushed in, and vice versa.

Naturally, the amount of movement of the thin sheet of metal is not much, but it is transmitted by a system of levers that magnifies the motion and passes it on to a coiled spring that in turn swings a pointer on the ouside of the box. The pointer marks the air pressure as so many inches (or millimetres) of mercury, according to the distance the air pressure has made it swing.

Since such a barometer does not involve the use of a liquid, it is called an aneroid barometer, from the Greek *a-* (general negative) and *neros* (wet). It is a 'not wet' barometer.

Since air pressure decreases with increasing height according to a known rule, an aneroid barometer will tell you how high above sea-level you are (if allowance is made for weather conditions, since air pressure is lower on stormy days than on fine days). In that case the instrument becomes an altimeter, from the Latin *altus* (high) and *metrum* (measure).

The suffix -meter can be used with either Latin or Greek prefixes, since *metrum* is Latin and *metron* Greek, both meaning measure. Many names for scientific instruments include it. Perhaps the most familiar example is the thermometer, which measures temperature (see CELSIUS). More glamorous examples are the instruments put into artificial satellites which make various measurements in the fringes of the atmosphere and make the results known to us by appropriately varying radio signals. These are examples of telemeters (from the Greek *tele*, meaning distant). They are devices that measure at a distance.

Aneurysm

THE HUMAN HEART beats constantly at a rate of 60 to 80 times a minute throughout a long life that may last over a century. In a hundred years of faithful labour, it will beat some 4 000 million times and pump approximately 600 000 tonnes of blood.

With each beat the heart ejects about 130 cubic centimetres of blood, and most of this goes into the aorta at a speed of 40 centimetres per second. The surge of blood subjects the aorta, the largest and most important artery in the body, to a periodic strain. Fortunately, the aorta and other arteries have thick, elastic walls which expand as the blood tumbles in and contract again as it passes by. To a certain extent the arteries mimic the heartbeat, keeping time and allowing for whatever delay is required for the blood surge to reach them. This arterial beat is called the pulse, from a Latin word meaning to beat, and it can be felt easily where arteries are near the skin, notably in the wrist.

If for any reason the wall of an artery is damaged or thinned so that it expands too much when the blood surge enters and contracts too little afterwards, the result is an aneurysm, from a Greek word meaning widening. Such a flaccid, too wide artery remains in momentary danger of rupture with possibly fatal results. Since the aorta gets the first brunt of the blood surge, it is most subject to aneurysms, and the condition is most dangerous there.

Little could be done about aneurysms until recent times. In 1948 the American surgeon Michael E. De Bakey, working at Baylor University in Houston, began to operate on aneurysms. He cut out the affected portions of the vessels and replaced them with less important and unaffected vessels from other parts of the body. Eventually he replaced them with Dacron tubes. Thousands of certain deaths have been prevented in this way.

Angiosperm

MOST OF THE plant life in the world belongs to the primitive Thallophyta (see PLANKTON), many of which are marine. The most advanced plants (and those most familiar to us) are the land-dwelling spermatophytes. These have roots, leaves, flowers, fruit, seeds—all the things we usually think of in connection with plants. In fact, the Greek *sperma* means seed, while *phyton* means plant, so that the spermatophytes are seed-plants, as opposed to thallophytes and certain of the more primitive land plants such as mosses and ferns which do not produce seeds.

The spermatophytes include first the gymnosperms. In these plants, the ovule (that is, the object which will, after pollination, become the seed—the word being derived from a Latin form *ovulum,* meaning little egg) is exposed on the surface of the organ that forms it. The Greek *gymnos* means naked, so a gymnosperm is a naked-seed plant. The gymnosperms include the various evergreens, with their needle-shaped leaves that are not shed in the autumn and their woody fruit, called cones because they are often conical in shape.

The remaining spermatophytes are the angiosperms, from the Greek *angeion,* meaning vessel, because the ovules when formed are enclosed in a vessel called the ovary (from the Latin *ovum,* meaning egg) into which the pollen must penetrate to achieve pollination. All the common flowering plants and deciduous trees belong in this group. A deciduous tree is one that sheds its leaves in the autumn, from the Latin *de-* (down) and *cadere* (to fall); the leaves fall down.

When the seed develops the first tiny leaves to appear lie in a small hollow in the seed. They are called cotyledons, a Greek word meaning a cup-shaped hollow (from *kotyle,* a kind of cup used by the Greeks). The angiosperms are divided into two groups, depending on whether they possess one cotyledon or two. The monocotyledons (the Greek *monos* means alone or solitary) include all the grains and grasses as well as some flowers such as lilies and orchids. Most angiosperms, however, have two such leaves and are dicotyledons (the Greek *di* means twice).

Angle

GEOMETRY (Greek *ge,* earth; *metron,* measure) is much concerned with the intersection of straight lines. Two lines intersecting form an angle, a word taken from the Latin *angulus,* meaning a corner. The most familiar angle is the right angle (see PERPENDICULAR) and the ordinary street corner often is a right angle.

When two lines meet more sharply than those in a right angle the resulting angle is acute. In Latin *acuere* means to sharpen, and the past participle, sharpened, is *acutus.* If the lines meet less sharply than in a right angle this is an obtuse angle. The word obtuse is taken from the Latin prefix *ob-,* meaning against, and *tundere,* meaning to strike. The past participle of the verb *obtundere* is *obtusus.* If you strike against a sharp edge, you see, you blunt or dull it; you make it obtuse.

If the two lines that form the angle become so obtuse that they meet in the same straight line there is no angle in the ordinary sense, but the geometricians, to be completely logical, find it convenient sometimes to call a straight line a straight angle. (The word straight comes from the Anglo-Saxon word *streccan,* which also gives us our word stretch. Stretch a cord, after all, and you make a straight line out of it.)

An angle can be still larger than a straight angle; one arm bends beneath the horizontal and is on its way back to the other arm, so to speak. This is a reflex angle, from the Latin *re-* meaning back and *flectere* meaning to bend. Bent back in Latin is, in fact, *reflexus.*

Aniline

IN 1826 the German chemist O. Unverdorben heated indigo strongly and broke its molecule into smaller pieces. One of these pieces made up a new nitrogen-containing organic liquid. In 1840 the method was improved and the name aniline proposed for the new compound, from *anil,* the Spanish name of the indigo plant (see INDIGO). Aniline could also be obtained from coal tar (a pitchy substance derived by heating soft coal in the absence of air).

In 1856 an eighteen-year-old British chemistry student, William Henry Perkin, was trying to make quinine (an anti-malarial medicine), out of simpler chemicals. The make-up of the quinine molecule was not known at the time, so the chances of success were very minute, and, of course, Perkin failed.

But what a failure! In the course of his experiments Perkin treated aniline with various chemicals (because he thought, mistakenly, that the aniline molecule resembled the quinine molecule) and got a black, gooey mess for his pains. Undoubtedly he ought to have dumped it, but it had a purple glint, and he began thinking. He sent some to a dye establishment, and they were interested.

Perkin dropped everything and concentrated on getting the purple dye out of the black mess and learning how to make more of it. He discovered a more economical way of making aniline from coal-tar chemicals and set up a factory. The dye was named Aniline Purple, but the French dyers who took up the new material invented the word mauve for the colour, because it resembled in colour the flower of the mallow *(Malvus sylvestris).* The dye was also known as mauveine.

Mauveine was the first of hundreds of synthetic dyes produced by the chemical industry. As a class these are called aniline dyes or coal-tar dyes. They have put the natural dyes out of business, including indigo, from which aniline was first obtained, and after which it was named. And Perkin lived an additional half-century, rich and famous, because he had the wit not to tip a mess down the sink and the drive to construct an industry out of it.

Anthracite

BITUMEN IS the Latin word for a kind of tarry material that was useful in that it was soft enough to be smeared on objects, sticky enough to remain and harden there, and capable of making the objects so treated waterproof. Bitumen is referred to in the Latin version of the Bible, and is translated as pitch or slime. Noah's ark was coated with bitumen, and so was the ark of bulrushes that saved the infant Moses.

Now, coal is mostly carbon, having been formed from plant life that grew in aeons past. During the course of ages most of the hydrogen, oxygen and nitrogen atoms in the plant tissues were lost, leaving the carbon. Most coal, if heated in the absence of air, will give up what is left of these other atoms in various molecular combinations as gases and vapours. Some of these vapours can be cooled into a black, bitumen-like material called coal tar, so the kind of coal which can be so treated is called bituminous coal. It yields bitumen, in other words.

A less common variety of coal (found particularly in South Wales) contains so much carbon (90 per cent or more), and so little of anything else, that it produces no bitumen to speak of. It burns with more heat and less smoke that does bituminous coal, and is therefore a favoured variety for home-heating. It is anthracite coal, from the Greek *anthrax,* which means coal, so that in a sense anthracite coal is coal coal.

One of the compounds obtained from coal tar has a molecule made up of 14 carbon atoms in a triple ring. It is called anthracene, again from *anthrax*. Anthrax itself is the name given to a fatal disease attacking domestic animals, and sometimes man. One of the symptoms in man are the coal-black pustules that appear —hence the name.

On similar principles, the word carbuncle (from the Latin *carbunculus,* meaning a little coal) can be applied to a red garnet or to a large boil, because both resemble a smouldering coal. I imagine that the boil makes the more convincing imitation to a person unfortunate enough to have one.

Anti-deuteron

IN THE 1930s physicists became convinced that every sub-atomic particle had a twin *(anti-particle)* that was opposite to itself in some key property. The first such anti-particle to be discovered was the positron, which was just like an electron but had a positive electric charge the precise size of the electron's negative electric charge. The positron is sometimes called the anti-electron.

Again, there is the proton, which has a positive electric charge, and the anti-proton, which has all the other properties of a proton, but has a negative electric charge. The neutron has no electric charge at all, but has a magnetic field pointing in a particular direction. The anti-neutron has a magnetic field pointing in the other way.

Our part of the universe is made up entirely of particles. Anti-particles made in the laboratory soon react with their opposite numbers and disappear. If there were a section of the universe in which only anti-particles existed it seems certain they would come together in the same way particles do to form whole atoms and molecules made up of anti-particles alone. They would make up anti-matter.

But is this just theory or can we find evidence for it? The simplest way to put anti-particles together is to join an anti-proton and an anti-neutron. A proton and a neutron in conjunction make up the nucleus of a type of hydrogen atom twice as massive as ordinary hydrogen (whose atomic nucleus contains only a proton and nothing more). The proton-neutron hydrogen is called deuterium, from a Greek word meaning second, because its nucleus contains a second particle. The nucleus alone, the proton-neutron combination, is a deuteron. An anti-proton/anti-neutron combination would therefore be an anti-deuteron. In 1965 American physicists actually produced such anti-deuterons, and for the first time anti-matter (the simplest possible variety) was formed.

Anti-particle

THE WORD particle is from a Latin word meaning little part, so that a particle is a very small piece of matter.

Until the 1890s, chemists thought that atoms were the smallest possible pieces of matter, incapable of being broken into anything smaller. The very word atom is from a Greek term meaning uncuttable.

Then in the 1890s it was found that the atom was made up of still smaller objects. The first of these to be discovered, carrying a negative electric charge, was called an electron. Another, considerably larger than the electron and carrying a positive electric charge, was named proton. In 1930 an object the size of a proton but with no electric charge at all was discovered and named neutron. Since then dozens of objects smaller than atoms have been discovered. To give them all a name in common, they are called sub-atomic particles.

In 1928 an English physicist, Paul Dirac, presented theoretical arguments for supposing that for every sub-atomic particle there should be another that was opposite in properties. For instance, there should be a particle opposite to the electron, with a positive electric charge but otherwise just like it. Such an opposite-of-the-electron particle was discovered in 1932 by the American physicist Carl D. Anderson, and was named positron because of its positive charge.

It was not till 1955, though, that the opposite of the proton (just like it, but with a negative electric charge) was discovered. Instead of being given a special name it was called an anti-proton, for the Greek prefix *anti-* means opposite to.

Physicists went on to discover opposites to all the sub-atomic particles that could have been in theory. The general name they gave these opposites was anti-particles.

Antibody

CITIZENS OF ancient Rome owed their governments certain 'gifts'. These might be of money (as we pay taxes) or of services (as we go off to war). Certain citizens would for one reason or another be exempt (just as today some organizations are tax-exempt).

The Latin word for these services or obligations is *munia,* and the Latin prefix *im-* means not. A person who was not expected to make a particular gift of money or services was said in Latin to be *immunis.* In English the word has come down as immunity.

We all know that a person who has had measles, or certain other diseases, will practically never get it a second time. He has become immune to that disease.

The body achieves immunity in its attempts to battle the first attack of a disease. The body then begins to form protein molecules in the blood that are especially designed to combine with the disease germ and render it harmless. Or the molecules might combine with a poisonous compound produced by a germ and neutralize its action. (Such germ-produced poisons are called toxins, from the Greek *toxon,* meaning bow, because arrows were often poisoned with what the Greeks called *toxikon pharmakon,* or poison of the bow.)

The germ-fighting proteins linger in the blood after the person recovers from a disease such as measles, and a second attempt at infection is met instantly by these ready defenders. The defence proteins are called antibodies, from the Greek *anti* (against), since they are bodies (i.e., substances) formed to act against germs. The germs or their poisons (or anything, in fact, that causes the production of antibodies in the first place) are antigens, the Greek suffix *-gen* meaning to produce.

Apollo, Project

THE SUCCESS of the American space effort in terms of the one-man orbital flights of Project Mercury and the two-man orbital flights of Project Gemini meant the next effort would surely be the 380000-kilometre leap across space to the moon. In 1961 President John F. Kennedy stated that it was an American aim to get a man to the moon and bring him back to earth safely by the end of the decade. By the time Project Gemini was concluded most of the decade was gone.

Since Projects Mercury and Gemini both had names drawn from mythology, it seemed reasonable to use mythology for the moon-landing project as well. The classical moon deities were all females, however, and a masculine name was desired.

The mythological beings associated with the sun were masculine, and the best known was Apollo. The moon-landing programme therefore became Project Apollo. It was fitting, for Apollo drove the golden, flaming chariot of the sun, while the three men who were to manoeuvre the space vessel to the moon were driving a fiery rocket exhaust. It was almost as though the myth were coming to life.

There was something ill-omened about driving the chariot of the sun. Phaethon, the son of the sun god, once tried to guide the chariot. In his inexperienced hands the chariot went off course and endangered the whole world, so that Zeus was forced to kill Phaethon.

In January 1967 three astronauts, testing an Apollo capsule on the ground, died tragically and Phaethon-like in an accidental fire. Nevertheless, on July 20, 1969, *Apollo 11* was guided safely to the moon, and Neil A. Armstrong became the first man to set foot on another world. He and his comrades came safely back to earth and President Kennedy's goal was met, though he himself had died at the hands of an assassin years before and was not present to witness the great day.

Project Apollo was terminated officially in December 1972 with *Apollo 17*. There were six moon landings in all, *Apollo 13* being unsuccessful. The entire cost of the Apollo programe was stated officially to be 25 000 000 000 dollars.

Appendix

LIVING CREATURES, including man, are virtual museums of structures that have no useful function, but which represent the remains of organs that once had some use. (Man has tiny bones once meant for a tail, and unworkable muscles once meant to move the ears.)

The small intestine, for instance, enters the large intestine about seven centimetres from its lower end. That lower end is therefore a kind of blind alley called the caecum, from the Latin *caecus* (blind). At the end of the caecum is a much narrower tube about eight to ten centimetres long. In general, structures that hang from a part of the body are called appendixes (singular, appendix) or appendages, from the Latin *ad-* (to) and *pendere* (to hang). They are structures that hang on to the body.

This narrow tube (also a blind alley) is just long and narrow enough to resemble a worm, so that it is called the vermiform appendix, from the Latin *vermis* (worm) and *forma* (form)—the appendix in the form of a worm. However, such is its notoriety that this particular organ is usually called simply the appendix.

In certain plant-eating animals the caecum is a large storage-place where food may remain to be broken down by bacteria so that the animal itself may more easily digest and absorb it. The appendix in man and the apes (it occurs in almost no other animal) is what remains of that large caecum. It indicates that the fairly near ancestors of man and the apes were plant-eaters. The appendix is thus the useless remainder of a once useful organ; it is a vestige, from the Latin *vestigium* (footprint). Just as a footprint is a sign that a man once passed that way, so a vestige is a sign that a useful organ once passed that way.

Every once in a while the appendix is worse than useless. It becomes inflamed, and must be cut out lest it rack or kill its owner. The condition is appendicitis; the operation an appendectomy. The suffix -itis is from the Greek, and is used to mean an inflammation of, while the suffix -ectomy comes from the Greek *ektome* meaning cut out.

Appestat

THERE ARE TWO chief factors to weight increase under normal conditions—the amount of calories that comes in as food and the amount that goes out by way of physical activity. A plump person may eat no more than a skinny one, but may be considerably less active.

Many people keep their weight steady year after year just by eating when they are hungry. If they have been very active they get hungry sooner and stay hungry longer. In that way they take in enough calories to balance the work or exercise they have been doing. If they have been taking it easy for quite a while they are less hungry, and are more quickly filled.

There is some device in the body which controls the appetite in such a way as to keep food intake matched to energy output. This is similar to the way in which a thermostat (from Greek words meaning heat-stationary) controls a furnace so as to keep the temperature steady. The appetite-controlling device is called, by analogy, the appestat.

It seems likely that in plump people, who gain weight easily, the appestat is set higher than it should be. They get hungry too soon and stay hungry too long, just as a house might have its thermostat set too high so that it overheats. Plump people find it agony to try to diet, since all the while they are cutting down on food the appestat is constantly signalling for more.

The appestat seems to be located in a portion of the brain called the hypothalamus (beneath the thalamus). Thalamus is from a Latin word for a kind of room, since the ancient Romans thought this brain portion was hollow like a room. When the hypothalamus of a laboratory animal is damaged the appestat seems to be shoved high. The animal begins to eat voraciously, and soon gets enormously fat.

Aqua Regia

THE ALCHEMISTS of the Middle Ages used colourful language to describe the materials they worked with. Liquids, for instance, were generally named as some descriptive type of *aqua* (which is Latin for water).

Thus when it was first learnèd how to distill wine and get alcohol, a watery solution containing enough alcohol to burn was obtained and this was called *aqua ardens,* which is Latin for burning water. A solution containing larger quantities of alcohol was called *aqua vitae* (water of life), probably because of the new life that drinking it appeared to give people who were feeling somewhat passé. The expression is still used for various forms of brandy and similar liquors.

Somewhere about 1300 the alchemists discovered the strong mineral acids. This was a milestone in chemical history, since acids can be used to dissolve many things that will not dissolve in water. The strongest known acid in ancient times was vinegar (see ACID), but the mineral acids were immensely stronger, and made many chemical reactions and processes possible that were not possible before.

Nitric acid, for instance (see RAYON), was discovered and named *aqua fortis* (strong water), and so it was, for it ate away almost any substance with which it came in contact, including all the metals then known, except gold. If hydrochloric acid (which was discovered three centuries later) or ammonium chloride were added to the nitric acid the mixture turned green and the acid became still stronger, for now it actually dissolved gold. (This was because the hydrochloric acid reacted with the nitric acid to form the green element chlorine, which was what attacked the gold.)

Since gold was the king of metals, an acid that would dissolve it must surely be the king of waters, and so it was named *aqua regia* (royal water). Though almost all the names of alchemy have died out, this one persists, and is used to this very day for a 1-to-4 mixture of nitric and hydrochloric acids. (It dissolves platinum, too.)

Arctic

THE EARTH'S AXIS of rotation is inclined 23½ degrees to its plane of revolution. On December 21, then, when the North Pole is inclined the full 23½ degrees away from the sun, all points within 23½ degrees of the Pole go at least one full day without seeing the sun. At the same time the South Pole is inclined the full 23½ degrees towards the sun, so that all points within 23½ degrees of that Pole will enjoy uninterrupted sun for at least one full day. On June 21, when the North Pole is inclined toward the sun and the South Pole away from it, the situation is reversed.

Now as you go farther north from the equator the stars of the northern sky seem to climb higher in the sky. Eventually the most prominent constellation of the northern sky, the Big Dipper or Great Bear, would be overhead at some time of the night. The Greeks therefore referred to the north as the arctic, from *arktos* (bear). It was the region where the bear was overhead.

Naturally, the southern regions were the antarctic, from the Greek *anti-*, meaning against or opposite to. The south, after all, was in the direction opposite to the north.

In modern geography the Arctic refers to that portion of the earth's surface lying within the 23½-degree limit of the North Pole, while the Antartic is the similar region about the South Pole. The imaginary circle that limits this region is the Arctic Circle in the north and the Antarctic Circle in the south. Furthermore, the ice-covered continent lying almost entirely within the Antartic Circle is Antarctica.

The Great Bear itself is known to astronomers by the Latin name Ursa Major (from the Latin *ursus*, meaning bear, and *major*, meaning larger), but the Greek word comes into play in connection with one of the stars, a bright one in the constellation Boötes. It is near Ursa Major, and seems to be a gleaming eye keeping perpetual guard over the bear. It is named Arcturus, from the Greek *arktos* (bear) and *ouros* (guard).

Argon

THERE IS an important difference between iron and gold. Iron rusts and crumbles away. Gold does not. The reason is that iron will combine with the oxygen and water vapour in air to form rust (from an Anglo-Saxon word meaning red). Gold, on the other hand, does not combine with oxygen, or with scarcely any substance except under extreme conditions. It is an inert element (see INERTIA).

To the ancients, however, there seemed another explanation. Standoffishness was characteristic of aristocrats. Noblemen, after all, didn't associate with just anyone, but only with their equals. The higher the nobleman the fewer his equals, and the more standoffish he was. Gold, obviously, was a noble metal.

Now in 1894 the Scottish chemist William Ramsay discovered a gaseous element that made up 1 per cent of the air. It would not combine with other elements under any conditions. Its atoms would not even pair up with each other. It is one of the group of inert gases for that reason (these are also called rare gases, because present in air in such small quantity).

Ramsay called this particular gas argon, from the Greek *a*- (no) and *ergon* (work). It was too lazy to do the work of combining with other substances. But the old notion of nobility persisted. Here was something even more standoffish than gold itself, and so argon was called a noble gas.

Ramsay proceeded to find four other inert gases in the air in the next four years. They were even rarer than argon. One was helium, found years before in the sun (see HELIUM). Then there was neon (from the Greek word for new, which is *neos*), krypton (from the Greek *kryptos* meaning hidden) and xenon (from the Greek *xenos* meaning stranger). After all, these gases were new and strange, and had been hidden in air a long time before their discovery. (A sixth and last member of the series was discovered later and named radon. See RADIOACTIVITY.)

Artery

IN ANCIENT TIMES only one type of blood-vessel was recognized. In Latin, the word for it was *vena,* and in English it is vein.

There are other vessels like the veins, but in general larger and with thicker walls. When anatomists examined these in dead bodies they found them to be empty. They assumed that they carried air to various parts of the body, and were therefore extensions of the windpipe.

The Greek name for the windpipe was *arteria,* perhaps from *aer,* air, and *terein,* to keep, or from *airein,* to raise; a vessel in which air was kept. This name was transferred to the empty vein-like vessels which are now called arteries. (The windpipe is no longer known by its old name but is called the trachea, from the Greek *trachys,* rough, because it has circular stiffenings of cartilage which make it rough to the touch, as you will find if you feel your throat.)

The Greek physician Galen, in the second century A.D., was the

35

first to find that the arteries carried blood as did the veins, but for many centuries afterwards no one understood properly why two sets of vessels were needed. In 1628 the English physician William Harvey set forth the principle of the circulation of the blood; that it did not rest or oscillate back and forth in the vessels, but flowed in a single direction, leaving the heart via the arteries and returning via the veins.

There was even then no known connection between arteries and veins, yet this was necessary if circulation were to take place. In 1661, four years after Harvey's death, the Italian physiologist Marcello Malpighi supplied the missing link. He was the first to use the microscope systematically on plants and animals, and, among other things, he saw (in frogs first) tiny vessels, thinner than hairs, through which blood flowed from arteries to veins, bathing all tissue cells as it did so. These tiny, thinner-than-hair vessels are capillaries, from the Latin *capillus,* meaning hair.

Ascorbic Acid

UNTIL FAIRLY RECENT times scurvy (medieval Latin name *scorbutus,* of uncertain origin) was a serious human affliction. It began with weakness and with muscular pains, then continued with sore gums and haemorrhages. Eventually teeth fell out, haemorrhage grew more severe and death came finally. Looking back from the greater knowledge of today, it seems people ought to have noticed the connection of the disease with diet earlier than they did.

For instance, it struck whenever diet was monotonous and lacked fresh fruit and vegetables—on long sea voyages particularly; in hard-pressed armies; in besieged cities; in prisons and poorhouses.

Scurvy will not strike people under any of these conditions if certain fruit juices are available. After several decades of sporadic experimentation, the Royal Navy in 1795 began to force its sailors to drink a daily ration of lime-juice. The sailors must have complained bitterly, but scurvy stopped. (And to this day, British sailors are referred to slangily as limeys.) The other citrus juices, tomato juice and various fresh vegetables are also effective in preventing scurvy.

By 1907 biochemists began suggesting in print that there might be a chemical in these foods which was needed by the body and that scurvy was the result of the absence of this substance. About that time chemicals called vitamins began to enter the thinking of nutritionists (see VITAMIN), so the hypothetical anti-scurvy, or antiscorbutic, chemical was eventually named vitamin C. (The letters A and B were already taken.)

The Hungarian biochemist Albert Szent-Györgyi isolated a chemical from cabbage in 1928 which the American biochemist Charles G. King showed in 1932 was the chemical needed by the body to prevent scurvy; it was named ascorbic acid, no-scurvy acid.

Asteroid

A PLANET IS an astronomical body which revolves about a sun, but there are other astronomical bodies which revolve about planets. The moon, for instance, revolves about the earth (and is carried by it about the sun, so that it revolves about the sun, too). Most of the other planets have smaller bodies circling them.

Such smaller bodies may be called moons in analogy to our own moon, but a more frequently used word is satellite, from the Latin word *satelles,* meaning attendant (since the moon attends its planet on the latter's journey about the sun). The Russian word is *sputnik,* meaning one who travels with another.

From 1801 onward, hundreds of tiny planets have been discovered circling the sun in orbits lying between those of Mars and Jupiter. They circled the sun, so they were planets. They were so small, however (even the largest is only 773 kilometres in diameter, compared with a minimum of 4800 kilometres for the other planets), that some new name seemed called for.

The most frequently used name is asteroid, from the Greek word *aster,* meaning star, and the Greek suffix *-oeides,* meaning having the form of. These small planets have the form of stars when seen through the telescope, instead of showing visible discs as the other planets do. In actual fact, however, they are completely unlike stars, so many people prefer the alternate name planetoid.

Even planetoid isn't quite a fair name. The planetoids do not merely have the form of planets; they are planets. To emphasize their small size, though, they are frequently called minor planets, and that is perhaps the best name of all.

Tiny minor planets that enter the atmosphere of the earth and burn up are called meteors, from the Greek word *meteoron,* meaning a heavenly phenomenon. If a meteor is not entirely consumed in the atmosphere the part that survives and hits the earth is a meteorite. While it is still in space, and before it enters the atmosphere, it is a meteoroid. If it is of microscopic size, as millions are, it is a micrometeor.

Astrochemistry

MANY GALAXIES, including our own, contain vast, thin clouds of gas and dust, which astronomers assume to be composed of the common varieties of atoms making up the universe —hydrogen, helium, neon, oxygen, carbon and nitrogen.

The atoms are spread out so thinly in these galactic clouds that in their random movements they would be expected to approach one another only rarely. For that reason astronomers expected to find little, if any, atom combinations. Yet in the 1930s astronomers could tell from the light-absorption of such clouds that there might be small quantities of carbon-hydrogen (CH) and carbon-nitrogen (CN) combinations.

Once radio-waves from the heavens began to be studied and analysed, in the 1940s and afterwards, however, a sharper tool was available. In 1963, for instance, radio waves from certain interstellar clouds were analysed and found to indicate the presence of surprising quantities of oxygen-hydrogen, or hydroxyl (OH) combinations.

Two-atom combinations were surprising enough, but then three-atom combinations began to turn up. From late 1968 on, radio-wave absorption was found to indicate the presence of such three-atom combinations as water (H_2O) and hydrogen cyanide (HCN), and the four-atom combinations of ammonia (NH_3).

Even more complicated molecules were detected: formaldehyde (HCHO), formic acid (HCOOH) and cyanoacetylene (HCCCN). The most complicated ones yet found are the six-atom combinations of methyl alcohol (CH_3OH) and formamide (NH_2COH). The last is the first to be detected containing all four major combining atoms: nitrogen, hydrogen, carbon and oxygen.

How these complicated molecules are formed astronomers don't know, but now the necessity arises of studying the chemistry of very thin gas clouds—a science called astrochemistry.

Astronaut

MAN HAS ALWAYS DREAMED of flying, of winging through the sky like a large bird. In ancient times it was felt the air continued indefinitely upward, even to the heavenly bodies. As late as 1638, an English bishop, Francis Godwin, wrote a fanciful tale of a man who was carried to the moon by large birds.

In 1643 an Italian physicist, Evangelista Torricelli, weighed the atmosphere and it became obvious it could only extend a few miles upward. The moon was separated from earth by nearly 400000 kilometres of vacuum, and other heavenly bodies were farther away still.

In 1657 a French poet, Cyrano de Bergerac, in a work published posthumously, first mentioned the possible use of the rocket principle in travelling through space. This is the one practical method of crossing a vacuum, and it was by rockets that men actually reached the moon in 1969.

Space is what separates two objects. People are more interested in objects than in the space between them, so space came to be thought of as nothingness. The ideal nothingness is a vacuum, and since a vacuum exists beyond the atmosphere and between the heavenly bodies, it is that region which came to be known as space.

In the twentieth century one could talk of space flight for journeys beyond the atmosphere and space travel for journeys to the moon or other outer-space destinations. (Rocket flight and rocket travel might also be used, since rockets were the means of propulsion.)

About 1930 the term astronautics, from Greek words meaning star voyaging, originated in France. It will be a long time before man will be able to journey between the stars, to be sure, and for now he must confine himself to the solar system, the neighbourhood of his own star. Nevertheless, the word has stuck, and in the West these men who make rocket flights are called astronauts —that is, star voyagers—while the Soviet Union calls them cosmonauts (see COSMONAUT).

Atmosphere

A SPHERE (from the Greek *sphaira,* a ball) is a solid, the surface of which curves equally in all directions. The earth's shape is roughly that of a sphere, but not quite. It is slightly flattened at the poles, so that the curvature is less sharp there than at the equator. It is a spheroid, but under ordinary circumstances it is still spoken of as a sphere.

The solid matter of the earth itself, for instance, is the lithosphere, from the Greek *lithos* (stone). Enclosing three-fourths of the lithosphere is a shell of liquid making up the earth's ocean. If the lithosphere were to disappear the oceans would be seen to form a sort of hollow (and incomplete) sphere. They are referred to, then, as the hydrosphere, from the Greek *hydor* (water). Outside both lithosphere and hydrosphere is a hollow sphere of gas, the atmosphere, from the Greek *atmos* (vapour).

The atmosphere grows rapidly thinner as it extends outward from the earth, but very thin wisps reach outward for hundreds of kilometres, and it has no definite end. As the atmosphere thins, various layers have different properties, and have received different names.

About 75 per cent of the atmosphere lies within eleven kilometres of the earth's surface. In this lowest layer are the clouds and storms and all the weather changes we witness. It is the troposphere, from the Greek *tropos* (change). The sixteen kilometres or so above the troposphere consists of a layer of air called the stratosphere. This is from the Latin *stratum,* which is the past participle of *sternere* (to spread). The theory was that in the absence of storm and change the air simply spread out quietly, perhaps in several sub-layers.

For several hundred kilometres above the stratosphere is the ionosphere, so called because layers of ionized gas (see ION) occur within it, the ions being formed through the action of the short-wave radiation of the sun. Finally, from a height of 480 kilometres or so, indefinitely upward, is the exosphere, from the Greek *exo* (out). It is the part of the atmosphere that is at the outer limit, you see.

Atom

THE ANCIENT GREEKS were much interested in speculating on the nature of the world about them, and succeeded in evolving many fascinating theories as a result. They were usually wrong, if we judge them by what we now believe to be true—but not always.

For instance, two Greek thinkers, Leukippos of Miletus and Demokritos of Abdera, decided that substances could not be broken up into smaller and smaller particles indefinitely. Eventually, they thought, particles would be obtained so small they could be divided no farther. There were a number of varieties of such particles, each making up a different substance. By combining them in different ways, still other substances would result. The Greek word for indivisible is *atomos,* and so these particles expressed the fact that they could not be divided. They were called atoms.

This theory did not win favour among the Greeks, but over two thousand years later it was resurrected. In 1803 the British chemist John Dalton decided that the facts uncovered by the still new science of chemistry coud best be explained by supposing every chemical element to be formed of tiny indivisible particles. Each element had its own characteristic type of particle, and by varying the manner of combination of these, all existing substances could be constructed.

Dalton, following the old Greek theory, named his indivisible particles atoms, and this time the atomic theory met with approval.

Strangely enough, however, it was discovered in 1896 that atoms are not indivisible after all. Certain complicated atoms, it was found. broke up spontaneously, liberating particles far smaller than atoms. Then scientists learned how to break up atoms in the laboratory. Now man's whole future hinges upon the manner in which atoms break up and fuse together, and on the behaviour of particles smaller than atoms. But still the name is atom —indivisible.

ATP

IN 1905 two English chemists, Sir Arthur Harden and W. J. Young, were studying the effect of phosphate ions on certain enzymes. To their surprise the phosphate ions seemed to disappear. A search revealed them to be attached to sugar molecules, and these were the first organic phosphates to be studied. By the 1920s it began to appear that organic phosphates were essential to the production of energy in living tissue.

In 1941 the German-American chemist Fritz Lipmann found that certain phosphate-containing compounds could release energy in higher than ordinary amounts. These were called high-energy phosphates.

Apparently when food molecules are broken down phosphate groups are added. At certain key points chemical changes are brought about that turn the organic phosphate into a high-energy phosphate. Some of the chemical energy in the food is concentrated in that phosphate group, and. could be made use of at will.

Of the high-energy phosphates, one in particular was lower in energy than the rest. It stood in an intermediate position. It was capable of accepting a phosphate group from higher-energy phosphates and then passing it on, along with its energy load, to ordinary molecules. Because of this phosphate's use in almost every reaction requiring energy, it was clearly of key importance to the body.

This compound had been discovered in muscle in 1929 by the German chemist K. Lohmann long before the nature of its role was suspected. It was named adenosine triphosphate because it consisted of adenosine, a well-known tissue component, to which were attached three phosphate groups in a row. When its importance in energy utilization was made clear it came to be referred to so often by chemists that a short cut for its heptasyllabic name seemed desirable. The initials ATP therefore came into use.

Aurora Borealis

EAST WAS probably the first of the directions to be recognized by primitive man, if only because it was in that direction that the sun rose. Huddled in the cold and danger of the long winter night, he must have waited eagerly for the first signs of the returning sun with its light and warmth; and he must have learned in which direction to watch for it.

The Sanskrit word *ushas* was used for that direction, apparently derived from another word meaning shining. And from that came the Greek *eos* and the Latin *aurora* (which may have originally been *ausosa*). The words mean both dawn and the direction of the dawn. Our word east obviously comes from *eos,* or at least the two have a common ancestor. (Similarly, orient, which also means east, comes from the Latin *orior,* meaning to rise. Latin for rising is *oriens*.)

There is one kind of night light or dawn, however, which did not seem to have any connection with the sun. (Actually it does, but the ancients couldn't know that.) Periodically an upheaval on the sun's surface sends a stream of electrons shooting into space. When this stream strikes the upper regions of the earth's atmosphere it pumps energy into the atoms and molecules of the air and causes them to glow. A beautiful many-coloured light in moving sheets and streamers is formed. The light derived its name, however, from that which seemed most remarkable of all to the early observers—its direction.

The electrons, being electrically charged, are deflected by the earth's magnetic field and strike the atmosphere in the polar regions mainly. From the Northern Hemisphere, the light is seen in the north, and is called the Aurora Borealis (Latin for northern dawn, since Boreas was the word in both Latin and Greek for the north wind). It is usually called the Northern Lights in English. How different from the much more usual eastern dawn!

In the Southern Hemisphere the light is seen in the south and is called the Aurora Australis (southern dawn, since Auster is the south wind in Latin).

Autoradiography

PROBABLY THE MOST IMPORTANT SET of chemical reactions on earth is that of photosynthesis, whereby green plants use the energy of sunlight to convert carbon dioxide and water into their own tissue substances. It is upon plant tissue that all animal life exists.

Scientists have tried to penetrate the details of the photosynthetic reaction. The number of different substances present in plants is, however, enormous and there is no way of telling by ordinary methods what was formed first, what next, and so on.

In 1948 the American biochemist Melvin Calvin began a series of experiments designed to work out the details. He developed a scheme for exposing green plant cells to carbon dioxide containing radioactive carbon atoms for just a few seconds, after which the cells were killed so that the photosynthetic reaction was stopped. In the few seconds of exposure, only the first few steps in the reaction had time to take place. If one then analysed the contents of the cells, and searched only for those compounds that contained radioactive carbon atoms, one could locate the places where the carbon dioxide entered the system.

To do this, Calvin mashed up the cells and dissolved the contents. He then let the solution seep up some porous filter paper. The various substances in the solution seeped upward, each at its own rate. In the end, he had small quantities of particular substances, each concentrated in a certain area of the filter paper.

Calvin then placed a photographic film under the filter paper. Those areas containing substances with radioactive carbon atoms gave off radiations that fogged the film. In this way he could identify the spots containing compounds to be studied further.

This process of having radioactive compounds indicate their own presence is called autoradiography, which means literally own-radiation marking.

Auxin

PLANTS ARE in general simpler than animals. Plants, for instance, lack the nerves and muscles which enable animals to move quickly. Yet plants do move, even though only slowly. A plant stem will turn upward and grow in the direction opposite to the pull of gravity; it will also turn to face the light.

That plants can do so is the result of differential growth. Suppose a stem is forced to remain horizontal, along the ground. The cells on the lower side of the stem will grow and multiply more quickly than those on the upper side. The lower side will elongate, and the stem will naturally curve upward. In the same way, if a stem is illuminated by light coming from one side only, cells on the shaded side will grow and multiply more quickly and the stem will turn towards the light.

But what stimulates the growth? Growth is stimulated by a hormone, an organic compound which is produced naturally in plants. The hormone concentrates in those places where growth is required and stimulates that growth. Because these particular hormones are found exclusively in plants, not animals, they are called plant hormones, or phytohormones. (The prefix phyto- comes from a Greek word meaning plant.)

The notion of plant hormones was first worked out in detail by a Dutch botanist, Fritz Went, in the 1920s. The plant hormones he studied were called auxins, from a Greek word meaning to increase.

Chemists have produced synthetic compounds which have auxin-like properties. These can be used in a variety of ways—to prevent flowering, to keep fruit hanging on the bough for longer periods, or occasionally to produce seedless fruits. They are most commonly used as weed-killers, however, since in high enough concentrations they seem to overstrain the growth mechanism of plants and in doing so kill them.

Avidin

IN 1936 two Dutch chemists isolated a substance that in small quantities was essential to life. It proved to be one of the family of vitamins that make up the B complex (so called because the first of the group was called vitamin B to distinguish it from two others labelled A and C, the letters being assigned arbitrarily). Because the new vitamin seemed to be found in all forms of life, it was named biotin, from the Greek word for life.

To test its functions the chemists tried to rear animals on diets lacking biotin, to see what would happen if it were not present. Almost any diet, however, had enough biotin for the needs of the animals.

It had earlier been discovered, though, that rats fed diets that included large quantities of raw egg white suffered a certain disorder called egg-white injury. This was prevented if certain other foods were also added to the diet. By 1940 biochemists showed that it was the biotin in the added food that prevented egg-white injury.

There was something in raw egg white, apparently, that combined with biotin, producing a biotin deficiency that was responsible for the egg-white injury. If enough biotin was added to the diet this was prevented.

Heating the egg white destroyed its ability to tie up biotin, showing the substance in question to be a protein (for proteins are heat-sensitive). The protein was finally isolated and named avidin because it combined so avidly with biotin.

Avidin is not a danger to human health, for few people eat so many raw eggs as to induce abnormal symptoms. Nor is it known what the function of avidin in egg white is. However, avidin has been useful to biochemists, who could with its help devise biotin-deficient diets through which to study the function of the vitamin.

Bacteriophage

VIRUSES,WHICH ARE organisms so small that they cannot be seen by the ordinary microscope (see VIRUS) are the cause of some of our most notorious diseases: colds, influenza, measles, mumps, smallpox, yellow fever and poliomyelitis. There are also the plant viruses, so called because they infect plants. One of these, the tobacco mosaic virus, was the first of all viruses subjected to experiment, and the first to be isolated.

There are even viruses which infect bacteria and are parasitic upon them. In 1915 F. W. Twort first noticed that certain colonies of the bacteria he was growing became translucent and seemed to melt away. If he made an extract of the colonies and filtered it the filtered extract, when added to normal colonies of bacteria, caused them to begin fading too. The Canadian investigator Félix Hubert d'Hérelle extended such studies in 1918, suspected the existence of a virus, and called it bacteriophage. The suffix -phage comes from the Greek *phagein* (to eat), so that a bacteriophage is a bacteria-eater, which is exactly right. Oddly enough, although they are parasites on simple one-celled creatures, the bacteriophages are larger and more complicated than most of the viruses that infest multicellular plants and animals.

There are creatures intermediate in size between bacteria and viruses. Like the viruses, they can grow only within the living cell and not on non-living media as bacteria can. They are large enough, however, to be made out by an ordinary microscope as small bodies within infected cells. These were called first rickettsial bodies, then simply rickettsia, after the American pathologist H. T. Ricketts who first discovered them in connection with Rocky Mountain spotted fever. They are now known to cause a number of diseases (including typhus), all of which are spread by such creatures as ticks and lice. The use of DDT to kill the vermin normally defeats these diseases.

Ballistics

WHEN AN OBJECT is thrown it follows a certain path under the influence of the force of the throwing arm and the force of gravity.

The path it follows is its trajectory, from the Latin *trans-* (across) and *jacere* (to throw); it is the path resulting, in other words, when an object is thrown across from here to there. The thrown object itself is a projectile. The Latin prefix *pro-* means before or forward, so that a projectile is something thrown forward. It may also be called a missile, from the Latin *mittere* (past participle, *missus*), meaning to send or to throw.

When projectiles are small and thrown weakly their paths may be estimated in advance quite well, even though that path is curved and not straight, and the thrower can make proper allowance. Thus a bowler may have excellent control and a boy may be fiendishly accurate with a catapult.

However, once cannon were invented and heavy lumps of stone and iron were hurled over long distances, mental estimates were no longer sufficient. Trajectories had to be studied mathematically, and the science was named ballistics from the Greek *ballein* (to throw)—which is possibly related to the word ball.

In a perfect vacuum a missile would follow a perfect parabola (see PARABOLA) as its trajectory, and this would be easy to compute. However, air resistance slows it in its flight and changes its trajectory. Since air resistance varies with the height of the missile above the earth, with wind velocity and other factors, this becomes difficult to calculate. The importance of the science is evident, though, when we think of the huge bomb-carrying rockets being developed by the USA and the USSR—the so-called ballistic missiles, which are simply thrown by rocket, as opposed to guided missiles, which are guided by radio-waves.

Barn

PHYSICISTS PRODUCE nuclear reactions by aiming energetic sub-atomic particles at pieces of matter. If such a particle hits an atomic nucleus it is likely to alter it somehow. If, on the other hand, it misses the atomic nucleus, nothing happens.

Naturally, physicists can't aim the particles because they are working with objects far too small to see. They can only make use of many particles and hope that some will just happen to score hits.

The number of hits scored by a particular volley of particles depends on the nature of the particles and of the target. Sometimes very few hits are made, as though the target nuclei are very small, and therefore easy to miss. Sometimes, under similar conditions, quite a few hits are made—and the nuclei seem bigger in that case.

Physicists would say that nuclei that were easy to hit under certain circumstances had a large nuclear cross-section; those that were hard to hit had a small one. Even an easy-to-hit nucleus has a cross-section of only about 10^{-24} square centimetres. This means that a quadrillion nuclei jammed closely together would cover an area of only one square centimetre.

In 1942 American physicists were working on the atomic bomb and they didn't want to talk too openly about what they were doing. They were bombarding uranium nuclei with neutrons under conditions designed to bring about as many hits as possible. Success was achieved to the point where two young physicists, Marshall G. Holloway and C. P. Baker, said the uranium nuclei seemed as big as a barn.

For that reason an area of 10^{-24} square centimetres was called a barn. For a while this was used only out of a desire for secrecy; people overhearing the phrase would not connect it with atomic bombs. Eventually, however, it was accepted as the legitimate unit of nuclear cross-section.

Barnard's Star

AS LONG AGO AS 1718, the English astronomer Edmund Halley was able to show that stars actually move. Stars are so distant, however, that it can take many centuries for the motion to become noticeable. The only stars that have motions that are relatively easy to spot are among those that are unusually close to us.

In 1916 the American astronomer Edward E. Barnard noted a dim star in the constellation Ophiuchus (too dim to see without a telescope) moving with a velocity faster than that of any other known star. It moves so quickly that in 180 years it traverses a distance equal to the diameter of the moon. So unusual was this property that the star was one of the few that came to be known by its discoverer's name—Barnard's Star.

The reason for the rapid motion of Barnard's Star is that it is very close to us—only six light-years away. Only Alpha and Proxima Centauri are closer.

Since 1943 it has been found that the motion of some of the stars nearest to us wavers slightly. The only reasonable cause for that would be the gravitational influence of a large planet revolving about a star. To be detected at a distance of light-years the wavering had to be considerable, which meant the star had to be smaller than the sun and the planet larger than Jupiter.

In 1963 wavering was detected in Barnard's Star. The Dutch American astronomer Peter Van de Kamp decided in 1969 that the wavering might best be explained in terms of two planets, one slightly larger than Jupiter, one slightly smaller. It was the first time any star other than our own sun has been found to show signs of having more than one planet. And the planets of Barnard's Star (if Van de Kamp is correct) are the smallest ever detected outside our solar system.

Barometer

IF YOU SUCK at a straw inserted in water the water-level rises within the straw until it reaches the top. If the straw were over ten metres long, however, and pointed straight up, no amount of suction by mouth or by any form of mechanical pump could pull the water to the top of the straw. The Italian physicist Galileo Galilei pondered over this failure of suction, but did not come to the correct conclusion.

Galileo's student Evangelista Torricelli thought, however, that what made the water move upward was not the pull of suction but the push of air pressure against the water-level in the well or container from which the water was being sucked or pumped. When a column of water reaches such a height that the pressure of its weight on the water-level of the well is equal to the pressure of the atmosphere's weight it can rise no higher.

Torricelli tested this in 1643 by using mercury, which is $13 \cdot 5$ times denser than water. He reasoned that a ¾-metre column of mercury would exert as much pressure as a 10-metre column of water, and that should be all the mercury the atmosphere would support. Torricelli filled a metre tube with mercury, placed his thumb over the open end, and upended the tube into a dish of mercury.

When he took away his thumb the mercury dropped until the column stood just 75 centimetres above the mercury-level in the dish, and there it continued to stay. Air pressure held it that high and no higher. (The 15 centimetres above the mercury column was a vacuum—from the Latin *vacuus,* meaning empty—the first decent vacuum produced by man. A vacuum produced in that manner is still called a Torricellian vacuum.)

In 1648 the French mathematician Blaise Pascal had such a mercury tube carried up a mountain-side. As altitude increased there was less and less air overhead. What was left had less weight and exerted less pressure than air at sea-level, so the mercury column should sink lower, and it did.

Such a mercury column is still used today to measure air pressure. It is called a barometer, from the Greek *baros* (weight or heaviness) and *metron* (measure). It measures the heaviness of air.

Baryon

IN THE EARLY 1890s physicists were studying radiation produced when an electric current was sent across a vacuum. It became clear eventually that the radiation was composed of tiny particles, much smaller than atoms. The particles, called electrons, carried a negative electric charge. All atoms were found to contain electrons.

But atoms are electrically neutral. If they contain electrons they must also contain particles with a positive electric charge to balance them. The search was on for such a positively charged particle.

These particles were found and studied in succeeding years, but they were much more massive than electrons. Even the smallest positively charged particle that was found was about 1836 times as massive as the electron. In 1914 the New Zealand-born physicist Ernest Rutherford decided this positively charged particle was responsible for almost all the mass of matter. It was of first importance to the mass, and he therefore called it a proton, from a Greek word meaning first. He maintained that the smallest atom, that of hydrogen, was composed of one proton and one electron with the opposite charges exactly balanced.

In 1930 another massive particle, just about as massive as the proton, was discovered by the English physicist Sir James Chadwick. He named it neutron because it was electrically neutral.

The fact that the proton and neutron are so much more massive than the electron is very important. The proton and neutron are in the atomic nucleus at the very centre of the atom, so that almost all the mass is there. The light electron remains in the atom's outer reaches. The massive particles interact in ways that are fundamentally different from the behaviour of the light electrons. For this reason the massive particles, including the proton, the neutron, and still more massive particles since discovered, are lumped together as baryons, from a Greek word meaning heavy or massive.

Base

THE ASHES of plants yield substances with alkaline properties (see POTASSIUM) which are capable of counteracting the properties of acids. Both acids and alkalis, if strong, can be corrosive and dangerous. However, a strong acid added to a strong alkali results in a mixture that may be very mild; in one case, at least (hydrochloric acid and sodium hydroxide) it will be nothing more than salt water. A substance neither acid nor alkali is neutral, from the Latin *ne* (not) and *uter* (either); it is not either (neither) one or the other.

The alkaline substances obtained from plant ashes can be made even more alkaline by heating them strongly. Part of the ash turns into vapour and disappears, and this is carbon dioxide. Part remains behind, this being sodium or potassium oxide which, on addition of water, becomes sodium hydroxide (caustic soda) or potassium hydroxide (caustic potash). The word caustic comes from the Greek *kaustikos,* which in turn comes from *kaiein* (to burn)—which is a good description of what happens if either compound comes in contact with your skin.

To the early chemists, the part that remained behind after heating naturally seemed the steadier and firmer part of the original ash. So it was called the base, from the Greek *basis* (pedestal). It formed the pedestal, in other words, upon which the rest of the compound could be built.

Of course, base quickly came to mean any compound that could neutralize an acid, and this led to a contradiction. Ammonia, for instance, could neutralize acids, and yet it was a gas that was given off as vapour on heating ammonium salts. For this reason ordinary bases were called fixed bases (fixed in the sense of tied down), from the Latin *figere* (to fasten), while ammonia was called a volatile base, from the Latin *volare* (to fly) and *volatilis* (flying from). After all, a gas will fly away, and now we have a flying pedestal, which is not the most sensible notion possible.

However, modern chemists don't worry about it. A base is any compound which will neutralize an acid, whether the compound be solid, liquid or gas. A strongly basic substance is still called an alkali but there is no special name for strongly acidic substances.

Bathypelagic Fauna

PLANT LIFE is found only in the topmost layer of the ocean, for it depends on light, and sunlight can only penetrate some dozens of metres at most into the water. It is only in this photic zone (from a Greek word meaning light) that plants grow.

Animal life penetrates more deeply, for animals feed on each other and on debris that sinks downward continually from the photic zone. Scientists first became aware of this in the 1840s when the English naturalist Edward Forbes dredged up a starfish from a depth of a half a kilometre. Then in 1860 a telegraph cable lifted from the bottom of the Mediterranean Sea, two kilometres down, was found encrusted with life-forms.

The beginning of the systematic study of underwater life came in

1872, when a team in the exploring vessel *Challenger,* under the British naturalist Charles Thomson, made a voyage totalling over 111 000 kilometres, in a thorough attempt to dredge up organisms from the depths.

Deep-sea fish were brought to the surface. Some had luminescent patches in their skins, so that those parts glowed in the dark. Anglerfish had fleshy growths on their noses that resembled wriggling worms and attracted smaller fish within reach of the angler's gulping mouth. Fish with extensible stomachs could swallow other fish larger than themselves. It was a strange new world that was revealed.

There were deep-sea creatures other than fish, too. There were Protozoa, jellyfish, worms, shellfish, and so on. These are now referred to as a whole as bathypelagic fauna (from Greek words meaning deep-sea animals). The term is properly applied to animals that float or swim in open water; but there are also animals attached to the deep-sea bottom. These latter are benthic fauna, from another Greek word referring to the deep sea. The true study of benthic fauna came in the 1960s, when men penetrated to the very deepest part of the ocean and found life on its floor.

Bathyscaphe

ONE DIFFICULTY that stands in the way of man's exploration of the ocean depths is the enormous pressure resulting from the weight of kilometres of water. A man in even the most elaborate diving suit cannot safely descend more than 100 metres or so.

In 1930 the American naturalist Charles W. Beebe designed a thick-walled spherical steel vessel with thick quartz windows into which a man could fit and be lowered to great depths. Beebe called his device a bathysphere. The prefix bathy- is from a Greek word meaning deep, so it was a ball of the deep.

In 1934 Beebe descended over 900 metres in his bathysphere, while a co-worker, Otis Barton, finally reached a depth of 1400 metres in 1949. Barton used a modified bathysphere he called a benthoscope (from Greek words meaning to see the sea depths).

Neither the bathysphere nor the benthoscope was manoeuvrable. In 1947 the Swiss physicist Auguste Piccard designed a bathysphere that was suspended from a dirigible-like bag containing petrol. Since petrol is lighter than water, this buoyed up the sphere and let it sink slowly. The vessel carried a ballast of iron pellets, which it could jettison. The vessel would then be lightened and lifted upward by its overhead petrol bag. The vessel also had electrically driven propellers to allow it to move horizontally. The device was called a bathyscaphe (ship of the deep).

In 1960 the inventor's son, Jacques Piccard, and an American naval officer, Don Walsh, stepped into the *Trieste,* a bathyscaphe built by the Piccards, who sold it to the US Navy. In it they prepared to explore the Marianas Trench in the western Pacific, the deepest spot anywhere in the ocean. They plumbed downward over 11 kilometres to the very bottom and then returned safely to the surface. The entire sea had been opened to human exploration.

Benzene

THERE IS A TREE in Indonesia called benzoin or benjoin. It comes from the Arabic phrase *luban jawi,* meaning incense of Java. A resinous material obtained from incisions in the bark of this tree is gum benzoin. An acid easily obtained from this resin is called benzoic acid.

In 1834 the German chemist Eilhart Mitscherlich converted benzoic acid into a hydrocarbon (a compound with a molecule made up of only carbon and hydrogen atoms) and named it benzin. Another German chemist, Justus Liebig, objected that the -in ending was being used for compounds that contained nitrogen, which this one did not. He suggested the name benzol, the -ol ending signifying *Öl* (German for oil).

Liebig was a most influential chemist, and his suggestion holds good to this day in Germany, but with all respect to Liebig, it is still a bad name. The ending -ol is used by chemists for alcohols, which benzol is not. In England, France, and America, therefore, the compound is known as benzene, which is best, since -ene is an ending reserved for certain hydrocarbons. Benzol or benzole is still used for commercial crude benzene, and benzine for a mixture of petroleum hydrocarbons.

But benzene had actually been discovered before the time of Mitscherlich. In 1825 the English electrochemist Michael Faraday isolated some out of an oily residue obtained from illuminating gas. He called it carburetted hydrogen. However, in 1837 the French chemist Auguste Laurent, leaving the Germans to fight over benzin and benzol, suggested the name pheno, from the Greek *phainein* (to shine), in recognition of the fact that the compound was first found in something that shone—i.e., illuminating gas.

This did not catch on for benzene itself, but when a benzene molecule is attached to other atom combinations the benzene portion of the overall molecule is referred to now as the phenyl group, so Laurent half won. Furthermore, a benzene molecule to which a hydroxyl group (one made up of a hydrogen and an oxygen atom) is attached, is called phenol. (Here the -ol ending is justified. Phenol is a kind of alcohol.)

Berylliosis

AS TECHNOLOGY DEVELOPS human beings are exposed to hazards that did not exist in a pre-technological era. For instance, prior to the twentieth century there was very little opportunity for human beings to encounter concentrated sources of radiation such as X-rays or gamma rays. Nor was there any chance at all of encountering various synthetic organic compounds which never existed in nature but which were produced in the chemical laboratory—some of which proved instrumental in causing cancer.

Even substances which do occur in nature may exist only in small concentrations or in relatively harmless form but be concentrated and made dangerous by the requirements of technology.

Thus, once fluorescent lights were developed in the mid-1930s there was need for phosphors (from a Greek word meaning to carry light), which give off visible light after absorbing ultra-violet light. These phosphors were used to coat the inner surfaces of the fluorescent tubes, and one of them was a powdered compound of the light metal beryllium.

In 1946 an increasing number of illnesses were noticed among the workers engaged in the manufacture of fluorescent lights. The symptoms, involving the lungs, sometimes appeared months or even years after the patient had ceased working on fluorescent lights, and the illness usually ended in death. The blame was soon pinpointed on the beryllium compound used and it was dropped from the list of phosphors.

This prevented a serious danger. Not only those working in the field ran risks, but also anyone who accidentally broke a fluorescent light, breathed its dust, or was cut by the jagged glass.

The disease, now no longer an acute threat, is called berylliosis, the suffix -osis being commonly used in medicine to signify a pathological state.

Big Bang

HOW DID THE UNIVERSE come into being? A hint at a plausible answer came in the 1920s, when the American astronomer Edwin P. Hubble showed that the distant galaxies were receding from us in a very systematic way, as though the entire universe were expanding.

In 1927 the Belgian mathematician Georges Lemaître pointed out that if we looked upon the movement of the universe in reverse, as though we were reversing a film, it would seem to be contracting—eventually into a hard mass. Lemaître suggested that all the mass of the universe was in the beginning squeezed into one relatively small, enormously dense cosmic egg, which exploded and gave birth to the universe as we know it.

Fragments of the original sphere of matter formed galaxies, still rushing out in all directions as a result of that unimaginably powerful ancient explosion.

The Russian-American physicist George Gamow went on to elaborate this notion. He calculated the temperatures in the fragments of that explosion; how quickly each temperature would drop; how the initial energy would be converted into sub-atomic particles, then simple atoms, then complicated ones. Thinking of that first tremendous explosion, Gamow called it the big bang theory of the universe's origin, and the name has stuck.

The big bang theory was just theory to begin with. How to gain evidence for it? In 1965 two American radio-astronomers, Arno A. Penzias and R. W. Wilson, showed that there was a general background of radio waves from every part of the sky. This seemed to be the remnant of the radiation of that vast explosion, still lingering after all those millions of years. This is at the moment the strongest piece of evidence in favour of the big bang.

Bile

THE SECRETION formed by the liver may be called either bile or gall. Bile comes from the Latin word *bilis,* which is what the Romans called the secretion, but gall is Anglo-Saxon in origin. Both words are used in combination with others, and the choice of which to use depends on the derivation of the other word.

For instance, the fluid is stored in a small pear-shaped bag called a gall bladder (never a bile bladder, since bladder, like gall, is of Anglo-Saxon derivation), and is led into the intestines by means of a bile duct (never a gall duct, since duct is from the Latin *ducere,* meaning to lead).

Sometimes material dissolved in the bile settles out in little crystals which may collect into hard masses. If these get stuck in the bile duct and form an obstruction the result is pain and possibly surgery. Such masses are called gallstones (never bilestones, since stone is from the Anglo-Saxon). On the other hand, the Latin word for a small stone is *calculus* (plural, *calculi*) and this name may be used for gallstones. But, because of the change to Latin, they are never called gall calculi, but always biliary calculi. (The mathematical calculus, frequently causing the heart to drop like a stone, derives from the old method of reckoning on the abacus, using small pebbles—see CALCIUM.)

When gallstones obstruct the way to the intestines the bile being formed by the liver backs up and is forced into the bloodstream. The material in bile consists in part of strongly coloured compounds called bile pigments (never gall pigments because pigment is from the Latin *pingere,* meaning to paint, and *pigmentum,* a paint). In general, these pigments are dark green or brownish red. In the blood, where the colour is mixed with the red of blood and is seen through the light yellow of the skin and its underlying fat, the pigments give the complexion a sickly green-yellow appearance.

For that reason any disorder involving bile in the blood is called jaundice (from the French *jaune,* meaning yellow), and when the disease is caused by a gallstone obstruction it is called obstructive jaundice.

Binary Digit

OUR USUAL SYSTEM for expressing numbers makes use of powers of 10. In the number 222, for instance, the first 2 is two hundred (2×10^2), the second 2 is twenty (2×10^1) and the third 2 is two (2×10^0); so 222 is really (2×10^2) + (2×10^1) + (2×10^0). It is important to remember that 10^0 or any other number to the power zero equals 1.

Any number other than 10 can be used as a base. In a five-based system 222 would be (2×5^2) + (2×5^1) + (2×5^0), or 62 in the ten-based system. In a three-based system 222 would be (2×3^2) + (2×3^1) + (2×3^0), or 26 in a ten-based system.

In any system the number of different digits, counting zero, is equal to the value of the base. There are ten digits in our ordinary system, but only five in a five-based system: 0, 1, 2, 3, 4. The digit 5, in the five-based system would be expressed 10, our 6 would be 11, and so on. In the three-based system, only 0, 1, and 2 would exist, and so on.

The simplest system is the two-based system, which would have as digits only 0 and 1. In the two-based system, a number like 1101 would be (1×2^3) + (1×2^2) + (0×2^1) + (1×2^0), or 13 in the ten-based system. Writing numbers in order in the two-based system, we would have 1, 10, 11, 100, 101, 110, 111, 1000, 1001, 1010, and so on.

Computing in the two-based system is very simple, especially for electronic computers. Every on-off switch within a computer can represent 0 in the off position and 1 in the on position. The switches can be designed to manoeuvre according to two-based arithmetic and give answers with the speed of the electric current. The two-based system is also called the binary system, from a Latin word meaning two at a time. Therefore 0 and 1 are binary digits, which can be abbreviated 'bits'. Zero and 1 are the smallest pieces of information that can be fed into a computer, so that we can speak of so many bits of information.

Bioastronautics

AS SOON AS man succeeded in launching satellites into orbit about earth the question naturally arose: how would man fare in space? If space was to be explored men would have to be sent beyond the atmosphere and to other worlds. Could this be done?

In some ways the experience of space travel would introduce entirely new stresses on living organisms. There was the high acceleration, for instance, that would have to be endured while the spacecraft was being placed into orbit. Then there was the experience of weightlessness once the orbit was established. And what of the charged particles and energetic radiation that are present in space—particles and radiation that are ordinarily filtered out by our atmosphere and never reach organisms on earth's surface?

To go further, there are questions as to how a human being can be kept alive inside the space capsule even when there are no outside dangers. How is the air to be kept pure; how is the occupant to be fed; how are wastes to be disposed of?

All this can be (and has been) called space biology or space medicine, but a more formal name for the study is bioastronautics (from Greek words meaning life aspects of space travel).

The science began on earth, for the effect of acceleration can be studied on life forms by whirling them rapidly in centrifuges. It entered space itself early, however. The second satellite to be placed in orbit (Sputnik II, launched by the Soviet Union in November 1957) carried a dog, whose reactions were monitored by instruments.

Since then many men have been placed in orbit; some have stayed in space for weeks without permanent harm; some have reached the moon and returned; and all seem well. The same is true of the one woman launched in space (by the Soviet Union). In fact she married a man who had been in space, and they have a seemingly normal child.

Bionics

MANY OF THE EFFECTS man is trying to achieve by means of his machinery have been obtained in living systems during millions of years of trial and error. The devices of life, however, are sometimes not suitable for machines. Much time, for instance, was wasted by inventors who imitated birds and tried to design flying machines with flapping wings.

On the other hand, consider dolphin skin. Dolphins swim at speeds that would require $2 \cdot 6$ horsepower if the water about them were as turbulent as it would be about a vessel of the same size. Water flows past the dolphin without turbulence, however, because of the nature of dolphin skin. If we could imitate the effect in ship hulls, the speed of an ocean liner could be increased while its fuel consumption was decreased.

Again, the American biophysicist Jerome Lettvin studied the frog's retina in detail by inserting tiny platinum electrodes into its optic nerve. It turned out that the retina had five different types of cells, which reacted (1) to sudden changes in illumination from point to point, (2) to dark, curved surfaces, (3) to rapid motion, (4) to dimming light, (5) to the blue of water. All combined to make seeing extraordinarily efficient for the frog. If man-made sensors could be made to use such tricks they would become far more versatile than they now are.

In short, engineers are now making conscious efforts to adapt biological systems to man-made electronic devices. In 1960 an American engineer, Jack Steele, shortened the descriptive term biological-electronics to bionics.

In a sense, the ultimate goal of bionics is to invent a device that would imitate the functioning of the human brain and give us that science fictional dream, the robot.

Biosphere

THE ANCIENT GREEKS considered the universe to be made up of a series of concentric shells. At the centre was the solid ball of earth itself, which we now call the lithosphere (from Greek words meaning stone ball). Surrounding that, though not quite entirely, was a shell of water, the hydrosphere (water ball), and outside that the atmosphere (vapour ball). Beyond the atmosphere the Greeks envisaged a sphere of fire and various planetary spheres, theories which we have abandoned. However, we have divided the atmosphere into various regions, using the same suffix—troposphere, stratosphere and so on.

In one respect, though, modern scientists have gone ahead and added another shell. In a thin shell around earth, including the hydrosphere, the outermost part of the lithosphere and the lower reaches of the atmosphere, are included all the life-forms of the planet, and all the living activity. This was first pointed out by the French zoologist Jean Baptiste de Lamarck at the beginning of the nineteenth century, and the concept was sharpened by the Austrian geologist Eduard Suess in 1875. This thin shell in which all of life lives is the biosphere (life ball).

The total mass of living things on earth is estimated to be about seventeen trillion tons. This is only about 1/300 the mass of the atmosphere, and only 1/70 000 the mass of the ocean. Nevertheless, the biosphere is so active chemically that it is responsible for much of the environment about us. For instance, it is thanks to the green plants that the air is full of free oxygen. Also, the great layers of limestone produced by corals make up large atolls and long reefs.

On the other hand, the biosphere is extraordinarily fragile, and even small changes in temperature or the addition of small quantities of poisons could change it drastically or wipe it out altogether. Man himself has been putting great strains on the biosphere in recent years, and this is of growing concern to scientists.

Black Hole

IN 1916 Albert Einstein, in his general theory of relativity, suggested a new way of looking at gravitation. He said that masses distorted space in such a way that objects found the shortest distance between two points to lie along a curved path. Objects passing by each other therefore followed this curved path. If they were close enough to each other the curvature would be so sharp that they would follow closed paths about each other.

Einstein differed from the older suggestions of Sir Isaac Newton in his belief that, in the relativistic view of the universe, light rays also followed a curved path in the presence of masses. Because the photons that made up the light moved so much more rapidly than ordinary astronomical bodies, they had a chance to curve only very slightly before moving on far beyond the space-distorting mass.

Still, the amount by which the path of light curves depends on the mount of mass present and its concentration. If a very large quantity of mass is concentrated into a very small volume, then space is enormously distorted in its immediate vicinity. Light passing near such a body would be trapped into a closed orbit and would never escape. This was pointed out in 1916, almost as soon as Einstein's paper appeared, by a German astronomer, Karl Schwarzschild.

In the vicinity of this massive compressed body, not only would light be trapped but so would everything else. It would be like a hole in space into which things could fall, but from which nothing, not even light, could emerge. It would be, therefore, a black hole.

For many years this was considered only theory, but now the newly discovered pulsars seem to be neutron stars. Neutron stars are almost massive enough and compressed enough to be black holes, and so astronomers are now searching avidly for any sign of the existence of the thing itself.

Borane

THE SIMPLEST HYDROCARBON (a substance made up of carbon and hydrogen atoms only) was named methane. As a result the -ane suffix was applied to many hydrocarbons, and then to compounds of hydrogen with one other element, even when that element was not carbon.

At the time this convention was established there were already compounds of hydrogen that had names of other sorts that could not be abandoned. Two hydrogen atoms with an oxygen were water, three hydrogens with a nitrogen were ammonia, three hydrogens with phosphorus, or with arsenic, or with antimony, were phosphine, arsine and stribine, respectively. (The prefix stib- is from *stibium*, the Latin name for an antimony compound used in ancient times.)

Silicon, which closely resembles carbon, was found later to form compounds with hydrogen, and these were called silanes. Like carbon, silicon atoms can form chains, so that we can have one silicon atom combined with four hydrogen atoms (monosilane), two connected silicon atoms combined with six hydrogen atoms (disilane) and so on.

Germanium, tin and lead (in the same family as carbon and silicon) also form compounds with hydrogen, which are germane, stannane and plumbane respectively. (The prefixes stann- and plumb- are from *stannum* and *plumbum*, Latin for tin and lead.)

One other element is of importance in this respect. Boron combines with hydrogen atoms to form boranes. A single boron atom should combine with three hydrogen atoms to form monoborane, or simply borane, but this has not been isolated. Chemists have, however, obtained diborane, with two boron atoms and six hydrogen atoms. Higher analogues with molecules containing up to twelve boron atoms (dodecaborane) are also formed. Boranes became important in the 1950s when they were added to rocket fuels to give them greater thrust.

Borazon

THE HARDEST SUBSTANCE KNOWN is diamond, which is made up exclusively of carbon atoms. The small carbon atoms can come unusually close to each other, so the attraction between them is great. Each carbon atom has four electrons in its outermost shell, which can hold eight. Each therefore shares two electrons with each of four neighbours, so that each has a share in eight electrons altogether. No other naturally occurring substance can form so many strong bonds in so many different symmetrically placed directions. (Carbon atoms can form bonds in an asymmetric pattern, as in graphite or coal, and the substance is then not particularly hard.)

A boron atom has one electron less than carbon, and has only three in its outermost shell. A nitrogen atom, with one electron more than carbon, has five. Both are small atoms, and when boron, in elemental form, has each atom bonded in three directions, it is pretty hard, though certainly not as hard as diamond. Nitrogen atoms combine in pairs and no more, so that the element is gaseous.

But imagine equal numbers of boron and nitrogen atoms bonded in alternation. A boron and nitrogen atom together would have a total of eight electrons in the outermost shells, just as a pair of carbon atoms would, and there should be similarities. Boron nitride, with molecules consisting of an atom of each element (BN), is indeed something like graphite in its properties. Graphite, under huge pressures, shifts to a symmetrical atomic arrangement and becomes diamond. What of boron nitride? At a pressure of 65 000 atmospheres and a temperature of 1500°C (a combination not available till the 1950s), the symmetrical form is produced—a kind of boron nitride as hard as diamond. It is called borazon (the second part of the name from azote, an old name for nitrogen). Borazon has the advantage over diamond of being much more resistant to heat.

Botulism

A VERY COMMON BACTERIUM is *Clostridium botulinum*. *Clostridium* is Latin for little spindle, which describes the bacterium's shape, and *botulinum* is from the Latin word for sausage, where it is sometimes detected.

C. botulinum is anaerobic (from Greek words meaning no air), since it can only live in the absence of oxygen. Under unfavourable living conditions, like many other bacteria, it forms a hard pellicle about itself and becomes a spore. Within the spore life ebbs low, and it can survive extreme conditions without actually dying. It remains ready to become actively alive again when conditions improve.

In canning food, or in making preserves, the product must be boiled after sealing. To make sure that the hardy spores are killed, the boiling must be continued for at least half an hour. Ordinary bacteria cannot live inside the tin, where oxygen is lacking, but *C. botulinum* can, for it requires no oxygen. If its spores survive they become active, grow and liberate botulinum toxin, which is the most poisonous substance known. (An ounce of the toxin, properly distributed, would be enough to kill every human being on earth.)

If food containing the toxin is eaten the latter is very slowly absorbed and affects certain nerves leading to muscles. There is a selective muscle paralysis affecting the eyes and throat first, so that those suffering from botulism find it difficult to focus their eyes and to speak. Then the chest muscles are paralysed, and it is the inability to breathe that kills.

Every once in a while an imperfectly heated batch of tins will lead to cases of botulism. Even one case is sufficient to set in motion a thorough search for any other possibly infected tins, so dreaded is the disease.

Bremsstrahlung

IN 1895 the German physicist Wilhelm Röntgen discovered a new kind of radiation so energetic it could pass through glass or cardboard. He did not know its nature, so he called it X-rays. The radiation turned out to be like ultra-violet radiation, but with smaller and more energetic waves.

Röntgen had obtained his X-rays through the action of cathode rays (speeding electrons) as they collided with the glass of the tube within which they were produced. What if the cathode rays were allowed to strike a piece of metal instead? Would the situation change? Metal pieces were inserted into the cathode ray tube, and when the stream of electrons struck X-rays were produced in greater quantity and possessing greater energy.

It was found by the English physicist Charles Barkla, in 1911, that the energy content of the X-rays depended on the nature of the metal used to stop the electrons. The English physicist Henry Moseley went on to deduce from those energies the structure of the atomic nuclei of the various metals. From this in 1914 he proceeded to work out the concept of the atomic number, which brought final order to the list of elements.

But how do the X-rays originate? The speeding electrons have a great deal of energy of motion. When they collide with some substance with which they interact they decelerate rapidly and lose this energy.

Energy, however, cannot be destroyed. If the electrons lose energy of motion this energy must reappear in some other form. As it happens, it appears in the form of radiation. The Germans, who first studied this, called the radiation *Bremsstrahlung* (deceleration radiation) and the phrase was accepted by English-speaking physicists without translation. X-rays, then, are a form of *Bremsstrahlung*.

Bromine

FEW CHEMICAL substances are named as the result of the sense of smell; very few in comparison with the names resulting from their colour. However, there are some familiar cases.

For instance, in 1824 a young French chemist Antoine-Jérôme Balard was studying some of the crystalline material obtained from the brine of a salt marsh. He noticed that when certain chemicals were added to a solution of this brine a brownish colour appeared. He investigated this, and found a new element, one of the few that is liquid at ordinary temperature. It is a deep red in colour, and has a strong odour, something like that of chlorine or iodine but stronger.

Balard suggested it be called muride, from the Latin *muria* (brine), but this did not gain favour. Instead the element was named bromine, from the Greek *bromos* (stench), because of its odour. (Actually, there are chemical compounds that smell infinitely worse than bromine, so it has always seemed unfair to me to pick on the poor element in this way.)

In 1839, to take another case, the German chemist Christian Schönbein discovered a gas which turned out to be a variety of oxygen. Whereas ordinary oxygen had a molecule made up of two oxygen atoms, the new gas had one made up of three oxygen atoms. The three-atom gas has a marked odour (something like that of weak bromine) and Schönbein named it ozone, from the Greek verb *ozein,* meaning to smell.

But prior to both those discoveries, the same situation had arisen with the same result. The British chemist Smithson Tennant in 1803 found when working with crude platinum that after the platinum had been dissolved in a mixture of acids a black metallic powder remained behind. Investigating that powder, he found two new elements. One of them formed a compound with oxygen that even in small quantities had a strong smell like that of chlorine. So he named that element osmium, from the Greek noun *osme*, meaning a smell.

Brownian Motion

IN 1827 the Scottish botanist Robert Brown was viewing a suspension of pollen grains under a microscope. He noted that the individual grains were moving about irregularly, even though the water in which they were suspended was quiet. Since the pollen grains had the potentiality of life within them, Brown thought at first the motion was a manifestation of this life. Then, however, he viewed a suspension of dye particles in water. These particles were definitely non-living and yet they moved about randomly just as the pollen grains had done.

Brown had no explanation for it, but because he reported it the phenomenon has been known as Brownian motion ever since.

In the 1860s the Scottish physicist J. Clerk Maxwell worked out the rules governing the behaviour of gases on the assumption that they consisted of small particles (atoms or molecules) in random motion. The analysis was very convincing, and it seemed that liquids too would consist of particles in random motion. A grain of pollen or a piece of dye would be bombarded on all sides by randomly moving water molecules, and since the force might be slightly greater from one side than another, the suspended grains would be moved first this way, then that.

The amount by which suspended grains of a certain size would move this way and that would depend in part on the size of the bombarding water molecules. In 1905 Albert Einstein worked out an equation describing this; and in 1908 the French physicist Jean Perrin set about making the necessary observations to determine the size of the values in Einstein's equation.

He succeeded, and was able to calculate the size of the water molecules. This was the first direct determination of molecular size, and was the final proof that atoms and molecules did indeed exist, and were not simply convenient chemical fictions.

Bubble Chamber

UNTIL WELL AFTER THE SECOND WORLD WAR, the best way of following the tracks of sub-atomic particles was to use the cloud chamber invented by Charles Wilson in 1911. It contained air saturated with water vapour. Particles, speeding through, chipped electrons away from the atoms they hit, forming electrically charged ions. Around these ions tiny water droplets formed, making a visible track from which much could be determined concerning the particles.

But air contains relatively few atoms. Particles scored hits only occasionally and the track was therefore thin and light. Events that took place very quickly were easily missed and the fine details of even comparatively slow events weren't clear.

Liquids are denser than gases. A speeding particle would strike many more atoms in a liquid than in a gas. But how to use a liquid? In 1952 the American physicist Donald A. Glaser, while talking over a glass of beer, began to watch the bubbles forming in the beer. It occurred to him that instead of following water droplets in air you could as easily follow gas bubbles in liquid.

Suppose you heated a liquid to the point where it was about to boil and put it under just enough pressure to keep it from boiling. A sub-atomic particle, speeding through the liquid, would produce ions, about each of which a tiny bit of boiling would take place. For an instant there would be a visible wake of bubbles left by the particles and this could be photographed. Before 1952 was over Glaser had built the first bubble chamber.

He used ether as the liquid at first, then switched to liquid hydrogen at extremely low temperatures. At low temperatures efficiency was greater and the tracks clearer. With bubble chambers very rapid particle interactions could be observed in great detail.

Caffeine

THERE IS A PLANT, native to Ethiopia, the seeds of which when roasted, ground and soaked in boiling water make an immensely popular drink. Its name is derived, according to one theory, from the Ethiopian province in which it may have been first grown, the province of Kaffa.

The drink spread from Ethiopia to Arabia, where it quickly grew popular. Islam forbade the drinking of intoxicating liquors, and while this was all very well, some sort of stimulating drink becomes popular with people despite all sorts of laws. The Ethiopian seed plus water offered a drink that, though not intoxicating, was stimulating enough to be an acceptable substitute for wine to the Arabs. So according to a second theory, it derives its name from an Arabic word for wine, *qahwah*.

From Arabia the drink spread through Europe in the 1600s, and there the French called it *café* and the English coffee. Little places which would specialize in selling coffee and other things to eat or drink on the side became so popular that café and cafeteria (the latter originally the Spanish word for coffee-house) are words for kinds of restaurants now.

The German chemist F. Runge in 1820 isolated from coffee seeds (or beans, as they are usually inaccurately called because of their shape) an alkaloid which was the stimulating ingredient of coffee. He named it, very naturally, caffeine.

The Chinese flavoured their boiled water with the leaves of a shrub they called *ch'a* or *tse* or *te* (depending on the province). This also swept Europe, later than coffee did, and won its outstanding victory in England, where it completely replaced coffee. The Chinese drink is called *chai* in Russian, *Thee* or *Tee* in German, *thé* in French, *Thee* in Dutch, and *tea* in English. Tea-leaves too contain caffeine, but when first isolated the caffeine from tea was thought to be a distinct substance, and named theine.

Caffeine is also found in seeds of a Brazilian shrub, named the guarana, used to make stimulating drinks, and the caffeine therein was named guaranine.

Calciferol

INFANTS SOMETIMES develop bones that are soft, and deform easily under weight or under muscle-pull, so that bandiness or curved spines or misshapen skulls result. This is called rickets or rachitis, possibly from the Greek *rhachis* (spine). A misshapen, soft skeleton results in weakness, and the common adjective rickety, meaning ready to fall, comes from the name of this disease.

In 1918 the English physiologist Edward Mellanby discovered a substance that when fed to infants seemed to prevent the development of rickets. This antirachitic factor was named vitamin D (see VITAMIN), in 1922 by the American biochemist Elmer Verner McCollum.

By 1935 the molecular structure of vitamin D was worked out, and it was found to resemble that of certain sterols (see CHOLESTEROL). In fact if these sterols are exposed to ultra-violet light the necessary change into vitamin D takes place. The

Latin word for to shed light upon is *irradiare,* from *radius* (ray). *Radiare* means to shed light and the prefix *in-* which means on has become here *ir-*. And so foods containing the necessary sterols may be irradiated until they contain the vitamin instead. The human skin contains some of the necessary sterols, and sunlight contains enough ultra-violet to irradiate us effectively. For that reason vitamin D is sometimes called the sunshine vitamin.

The action of vitamin D is to encourage in some way the transfer of the calcium ion floating freely in blood to the bone, where it is firmly bound into a crystal. The chemical name for the vitamin is therefore calciferol. The Latin *ferre* means to bear, so vitamin D is the calcium-bearer.

Different varieties of the vitamin, each about equally effective, may be formed from different sterols. From ergosterol (a sterol occuring in a type of fungus called ergot), ergocalciferol (vitamin D_2) is formed; and from 7-dehydrocholesterol (the sterol in our skins) cholecalciferol (vitamin D_3) is formed. (Vitamin D_1 does not exist. The name was originally given to a preparation that turned out to be a mixture of different substances.)

Calcitonin

A HORMONE generally has a powerful effect on some aspect of body chemistry, even when it is present in only tiny quantities. The presence of insulin, for instance, causes the blood content of glucose to go down. Since there is a desirable level for blood glucose, with too low a level being as bad as one too high, one doesn't want insulin to be present in too great a quantity. Feedback is therefore used. If the level of blood glucose drops insulin production is inhibited. As the insulin vanishes the level of blood glucose rises, and insulin is produced again.

Feedback does best when there are two hormones, working in opposite directions. Thus a second hormone, called glucagon, is also formed by the same gland that forms insulin. Glucagon's action is the reverse of insulin's.

Such double action from opposite directions was searched for elsewhere. The parathyroid glands produce a hormone (called parathormone after its organ of origin) which raises the level of calcium ions in the blood. Was there an opposition hormone to that?

In 1963 the Canadian physiologist D. Harold Copp worked with glands from 30 000 pigs and managed to extract about a tenth of a gramme of a substance that did indeed have an effect opposed to that of parathormone. It lowered the level of calcium ions in the blood. Copp named it calcitonin because it participated in regulating the tone (that is, the concentration) of calcium in the blood.

At first it was assumed that the calcitonin was formed in the same gland as its opposite number (parathormone), just as was the case for those other two opposites, insulin and glucagon. This proved not to be so. By 1967 it was clear that calcitonin was formed not in the parathyroid gland, but in the near-by thyroid gland. To emphasize this, the hormone is sometimes called thyrocalcitonin.

Calcium

ONE OF THE LATIN WORDS for stone is *calx* (genitive form, *calcis*), and this was applied particularly to a certain common type of stone which in English we call chalk (from *calx,* you see). In its most beautiful form this stone is called marble, from the Greek *marmaros,* which means sparkling stone. And it is true that much of the beauty of marble comes from a sort of fine-grained sparkle.

The Anglo-Saxon name for the material, by the way, is limestone. When it is heated it gives off carbon dioxide, and what is left is called lime. However, when in 1808 the British chemist Sir Humphry Davy first prepared a new element out of lime he went back to the Latin for the name. He added the -ium suffix common for metals and named the new element calcium.

(Calcium is not the only element to get its name from a common stone. The Latin word for flint (itself of Anglo-Saxon derivation), a rock even more common than chalk, is *silex* (genitive form, *silicis*). Consequently, the early chemists called flint and similar rocks silica. Then when in 1824 the Swedish chemist Jöns J. Berzelius found a new element in silica he simply added the -on non-metal suffix and the result was silicon.)

But stones achieve fame in another direction also. The Latin language generally indicates smallness by adding -ul to the roots of words. Since *calcis* means stone, *calculus* is a small stone or pebble.

Pebbles were useful in solving arithmetical problems. The earliest mechanical device for solving such problems was the abacus (a Latin word from the Greek *abax* (genitive, *abakos*), meaning a board on which one solved problems). This consisted of pebbles strung on wires, or placed in grooves, in a wooden framework. By moving the pebbles properly, problems in arithmetic could be solved.

Hence to work out an arithmetical problem is to calculate. Furthermore, when in 1666 Isaac Newton devised new mathematical methods for solving problems that could not be handled by the old methods the new methods were eventually called calculus.

Calendar

THE EARLIEST way of measuring periods of time longer than the day involved the moon. It is impossible to avoid noticing how night by night the moon changes from a crescent to a full circle and then back to a crescent. (The very word crescent comes from the Latin *crescere,* meaning to increase. The final shrinking form of the moon should, strictly speaking, be called a decrescent but it isn't.)

To primitive peoples it would seem that periodically a new moon was created and we still call the early crescent moon a new moon. It was convenient to measure time by the number of new moons that had appeared in the sky. The period from one new moon to the next is just about 29½ days and that period is the original month. In English, at least, the connection between month and moon is obvious.

The early Romans kept time strictly by the moon. They even had a ritual whereby the high priest would watch every month for the first appearance of the new moon's crescent. Once it appeared he would officially proclaim the beginning of the new month. The first day of the month was therefore called the calends, from the Latin *calare* (to proclaim). It was easy for this word to spread its meaning over the entire month, and now we call a table of months, or a system for telling time by months, the calendar.

For periods less than a month it is convenient to use the half-moon. From new moon to half-moon is just under 7½ days. The same period of time exists from half-moon to full moon, from full moon back to half-moon; and from half-moon back to crescent.

Consequently the Babylonians divided the month into seven-day periods. The Jews picked up the habit during the Babylonian Captivity and spread it to the rest of the world via Christianity. The English word week for this seven-day period is from the Anglo-Saxon, but may hark back to an old Gothic word meaning change (of the moon, you see).

Calorie

BEFORE THE 1850s chemists and physicists thought that heat was some kind of separate substance in matter that would flow from a hotter body to a colder one. It would flow into water to make it steam, and out of burning coal to heat the air. This substance was called caloric, from the Latin *calor,* meaning heat.

Caloric ran into trouble, however, when the American-born scientist Benjamin Thompson (he was a Tory during the Revolution and had to emigrate, eventually getting a title and becoming Count Rumford) noticed in 1798 that drilling brass cannon produced quantities of heat. He reasoned that both drill and cannon were cold to begin with and contained little caloric. Where did the unlimited supply of caloric come from once drilling began? The caloric theory was never quite the same again.

In 1857 the German physicist Rudolf Clausius advanced the theory that heat was not a material substance but energy—the energy of vibrating molecules. The heat of drilling thus came from the mechanical energy involved in forcing the drill through the brass. This view has been accepted ever since.

But an echo of the dead caloric still remains. To measure the quantity of heat physicists decided to call the heat required to raise one gramme of water from a temperature of 14·5 °C to 15·5 °C one gramme-calorie or, simply, one calorie. (Since heat is a form of energy, it can also be measured in ergs and joules (see ENERGY). One calorie is equal to 4·185 joules or to 41850000 ergs.)

A larger unit is frequently used. It is the kilogramme-calorie, or kilocalorie, which is equal to 1000 calories. Unfortunately, it has become too customary to call the kilocalorie simply Calorie (with a capital C). The capital cannot be heard by the ear, and this is a source of great confusion. People who talk of the calorie content of various foods are almost always talking about the kilocalorie content. In scientific terminology, the word calorie is now replaced by joule.

Capillarity

MOLECULES ATTRACT neighbouring molecules. They will attract others like themselves, so that a piece of steel, for instance, doesn't fall apart into dust even under thundrous blows, and a drop of water may hang from a tap for quite a while before dropping off. This is cohesion, from the Latin co- (together) and haerere (to stick). A substance will stick together.

Molecules will also be attracted to other molecules unlike themselves. Paint will stick to wood, mortar to brick, glue to almost anything. This is adhesion, the Latin prefix ad- meaning to. Paint sticks to wood.

Water in a tube will cohere. The water molecules near the tube itself will adhere to the glass or cellulose of the tube. The adherence is actually stronger than the coherence, so that where water meets tube the level curves up in order that as much water as possible be next the tube. In a narrow tube—say a centimetre in diameter—the upward curve of the water-level forms a crescent that is called a meniscus, from the Greek *meniskos* meaning a little moon. (If the tube were made of wax, water cohesion would be stronger than the water-wax adhesion and the meniscus would curve downward all round. The same is true for mercury in a glass tube, since mercury cohesion is unusually high for a liquid.)

If tubes are fine enough the adhesive forces are enough to pull the water-level a considerable way upward against gravity. Stick a blotter into water, for instance, and the water will soak upward through the fine spaces between the matted cellulose fibres of the blotter. Water will also soak upward through the very fine tubes leading up tree-trunks, and this is one of the mechanisms by which trees lift water hundreds of feet without a pumping mechanism like the animal heart. Because this works best in tubes that are as fine as hairs, or finer, this lifting of fluids is termed capillary action, or capillarity, from the Latin *capillus* (hair).

Carbonaceous Chondrites

UNTIL ROCKS AND SOIL were brought back from the moon in 1969 meteorites were the only matter known that was not originally part of earth.

Meteorites can be classified by their chemical structure. Some, for instance, are almost purely metallic, made up of a mixture of iron and nickel in proportions of about 9 to 1. Others are stony in nature, made up of minerals very like those making up the deeper sections of earth's rocky portion.

There are, however, other ways of classifying the meteorites. Of the stony meteorites, over 90 per cent include small, compact spheres of substance within them. These spheres are called chondrules, from a Greek word meaning grain of wheat, because they are frequently about the size of such grains. Meteorites containing these inclusions are called chondrites; those that don't are called achondrites.

The chondrites aren't ordinarily very different from the material surrounding them. They too are rocky in nature and resemble the kind of substances in earth rocks.

A small proportion of the chondrites, however, are black in colour, and the chondrules they contain have a considerable percentage of carbon. These are called carbonaceous chondrites in consequence, and they are by far the most fascinating of all the meteorites, since carbon is the basic element of life.

Can the carbonaceous chondrites represent traces of life stemming from another planet? In 1961 some American scientists actually reported finding shapes in the chondrules that might once have been part of life-forms. The furore soon died down, however, as it turned out the suspected shapes were contamination from the earth around them. Nevertheless, the presence of carbon itself remains intriguing.

Carborane

THE BORANES ARE COMPOUNDS of boron atoms and hydrogen atoms which came to be of increased interest in the 1950s because they were used as rocket fuel additives to give more thrust. As the boranes were investigated, the boron atoms in their molecules were found to be arranged in geometric patterns—in octahedrons and icosahedrons. An octahedron is a solid with eight faces and six vertices. The six boron atoms in hexaborane would be distributed, one at each vertex. An icosahedron has twenty faces and it has twelve vertices. The twelve boron atoms of dodecaborane are distributed one at each vertex.

The carbon atom is very like the boron atom in size, though it contains one additional electron. In 1963 it was found that one or more carbon atoms could substitute for boron atoms in these geometric patterns. This meant the discovery of an entirely new class of materials in which hydrogen atoms are bound to an interconnected web of boron and carbon atoms. These are the carboranes.

Even when for most of the boron atoms there are substituted carbon atoms, the borane structure is retained. The presence of the carbon atoms, however, gives the molecule much greater stability. Indeed, the more carbon atoms present the more stable the compound is to heat and the more inert to chemical reaction. There are uncounted numbers of possible carboranes, since one or more of the hydrogen atoms can be replaced by other atoms or groups of atoms.

As yet the compounds exist only in small quantities and are merely laboratory curiosities, but as knowledge concerning them increases, and as better methods of production are worked out, they will be formed in larger quantities. We may expect to find uses for them that will take advantage of their particular properties, so that whole new groups of polymers and plastics may become available.

Carcinoma

PERHAPS THE MOST frightening word in the medical vocabulary is cancer, and yet in Latin it is an innocent word meaning crab. We still use the word in its innocent sense as a sign of the zodiac (Cancer, the Crab; Tropic of Cancer. See SOLSTICE). The reason for the name is uncertain. It dates back to ancient times, but no one suggests that people connected the disease with the influence of the constellation (see INFLUENZA). Perhaps it's just that cancer clings to what it attacks as though by crabs' claws, or that it sometimes seems to take on a crab-like shape.

The word tumour is not synonymous with cancer, although the average man sometimes uses it in that way. A tumour is any abnormal growth on the body, from the Latin *tumere* (to swell). A tumour that is of limited growth, like a wart, may be unsightly, but it is no danger, so it is a benign tumour (from the Latin *bene*, meaning good, and *genus,* meaning sort; a good sort of tumour).

If, on the other hand, the tumour is of the type that grows without limit until it kills, it is a malignant tumour. Since the Latin *malus* is bad, this signifies a bad sort of tumour, and it is the malignant tumour that is cancer.

The most common types of cancer are those which strike the epithelial (Greek *epi,* upon, *thele,* nipple) cells, those which form the skin or line the alimentary canal. These are carcinomas, from the Greek *karkinos* (crab), so that, except for the switch in language, it is just another way of saying cancer.

The suffix -oma is used in modern medical terminology to mean a tumour—usually, but not always, a cancerous one. A cancer within the body attacking the bone or other connective tissue is a sarcoma, from the Greek *sarx* (flesh). An adenoma is a tumour of glandular tissue, and a hepatoma is cancer of the liver. (The Greek *aden* means gland and *hepar* means liver.)

Leukaemia is an exception. It is a condition in which the white cells of the blood multiply in cancerous fashion. The ending in this case is from the Greek *haima* (blood), while *leukos* means white and refers to the white cells (see HAEMOGLOBIN), so it means white cells in the blood.

Carnivore

DIET IS NOT necessarily a matter of choice. A cat, for instance, has teeth suitable for tearing food, but not for grinding it. It lacks our copious saliva with which to moisten food. It has a relatively short digestive tract, so that it can't indulge in prolonged handling of the food it eats. For all these reasons, the main items in its diet must be food that can be quickly digested even when swallowed in chunks. So it lives on other animals and is a meat-eater or carnivore, from the Latin *caro* (genitive *carnis*), meaning meat, and *vorare,* to eat or devour.

Although there are carnivorous birds, fish, insects and even plants, the most familiar carnivores are mammals, and many of these are included in a mammalian order entitled Carnivora.

Only one other mammalian order (the one including shrews and moles) is named after its feeding habits, and it involves meat-eaters too. The meat, however, is specialized, consisting mostly of insects (although the mole eats a great many earthworms). The order is therefore called Insectivora, or insect-eaters.

There are also a number of familiar animals with teeth designed to grind coarse food, and with long, complicated digestive systems to handle that food. (Coarse food like grass may be hard to digest, but at least it doesn't have to be chased, and doesn't fight back.) The grass-eaters, such as cattle, sheep, and horses, are herbivores, from the Latin *herba* (grass). These are spread over a number of mammalian orders.

Then there are specialized herbivores. Although many bats eat insects, some of the large ones, notably the flying foxes (so called because they have heads like those of small foxes), are fruit bats (so called because they live on fruits). Fruit-eaters are frugivores, from the Latin *frux* (fruit).

Of course, there are creatures who can eat almost anything, plant or animal. Such animals are generally successful in life, and include bears, pigs, rats, crows and, most particularly, man. These are the omnivores, from the Latin *omnis* (all); they eat all.

Carotid

THE NAMES OF arteries and veins usually come from the Greek or Latin names of the organs they serve. For instance, the hepatic artery leads blood to the liver, the Greek word for liver being *hepar.* In the same way the renal artery leads blood to the kidney and the pulmonary artery leads it to the lungs. The Latin word for kidney is *ren* and for lung is *pulmo* (genitive, *pulmonis*).

There are exceptions, though. There is the jugular vein, which passes through the neck. It is named, not after the organ it serves, but after the part of the body through which it passes. The Latin *jugulum* means collar-bone or, more generally, throat.

The arteries that serve the heart surround and encircle it like a crown. The Latin word for crown is *corona,* so the arteries to the heart are the coronary arteries. Any interference with those arteries means quick and painful trouble, since the heart can't stop working and must have nourishment. If a clot forms in the arteries, cutting off some of the blood-supply, the result is one form of heart attack. The Greek word for a clot is *thrombos,* so such a heart attack is called a coronary thrombosis. For this reason the word coronary by itself is a popular term for heart attack.

The brain, like the heart, is sensitive to lack of blood, but this shows up in another way. If the supply of blood to the brain is cut down the result is not pain, then death; but sleep, then death. Sideshow performers in ancient Greece, for instance, used to amaze their audiences by pressing a spot on a goat's neck (pinching off the artery leading blood to the brain) and causing it to go to sleep. Releasing the pressure would allow it to wake again. The Greek word for stupefy is *karoun,* and the arteries leading to the brain are therefore called the carotid arteries.

Catalysis

THE ANCIENT philosophers speculated on the possible existence of some substance which by its mere presence, and without being used up in the process, could change base metals into gold. This was called the philosopher's stone, and, of course, does not exist. However, other and infinitely more valuable philosopher's stones have been discovered.

As early as 1750, sulphuric acid was being formed in large quantity from sulphur dioxide and water by the use of nitrogen oxides which somehow made the appropriate chemical changes take place without being used up in the process. (And sulphuric acid is incomparably more valuable, though much less costly, than gold.)

In 1823 the German chemist Johann Wolfgang Döbereiner invented a kind of lighter in which a jet of hydrogen was directed on a piece of platinum. The hydrogen at once caught fire, because the platinum brought about its combination with the oxygen in the air without itself being used up.

Ordinary substances, too—simple acids, for instance—could bring about changes in this way, as in breaking down starch to sugar.

In 1836 the Swedish chemist Jöns Jakob Berzelius reviewed the whole subject. Though he could not explain it, he suggested a name. He called the philosopher's stone process catalysis, from the Greek *katalysis* meaning dissolution or destruction (*kata,* down, and *lysis,* breaking). The substance that brought about the catalysis was the catalyst, which by its mere presence destroyed sulphur dioxide, or hydrogen or starch.

We have since learned that it is more than a case of mere presence; that catalysts take part in the reactions they catalyse, but are reformed before the reaction is over, so that they seem not to have changed. The chemical industry today depends almost entirely on the use of appropriate catalysts, and so does all living tissue, including our own (see ENZYME).

Celluloid

ALTHOUGH BILLIARD-ROOMS are frequently pictured as places where idlers waste their time, and from which little good can be expected, the game of billiards was nevertheless the cause of an important chemical discovery. The billiard ball, it seems, must be hard, elastic, uniform in composition, and take a high polish. The ideal material for the purpose is ivory, which is the substance forming the tusks of elephants (and some other animals). These are a pair of enormously large teeth.

Now, there are things in this world easier to procure than elephants' teeth, and in the 1860s a prize was offered for the discovery of an adequate ivory-substitute for billiard balls. An Englishman, A. Parkes, had already discovered that if camphor is added to pyroxylin (see RAYON), the mixture becomes plastic; that is, it can be moulded into any desired shape. Camphor is an example, therefore, of a plasticizer.

The American inventor John Wesley Hyatt studied the camphor-pyroxylin product (which the British called Xylonite, from pyroxylin) and worked out practical mechanical devices for turning out Xylonite billiard balls, winning the prize in 1870. Hyatt called the material celluloid, since pyroxylin is derived from cellulose and the -oid suffix is from the Greek -oeides (having the form of). And thus billiards was responsible for the commercialization of the first artificial plastic.

The biggest shortcoming of celluloid was its inflammability. What was really needed for most purposes was a hard plastic that was also stable. The first thermosetting plastic (see POLYMER), one which after all these years can scarcely be improved on for hardness and stability, was formed by the polymerization of phenol and formaldehyde. This was accomplished in 1906 by the Belgian-born American chemist Leo Hendrik Baekeland who, more self-assertive than Hyatt, named his product after himself. He called it Bakelite.

Celsius

IN GENERAL, substances expand slightly when heated, and shrink when cooled. This fact gave mankind its first tool for measuring temperature accurately—the mercury thermometer. Thermometer comes from the Greek words *therme* (heat) and *metron* (measure); it is an instrument to measure heat.

The mercury thermometer was invented in 1714 by the German physicist Gabriel Daniel Fahrenheit, who filled a hollow bulb with mercury and allowed it to expand when heated up a very fine enclosed and evacuated tube. The amount by which the mercury thread crept up the tube was proportional to the temperature. (The glass also expanded, but nowhere near as much.)

Fahrenheit placed the mercury-filled bulb in a mixture of equal parts of salt and snow at the melting-point, and marked the height of the mercury column as 0. He next let it warm to the temperature of the human body, and marked the new height as 100. (The man he used must have been slightly feverish, or else Fahrenheit adjusted the level to allow the freezing-point and boiling-point of water to come out at whole values.) By drawing a hundred equal divisions between the two marks, he invented the Fahrenheit scale. On this scale the melting-point of pure ice is 32 degrees and that of the boiling-point of pure water is 212 degrees. The word degree comes from the Latin *de-* (down) and *gradus* (step). In marking off small divisions from 100 to 0 you go down steps.

In 1742 the Swedish astronomer Anders Celsius suggested that the temperature of melting ice be set at 100 degrees and that of boiling water at 0 degrees. (The 0 and 100 were later reversed.) This hundred-degree interval gave the new scale the name Centigrade, from the Latin *centum* (hundred) and *gradus* (step). It is the scale of a hundred steps from melting to boiling. It is now called the Celsius scale after the inventor.

The Fahrenheit scale is still sometimes used among the general public in England and America, but it is the Celsius scale that is used among scientists everywhere.

Centipede

MAN MUST early have realized that one outstanding difference between himself and the other animals was that he walked on two legs, the others on four. The various mammals and reptiles (but especially mammals) were therefore lumped together as quadrupeds, from the Latin *quattuor* (four) and *pes* (foot)—the four-footed ones.

Birds also had four limbs, but two were used as wings and only two as legs. They were bipeds, from the Latin *bis* (twice)—the two-footed ones. (Hence the inadequate ancient definition of man as *bipes implume*, a two-legged thing without feathers.) Man, of course (also the kangaroo and the frog), is a biped.

The insects, one and all, are blessed with six legs. This is taken note of, and the class name Insecta is sometimes referred to (but much less frequently) as Hexapoda, from the Greek *hex* (six)—the six-footed ones. (The Greek word for foot is *pous,* genitive, *podos.*)

Spiders are not insects, despite the popular misconception. Spiders lack wings, have no metamorphosis or change of form, two main body divisions rather than three, and, in common with their relatives the scorpions and king crabs, eight legs. Nevertheless, the name Octopoda is reserved for an order of sea creatures that don't have feet at all, but do have eight tentacles. (The word tentacle comes from the Latin *tentere,* meaning to touch.) The most common of these is the octopus. In both Latin and Greek *octo* means eight, so the octopus is the eight-footed one.

And there are crawling creatures with many more than eight legs. They are the centipedes and the millipedes, who were once grouped under the name Myriapoda. The Latin *centum* means a hundred and *mille* means a thousand, while the Greek *myrios* means ten thousand, so you get the idea. Both centipedes and millipedes are composed of many body segments. From each segment in the centipede are produced one pair of legs; from each segment in the millipede, two pairs of legs. The millipedes are therefore put in the class Diplopoda, from the Greek *diploos* (double), hence the double-footed ones, while the centipedes are the Chilopoda, from the Greek *chilias,* a thousand. The names are arbitrary of course: millipedes have nothing like a thousand legs and centipedes may have more or less than a hundred according to species.

Cepheid

IN THE GREEK MYTHS Cepheus was a king of Ethiopia whose daughter, Andromeda, was saved from a sea monster by the hero Perseus. The Greeks pictured the stars in constellations representing mythical characters and one of these was labelled Cepheus. The brighter stars in each constellation are named in order of brightness by Greek letters. The fourth letter of the Greek alphabet is delta, so the fourth brightest star of the constellation Cepheus is Delta Cephei.

77

Delta Cephei was found to be an unusual star in that its brightness varied regularly (it is a variable star). The star brightens and dims in a fixed pattern every 5·37 days. Eventually, other stars were found that brightened and dimmed in similar fashion, but with periods of anywhere from two to forty-five days. All variable stars with this sort of pattern were lumped together as Cepheids.

In 1912 Cepheids took on an unexpected importance, when the American astronomer Henrietta S. Leavitt was able to show that the period of variation was closely related to the amount of light given off by the star (its luminosity).

This meant that from its period alone, one could tell how luminous a Cepheid was. Then, from its apparent brightness, one could tell how far away it must be. It was by studying Cepheids that the American astronomer Harlow Shapley was first able to show the size of our galaxy—to demonstrate that it had to be 100 000 light-years across. He was also able to show that the solar system was nowhere near the centre of the Galaxy (as had been earlier thought) but was far out on the rim.

By using the Cepheid yardstick, it also became possible, for the first time, to measure the distance of some objects beyond the Galaxy. The Magellanic Clouds, for instance, were found to be about 150 000 light-years from us.

Cerebrum

UNTIL MODERN TIMES, the brain was little regarded (see PITUITARY). In fact to the ancients the most remarkable thing about heads were horns, and if the brain got a name it was apt to come from the word for horn, since the brain was something that was in the same part of the animal as were its horns (if it had any).

The word horn itself is related to the Latin word for it, *cornu.* (We keep the older pronunciation, by the way, when we speak of a horny callous on a toe as a corn.) The German word for brain is *Hirn,* and the similarity to horn is obvious. The similarity of brain itself is less obvious, and may not exist. An alternative theory is that it comes from the Greek *bregma,* meaning the top portion of the head, or from a common ancestor.

However, we are not through with horn. The Greek word for it is *keras,* and when the Romans called the chief portion of the brain the cerebrum (remember that in Latin c is always pronounced k) the relationship to horn seems clear. Behind and under the cerebrum is a smaller structure called the cerebellum. This is a Latin diminutive of cerebrum, so that it means literally little brain. Both cerebrum and cerebellum have come into English unchanged.

A word sometimes used to mean the whole brain is encephalon, which comes from the Greek *en-* (in) and *kephale* (head), so that it simply means that which is in the head. The word is better known in connection with virus infections leading to inflammations of the brain. This is called encephalitis, the Greek suffix *-itis* being commonly used in medical terms to mean inflammation.

A particular variety of encephalitis, which brings on a coma (from the Greek *koma,* meaning sleep) is encephalitis lethargica (lethargica is the feminine form of the Latin word meaning drowsy, which in turn comes from the Greek *lethe,* meaning forgetfulness, and *argos,* meaning idle). In everyday English the disease is called sleepy sickness (*not* sleeping sickness, which is caused by a protozoon parasite).

Cerenkov Radiation

THE SPEED OF LIGHT in a vacuum (299 792·5 kilometres per second) is considered the limit of possible velocity for any particle possessing mass. However, when light passes through some medium other than a vacuum it moves more slowly. Through water light travels at a velocity of only 224 140 kilometres per second; through glass at only 177 000 kilometres per second; and through diamond at only 123 970 kilometres per second.

It is possible for sub-atomic particles, as they move along at nearly the speed of light in a vacuum, to smash through water or glass at nearly their speed in air. Their speed in water or glass is therefore far greater than the speed of light in water or glass.

As the particles race through the medium they slow up somewhat and the energy they lose appears as radiation. A bluish light is emitted, but it cannot keep up with the speeding particles. It trails off behind like the wake behind a motor-boat, and the angle it makes with the line of motion of the particle depends on how much faster than the speed of light the particle is going.

The first to observe this blue light emitted by such very fast particles was a Soviet physicist named Pavel Cerenkov, who reported it in 1934. The light is therefore called Cerenkov radiation. In 1937 two other Soviet physicists, Ilya Frank and Igor Tamm, explained the existence of this light by pointing out the faster-than-light aspect of the particle's motion.

Special instruments, Cerenkov counters, have been designed to detect such radiation, measure its intensity and the direction in which it is given off. These are particularly useful in studying very energetic cosmic ray particles which move within eighty kilometres per second of the speed of light in a vacuum. Such

79

particles move faster than light does in air, so that they produce Cerenkov radiation even in air.

Cermet

AMONG THE FIRST MATERIALS used by mankind were certain earthy substances such as clay. If this was baked hard in a fire the result was something which was hard, insoluble in water and resistant to fire. Clay could be used to make pottery and bricks.

The Greeks called such baked clay *keramos*. As a result substances such as clay, which are hard, insoluble in water and resistant to heat are called ceramics.

Most ceramics are oxides, combinations of other elements with oxygen. Thus we may have silicon dioxide, aluminium oxide, chromium oxide, magnesium oxide and so on. Silicon dioxide (the most common) is often found in combination with one or more of the others, forming what are called silicates. Clay is an aluminum silicate. Glass, the various glazes and porcelain are examples of other silicates. (Glass is not sufficiently resistant to heat to be a good ceramic, but it has the virtue of being transparent.)

The greatest shortcoming of ceramics generally is that they are brittle. They will not bend without cracking, and a sharp blow will break them altogether. Once metals were discovered, therefore, they replaced stone and ceramics where toughness was needed. Metal knives could hold a sharper edge than could flint, and wouldn't blunt as quickly with use. Metal pots could be dropped without breaking. Metal objects could be bent and shaped in a variety of ways.

However, metal would rust, be damaged by water, and be more affected by heat than would ceramics. In recent years techniques have been devised to mix ceramic and metallic powders and compact them together under heat to form a kind of combination substance with a collection of virtues, as heat resistant as ceramics and as tough as metals. This combination is a ceramic-metal, or, in abbreviated form, a cermet.

Chiroptera

WHEN CREATURES have wings the Greek *pteron* (feather) and *pteryx* (wing) are apt to show up in their scientific names. This is true of reptiles (see PTERODACTYL) and insects (see LEPIDOPTERA). It is also true of mammals.

The one group of true flying mammals comprises, of course, the bats. This word can be traced back to an old Scandinavian word meaning to flutter, and the flight of the bat is indeed much more a fluttery thing than is the flight of birds. We still have this old meaning of bat when we bat our eyelids.

This notion is used most directly in connection with the additional fact that the most common English bats are just about the size and general appearance of a mouse (if you ignore the wings), and a very descriptive alternate name for the creature is flittermouse. The German for bat, similarly, is *Fledermaus,* as all operetta lovers will know.

It is interesting that the animal kingdom has developed wings four different times, and each time in a different style. First the insects developed ordinary membranes with no bony stiffening. Then reptiles such as the pterodactyl developed membranous wings with a single finger bone as stiffening, the others remaining outside as separate claws. Then birds developed feather-covered wings with all finger bones, fused together, as stiffening. Finally, the bat has a naked wing strung along four enormously lengthened fingers, so that only its wing of all those developed can flex like a hand. The order that includes the bats has therefore been given the name of Chiroptera, from the Greek *cheir* (hand) and *pteryx* (wing). They are the hand-winged creatures.

In the case of birds, however, there is little mention of wings in the scientific names. Two examples come to mind, though. There is an extinct bird, halfway from the lizard, called the Archaeopteryx, from the Greek *archaios* (ancient); it is the ancient winged creature. And there is the New Zealand kiwi (so called from the sound it utters). It is the one bird that is completely without exterior wings, so it is called Apteryx or no-wings.

Chlorine

THE GERMAN CHEMIST Johann Rudolf Glauber treated ordinary salt with sulphuric acid in 1658 and got a solution which gave off a choking vapour or 'spirit' (see GAS). He called the new substance spirit of salt.

The substance had acid properties in solution, and since it was prepared from salt, and salt was most easily prepared from seawater, the new substance was eventually named marine acid or muriatic acid. (The Latin word *mare* means sea, while *muria* means brine.)

Now, the so-called muriatic acid happens to be the most common example of an acid that does not contain oxygen. But in the

81

late 1700s there was a theory that all acids must contain oxygen, and, in fact, that is how oxygen received its name (see OXYGEN). It seemed certain, therefore, that the muriatic acid molecule must be made up of atoms of oxygen and of some unknown element which was named murium.

In 1774 the Swedish chemist Karl Wilhelm Scheele treated muriatic acid with manganese dioxide and obtained a greenish, chemically active gas with an unpleasant odour. He didn't realize he had a new element because the false theory about acids misled him. He considered it just muriatic acid with additional oxygen (from the manganese dioxide). In fact, the French chemist Count Claude Louis de Berthollet suggested in 1785 that the greenish gas be called oxymuriatic acid, and others suggested the name murium oxide.

It wasn't until 1810 that Davy (see POTASSIUM) first realized the truth and got the credit of finding a new element. He could find no oxygen in 'murium oxide', and came to the daring conclusion that there was no oxygen in muriatic acid in the first place. He decided that the greenish gas was a new element, and he overthrew all the previous misleading names and started afresh by calling it chlorine, from the Greek *chloros,* meaning green.

As for muriatic acid, its molecule is made up of one atom of hydrogen and one of chlorine. It is now commonly known as hydrochloric acid.

Chloroform

SCIENTISTS OBTAIN new chemicals in odd places. For instance, the English naturalist John Ray, back in the seventeenth century, cooked a batch of red ants and obtained a liquid which he (or someone after him) called formic acid, from the Latin *formica,* meaning ant. At that, it was a good name, because the pain of the bite of the red ant is due to a small quantity of formic acid which it injects into the flesh as it bites. Formic acid is strong enough as an acid to sting badly. It is also, by the way, supposed to occur in nettles, which explains a good deal.

The molecule of formic acid is made up of a single carbon atom to which are attached (1) a hydrogen atom, (2) an oxygen atom and (3) the hydrogen-oxygen combination called a hydroxyl group. Minor changes in these attachments result in other compounds which often retain the stem 'form' in their names. (It often happens in chemical terminology that one substance names another until the original derivation is lost sight of.)

For instance, replace the hydroxyl group with a hydrogen atom and the resulting molecule contains the aldehyde combination of elements (see KETONE). This particular aldehyde would, naturally, be formaldehyde. A 40 per cent solution of formaldehyde (which is a gas in the pure state) in water is formalin, and it is formalin which is used to preserve tissues against decay in zoology and anatomy laboratories. Students of either subject are well acquainted with the odour.

Again, if the hydroxyl group is replaced by a chlorine atom and the oxygen atom is replaced by two more chlorine atoms, the result is a kind of chlorinated formic acid. The name of the compound is rather in the nature of an abbreviation of this, for it is chloroform—a familiar name to most people, yet only a chemist would see the connection with ants.

Naturally, if bromine or iodine atoms are present instead of chlorine (and they can be), the results are, respectively, bromoform and iodoform.

Chlorophyll

USUALLY WHEN the prefix chlor- appears in the name of a chemical substance it is a pretty good sign that the molecule contains one or more atoms of chlorine (see CHLOROFORM). However, chlorine itself was so named for its green colour (see CHLORINE), and sometimes the prefix refers only to the colour of a substance and not to its chemical make-up.

The most important green in nature is, of course, the green of plants. The substance responsible for this colour was first isolated in 1817 by the French chemists Pierre Pelletier and Joseph Caventou, who named it chlorophyll, from the Greek *chloros* (green) and *phyllon* (leaf). Here the chloro- refers only to the colour, for there are no chlorine atoms in chlorophyll.

Chlorophyll acts to trap sunlight and convert carbon dioxide of the air into the food that all animals live on, so it is without comparison the most important coloured compound on earth. There are, however, other coloured compounds in plants which are infinitely less important and yet which we would hate to have to do without.

For instance, there is a type of coloured compound called anthocyanin that appears in flowers of many types. The name is derived from the Greek *anthos* (flower) and *kyanos* (blue). From this you would think it appears in blue flowers, and so it does in some, such as the delphinium. However, under the proper circumstances anthocyanins can be violet or red, and are responsible for the colour of the violet, dahlia, poppy and rose.

There are yellow pigments in plants, too. Some are called flavones (from the Latin *flavus,* meaning yellow) and some xanthones (from the Greek *xanthos,* also meaning yellow). However, the most common yellow-orange pigment in plants is carotene, so named because it was first isolated from carrots. It also occurs in many plant and animal fats, including that of man. Some peoples of East Asia have enough carotene in the fatty layers under the skin to cause people to speak of a Yellow Race.

Chloroplast

THE GREEN OF GREEN PLANTS is due to chlorophyll (from Greek words meaning green leaf). By means of chlorophyll the energy of sunlight is used to turn carbon dioxide and water into plant tissue and oxygen. Animal life, including ourselves, depends on those tissues and oxygen to eat and breathe.

In 1865 the German botanist Julius von Sachs showed chlorophyll in plant cells to be confined to small bodies in the cytoplasm. These small bodies were eventually named chloroplasts, the suffix coming from a Greek word referring to something with a definite form.

The interior of chloroplasts is divided by many thin membranes called lamellae, from a Latin word meaning little plates. The lamellae thicken and darken in places to form grana (from a Latin word meaning grains), and within each of these are 250 to 300 chlorophyll molecules.

Some chloroplasts are quite large, so large that there is only one to a cell. There are indications that they possess DNA, something characteristic of cell nuclei. Could it be that chloroplasts were once, in the dim past, independent organisms?

That this may be so is indicated by the fact that the very simplest plant cells do not have distinct chloroplasts or nuclei. The chlorophyll systems and the nuclear material are distributed throughout the cell. Such cells are called blue-green algae, where algae is the general name for unicellular plants living in water (from a Latin word for seaweed). Not all blue-green algae are blue-green in colour, but the first ones studied were.

Blue-green algae might almost be considered free-living chloroplasts. Bacteria very much resemble blue-green algae except for the lack of chlorophyll. Could these two classes of organisms represent the kind of life on earth before true cells with separate nuclei evolved?

Cholesterol

IN 1769 a French chemist, Poulletier de la Salle, working with gallstones (see BILE), obtained from them a white solid substance with a fatty feel. In 1815 another French chemist, Michel Eugène Chevreul, decided the substance was a kind of fat, and named it cholesterine. The word came from the Greek *chole* (gall) and *stear* (a hard fat).

In the early days of chemistry the suffix -ine or -in was a common one for the organic compounds in living tissue. As more and more came to be known about the chemical constitution of such compounds it became customary to reserve the -ine suffix for those that contained one or more nitrogen atoms in their molecules. Cholesterine did not have such nitrogen atoms. In 1859, moreover, still another French chemist, Pierre Berthelot, showed that cholesterine did have a hydroxyl group in its molecule, so that it was actually an alcohol (see ALCOHOL). The chemical names of alcohols have an -ol suffix, but it was not until 1900 or thereabouts that the switch was made and the compound was finally named cholesterol.

Slowly more facts were learned about its chemical structure. It wasn't until the 1930s that the last details were worked out, but by 1910 chemists were reasonably certain that the carbon atoms of cholesterol were arranged in a number of connected rings to which other carbon atoms (side chains) were attached. Other compounds were known by then which had the same ring system but slight differences in the side chains. By 1911 such compounds were given the general name sterols.

Then additional compounds were discovered with the same ring system, but lacking the hydroxyl group that made cholesterol an alcohol (though containing oxygen atoms in other combinations) and therefore not deserving of the -ol suffix. A still more general name was therefore invented in 1936: steroid, the suffix -oid being the usual Greek-derived ending meaning in the shape of. The ring combination found in all such compounds is now called the steroid nucleus.

Chromatography

ONE OF THE DIFFICULTIES in studying the chemistry of living tissue (a science called biochemistry, from the Greek *bios*, meaning life, hence life-chemistry) is that tissues consist of a mixture of a vast number of compounds, groups of which are so similar that they are almost impossible to separate. For instance, there are a number of very similar coloured compounds in leaves, and finding out anything about them was next to hopeless unless they could be separated, and separating them was almost hopeless, too.

In 1906, however, the Russian botanist Mikhail Tswett found the answer. He took the mixture of leaf pigments, dissolved it in petroleum ether, and poured the solution through a glass tube packed tightly with powdered limestone. The petroleum ether went through, but the pigment clung to the tiny limestone particles and remained behind (see ADSORPTION).

However, as he continued to pour fresh petroleum ether through the column the pigment was slowly washed downward. Each separate compound in the mixture was washed down at a slightly different rate, depending on how firmly it was attached to the limestone particle, and how easily it dissolved in petroleum ether. The result, as one particular pigment moved down quickly, another slowly and another still more slowly, was that separate bands appeared in the column. Each band contained one particular pigment separated from the rest, and each was its own shade of red, orange or yellow.

Tswett called this technique chromatography, from the Greek *chroma* (colour) and *graphein* (to write), because the solution to the mixture problem was 'written in colour' for everyone to see on the limestone column. It took twenty-five years for Tswett's discovery to be adopted by biochemists (no one seems to listen to Russian scientists), but today it is one of biochemistry's most powerful tools. Limestone has been replaced by more efficient powders, and very often simply by a sheet of absorbent paper. Furthermore, the technique is used mostly for colourless compounds now, but it is still called chromatography.

Chromium

SOME ELEMENTS are more apt to form coloured compounds than are others. A particular case in point is a silvery metal first isolated by the French chemist Nicolas-Louis Vauquelin in 1797. In working with this metal he obtained several of its compounds in red, yellow and green. As a result of this the name suggested for the new element was chromium, from the Greek *chroma* (colour).

This has had one odd result. Chromium is one of those metals which can be electroplated on to steel. The chromium skin not only takes a high and good-looking polish, but it is also resistant to rust, so that it protects the steel beneath, which would otherwise quickly rust and crumble. Chromium-plated steel is used commonly in the decorative metalwork on motor-cars, and it has become customary to speak of such metal as chrome, so that the Greek word for colour ends by being applied to a handsome but entirely colourless material.

More appropriately, the word chrome also occurs in the names of the coloured chromium compounds which are now used as pigments in paint. A compound of chromium and oxygen, for instance (called chromic oxide, logically enough), is also called chrome green because of its colour. If chromium and oxygen are combined with lead in various ways lead chromate is formed, which may appear in different colours and is called chrome red, chrome orange and chrome yellow.

A similar discovery occurred five years after Vauquelin's. The English chemist Smithson Tennant discovered a new element in crude platinum. This new element was also unusual for the number of differently coloured compounds it formed, and Tennant named it iridium, from the Greek *iris* or rainbow (genitive, *iridos*). It is interesting to note that to the Greeks Iris was the messenger of the gods, and it was only natural to suppose that the rainbow, which seems to stretch between earth and heaven, was the natural bridge by which Iris travelled, so her name was given to the rainbow. The coloured portion of the eye, which is differently coloured in different people, is also the iris, from the word rainbow.

Circadian Rhythm

ALTHOUGH WESTERN MAN is wedded to clocks, and is constantly aware of the importance of knowing the time, he isn't entirely dependent on them. Even without clocks he would know it was mealtime when he got hungry, and bedtime when he got sleepy. There are many people who can wake up at some desired time without having to set an alarm.

There are cyclic changes inside you that make you feel hungry or sleepy every so often and that keep you roughly aware of the passing of time. Such cycles are examples of biological clocks.

What set the biological clocks? There are steady cycles in the world outside the organism: light and dark alternate; the tides move in and out; the seasons bring rain and drought, or warmth and cold.

It is useful for an organism to respond to these changes. If its food is to be found only by night or only in the warm season it might as well go to sleep during the day or hibernate during the winter. If it is going to lay its eggs on the shore it can do so best at the highest high tide that comes with the full moon. Even plants respond to these rhythms, so that leaves curl at sunset, flowers or fruits come at particular seasons, and so on.

Undoubtedly this happens because of basic molecular changes within organisms developed by evolution, and in recent years biologists have been studying the details of these inner clocks with great interest.

The strongest rhythm is, of course, that of day and night, with its alternation of warm light and cool dark. There are many cycles that vary over a period of about a day. The Latin phrase for about a day is *circa dies,* so the daily rise and fall is called circadian rhythm.

These circadian rhythms are hard to ignore. People who make long jet trips find their circadian rhythm to be out of tune with the new position of the sun, and it can be very troublesome to try to adjust.

Cistron

THE PHYSICAL CHARACTERISTICS of an individual organism are determined by the genes it possesses. These genes are DNA molecules in the chromosomes within the nucleus of each cell.

Genes have enormous possibilities of variation among themselves because of their complicated molecular structure. A gene governing a particular characteristic always exists in a particular section of a chromosome, but it may exist there in any of a number of varieties. The gene governing eye-colour may be of a type that produces brown eyes, or one that produces blue eyes. Such varieties of a particular gene are called alleles, from a Greek word meaning another.

Chromosomes occur in pairs, so that a gene governing a particular characteristic will be present in each of two chromosomes. The gene may be the same allele in each of the two, or different alleles.

A chromosome can intertwine with its pair in such a way that portions are interchanged (a crossing over). The crossover may take place at any point between any two neighbouring genes. You might have two genes, a and b, next to each other in one chromosome and two alleles of those genes, a' and b', in its pair. Crossing over may take place between a and b. As a result the first chromosome may now have a and b' as neighbours, while the second has a' and b.

When a and b are in the same chromosome, that is called the cis configuration, from a Latin word meaning on the same side. With a and b on opposite chromosomes, we speak of the trans configuration, from a Latin word meaning across.

The a and b genes sometimes act as a unit in producing a

particular physical characteristic, even though they can be separated by crossing over. Such an action unit of separable genes is called a cistron, because they must be in the cis configuration to act normally.

Clock Paradox

ALBERT EINSTEIN'S special theory of relativity, published in 1905, showed that the measurements of mass and length weren't absolute, but depended on the velocity of the object being measured relative to the instrument doing the measurement.

An object in motion relative to an observer is a little more massive and a little shorter than that same object at rest. At ordinary velocities the change is extremely tiny, but at velocities comparable to that of light in a vacuum the changes mount up. An object moving 260 000 kilometres a second would be measured as half as long and twice as massive as it would be if at rest. If two objects, A and B, equal in length and mass move past each other at a velocity of 260 000 kilometres a second (each relative to the other). A would measure B as half as long and twice as massive as itself. B would measure A as half as long and twice as massive as itself also. When the two objects finish their trip and come together again each would measure the other as normally long and massive, for they are now at rest with respect to each other.

Einstein said this was also true of time. Time on a moving object passes more slowly than on an object at rest. If A and B are moving at 260 000 kilometres a second relative to each other, each would observe a clock on the other to be moving at a half-normal rate.

But changing time leaves a permanent mark. When A and B come together after their trip, A would expect B's clock to be slow and B would expect the same of A's clock. Rigorous logic appears to bring us to a contradiction here, and that is what we call a paradox. Since in the usual account scientists talk of observing clocks, this particular paradox is the clock paradox.

Fortunately, by taking acceleration into account, the clock paradox can be explained away, leaving the theory of relativity intact.

Clone

PLANTS ARE MORE VERSATILE than animals in some ways. A twig from one tree can be grafted to a plant of a completely different variety and that twig will continue to grow and flourish.

Or a twig can be planted and develop into a whole plant.

The Greek word for twig is *klon,* or as we spell it, *clone.* The word has come to refer to any group of cells grown from a single body cell—or to any whole organism developed from a part.

All animals have the power to regenerate lost parts of the body to some extent. Some simple animals can easily regrow entire limbs. Complex animals are less talented, but we can regrow hair and fingernails. We can patch damage to skin or liver, but neither we nor any other mammal can regrow a limb if one is lost.

And yet somehow it would seem that the cells of the body ought to be able to do more than they do. The fertilized ovum is a single cell that has the capacity to divide and grow into an entire complicated organism. In the process billions of cells develop from that original cell and those billions divide into dozens of varieties, none of which can by themselves develop into an organism. In some cases they can no longer even divide. Yet all the body cells contain the same genes present in the original fertilized ovum.

In each different type of cell different combinations of genes are blocked and put out of action; the pattern of those that remain determines the cell specialization. But what if the blocked genes are unblocked? One way of doing this might be to expose the cell nucleus to the cytoplasm of a fertilized ovum. In the late 1960s the nucleus of a frog's fertilized ovum was replaced with nucleus from a frog's body cell. The new cell developed normally, and a full-grown frog genetically identical to the frog of the body cell was formed. It was an animal clone.

Cloud-seeding

CLOUDS ARE COLLECTIONS of tiny water droplets or tiny ice crystals. Sometimes these collect into conglomerates large enough to fall as rain or snow. At other times the tiny fragments, too small to fall against air resistance, remain as they are, and there is no precipitation. When there is drought the presence of clouds which do not release their load of moisture is most frustrating.

One way of encouraging the necessary large conglomerates to form is by providing some sort of nuclei about which they can gather. The nuclei can be dust particles of the proper size and shape, or tiny crystals of certain chemicals, or electrically charged particles.

In the 1940s the American physicist Vincent J. Schaefer was carrying on a programme of experimentation with water vapour in a closed container kept at very low temperatures in order to duplicate cloud behaviour in the laboratory. In July 1946, during a hot spell, he found it difficult to keep the temperature low enough. He placed some solid carbon dioxide (dry ice) in the box, in order to force down the temperature. At once the water vapour condensed into large ice crystals and the box was filled with a miniature snowstorm. Would this work on a large scale, too?

On November 13, 1946, Schaefer flew in an aeroplane over a cloud bank and dropped six pounds of dry ice into it, starting the first man-made precipitation in history. Each particle of dry ice was a seed about which a conglomerate could grow. What Schaefer had done was called cloud-seeding. Schaefer's co-worker, Bernard Vonnegut, used tiny crystals of silver iodide. These were more effective seeds than dry ice, and could be blown into the clouds from the ground. Beginning with the 1950s, cloud-seeding has been used often in attempts to break droughts or to disrupt hurricanes before they can build up to dangerous levels.

Cloud Chamber

WHEN SUB-ATOMIC PARTICLES smash through the atmosphere they chip electrons away from the atoms they encounter. The electron-missing atoms (called ions) can carry away electric charge from metal foil and through this action the presence of sub-atomic particles and their approximate quantity can be determined. Another method is to have the sub-atomic particles strike certain chemicals, producing tiny scintillations of light which can be viewed under magnification.

In 1895 a Scottish physicist, Charles Wilson, was interested in cloud-formation. He came to the conclusion that water vapour turned into the tiny droplets that made up clouds through condensation around appropriate nuclei—dust particles or ions of some suitable type. Since sub-atomic particles formed ions, it occurred to him that water droplets might appear in their tracks.

In 1911 he pumped water vapour into dust-free air in a small closed chamber fitted with a piston. He pulled the piston upward, expanding and cooling the air. Cold air can't hold as much water vapour as warm air, but nothing happened. If there had been dust particles or ions to serve as nuclei, however, water droplets would have formed a tiny cloud.

Repeating the experiment, Wilson sent sub-atomic particles through the chamber and then expanded the air. This time a trail of water droplets formed about the ions, making the track of the particle visible. In the presence of a magnet the sub-atomic particle's path curved in a direction that depended on whether its charge was positive or negative and by an amount that depended on its mass. The track of water droplets curved too, and from that a skilled observer could tell a great deal about the sub-atomic particle. For obvious reasons, this detecting device was named a cloud chamber.

Coacervate

SCIENTISTS STUDYING the possible beginnings of life have duplicated conditions as they were thought to have existed in the early days of the earth. They have shown that from the very simple compounds present in the original ocean more complicated compounds could surely have been built up. Eventually compounds similar to those that now exist in living tissue could have formed.

But once such complicated compounds form how do they come together to form living systems? Somehow they must have come together to form organized cells; cells simpler than any now existing, perhaps.

It happens, though, that in solution complicated molecules do not invariably remain evenly spread out. Under certain conditions such a solution divides into two parts. In one part the complicated molecules are concentrated, while in the other few or none are to be found. (This is because the complicated molecules often have some sort of mutual attraction due to small electric charges here and there on the molecular structure.)

Such a separation into a part rich in large molecules and a part poor in them is called coacervation, from a Latin word meaning to heap up. The portion rich in the large molecules (where they are heaped up, so to speak) is the coacervate.

The Russian biochemist Alexander Oparin as long ago as 1935 suggested that the first cells might have formed as small droplets of coacervates. In 1958 the American biochemist Sidney W. Fox formed protein-like compounds he called proteinoids by processes not involving life. He dissolved them in hot water and let the solution cool. The proteinoids formed tiny spheres of coacervates about the size of small bacteria. Fox called these microspheres.

These microspheres turned out to have some properties reminiscent of cells, which made the coacervate theory more attractive.

Cobalt

ONLY SEVEN metals were known in ancient times: gold, silver, copper, iron, tin, lead and mercury. Medieval miners who came across ores of other metals were usually nonplussed and at a loss for methods of handling them.

About 1500, for instance, miners in Saxony came across ores which did not smelt properly, and which spoiled batches of ordinary ores. The miners had no real understanding of why this should be so, no concept of new metals that required new treatment for isolation. Their explanation was simpler and more direct: earth spirits had bewitched the ore just to be annoying.

One of the German earth spirits is the kobold. This comes from the primitive Germanic but may have some kinship with the Greek *kobalos,* a term used for a mischievous person, from which we ourselves probably get the word goblin. In any case, the Saxon miners called the annoying ore kobold.

About 1735 the Swedish mineralogist Georg Brandt, after years of interest in this ore (which had gained some importance in the manufacture of a deeply-coloured blue glass), isolated a new metal from it, to which he attached the name given it originally by the exasperated miners, and which has now become *Kobalt* in German and cobalt in English and French.

A second ore that also annoyed the long-suffering miners was called by them *Kupfernickel.* In German *Kupfer* means copper, and *Nickel* (like kobold) means a mischievous imp. (We ourselves sometimes refer to the devil as Old Nick.) *Kupfernickel,* therefore, means devil's copper or false copper.

In 1751 another Swedish mineralogist, Axel Fredrik Cronstedt, isolated a new metal from this second ore, and he too kept the miners' name, which was shortened eventually to nickel.

Codon

IN THE 1940s it was discovered that nucleic acids served as models for the synthesis of particular enzyme molecules in the cell. Nucleic acid molecules are made up of long strings of units called nucleotides, while enzyme molecules are made up of long strings of units called amino acids.

If each nucleotide in the nucleic acid chain corresponded to some particular amino acid you could imagine a neat transference of structure from nucleic acid to enzyme. The trouble was, though, that there were only four different nucleotides in nucleic acid molecules, but there were twenty different amino acids in enzyme molecules.

This, however, is not as puzzling as it sounds. There are only nine digits, but used in combinations, they can represent an enormous number of integers. Suppose a different combination of neighbouring nucleotides stood for each different amino acid.

If four different nucleotides occur in any combination along a chain there are 4 × 4 or 16 possible different neighbouring pairs of nucleotides, and 4 × 4 × 4 or 64 possible different neighbouring trios. It would have to be combinations of three nucleotides that corresponded to the twenty different amino acids.

During the 1960s biochemists determined which amino acid each of the sixty-four possible trios stood for, and this was called the genetic code. (Two or sometimes three or four different trios might all stand for the same amino acid.)

What does one call a particular combination of three nucleotides? The influence of the phrase genetic code made itself felt, and the three-nucleotide unit, corresponding to a particular amino acid, was called a codon. The -on ending was borrowed from the sub-atomic particles which were fundamental parts of an atom, as the codon was a fundamental part of the nucleic acid molecule.

Coelacanth

ABOUT 450 MILLION YEARS AGO, fish evolved and gradually became the dominant life-form of the oceans. One large division of fish had four stubby, fleshy appendages, fringed with fins. This group of fish were therefore named Crossopterygii from Greek words meaning fringe fins. The Crossopterygii were not as efficient in swimming as were those other fish which developed appendages that were less fleshy and bore longer fins.

Thanks to their more muscular appendages, however, the Crossopterygii could manage to stump their way across land when they had to. It is thought that land animals are descended from them, but they themselves were rapidly dwindling in numbers 250 million years ago.

The Crossopterygii were freshwater fish to begin with, but some forms colonized the salt sea. Certain of these are called coelacanths, from Greek words meaning hollow spines, because this was one of their features. No coelacanth fossils were found to be less than 70 million years old, and it was thought they had become extinct.

On December 25, 1938, however, a trawler fishing off South Africa brought up an odd fish about five feet long. A South African zoologist, J. L. B. Smith, who had a chance to examine it, recognized it at once as a living coelacanth. The Second World War halted the hunt for more coelacanths, but in 1952 another, of a different genus, was fished up off Madagascar. Before long numbers of this kind of fish were found. Since they are adapted to fairly deep waters, they die soon after being brought to the surface.

The coelacanths have a double interest. First, they are living examples of a type of creature that was thought to have grown extinct with the dinosaurs. Second, they are the only known direct descendants of a type of fish that seem to have been the ancestors of land vertebrates, and therefore of ourselves.

Coenzyme

IN 1904 two English biochemists, Sir Arthur Harden and W. J. Young, studying a yeast enzyme, zymase, disrupted it into two parts. Separated, the two parts of the enzyme could not carry through the normal task of bringing about the fermentation of sugar. When the two were combined, however, enzyme activity was restored.

One of the fractions was a protein very much like the original enzyme. It was the main portion of the molecule. The other was not protein, but was instead a relatively small molecule that was stable to heat. It was a substance that worked with the main portion of the enzyme to help it carry out its function, and so Harden and Young called it cozymase, where the prefix is a Latin one meaning together with.

In general a small molecule, necessary to enzyme function but radically different in chemical structure from the rest of the enzyme, is called a coenzyme.

It was soon found that many coenzymes contained atom groupings that were unusual, and were not found in other tissue components. Just as enzymes need be present in only small quantities to do their work, so with coenzymes and these atom groupings. It happens that the tissues of complex animals (and those of some simple organisms, too) do not maintain a complex system of chemical machinery in order to make these atom groupings that are needed in such small quantity. Instead it seems reasonable to count on finding sufficient amounts ready-made in food.

It is these relatively rare but necessary atom groupings which the body must find in food that are included among vitamins.

Some enzymes have associated with them atom groupings containing metal atoms—such as cobalt, copper, zinc or molybdenum—not often found elsewhere in tissues. They are present in organisms only in traces, and yet these trace minerals are essential to life.

Colloid

IN 1861 the Scottish chemist Thomas Graham placed various solutions in a cylinder in which the open-ended bottom was blocked off by a thin sheet of parchment paper. He then placed the cylinder in a bucket of pure water.

If the solution in the cylinder contained any of a number of substances, such as, for example, ordinary salt or sugar, these found their way through the parchment and could be detected in the water in the bucket by some appropriate chemical test.

There were other substances, however, which if present in the solution within the cylinder would never pass through the parchment, no matter how long one waited.

The first set (salt, sugar, etc.) formed thin, watery solutions and happened to exist in crystalline form when not in solution (see CRYSTAL). These, therefore, Graham called crystalloids. The second set, which included various proteins, gums, and so on, formed thick glue-like solutions and did not form crystals when not in solution. He called these colloids, from the Greek *kolla* (glue).

But later events spoiled the logic of the names, since biochemists have now managed to crystallize many substances that are colloids in solution. The difference between crystalloids and colloids has nothing to do with crystals, actually. It is just that crystalloids have small molecules that can get through the submicroscopic holes in parchment, while the colloids have large molecules (or large aggregates of small molecules) that cannot.

Many of the most important substances in living tissue are colloids in solution, and to purify them biochemists still make use of Graham's original experiment. They place a solution inside a membrane (improved over Graham's parchment, of course) and bathe the membranous bag in water. The small molecules escape through the membrane and the large do not, so that the two classes of substance are separated. Graham called this process dialysis, from the Greek *dialyein* (to separate), and the name remains in use to this day.

Comet

ANCIENT MAN was well aware of the regular movements of the heavenly bodies and knew how they marked off the seasons on earth (see SOLSTICE, EQUINOX) and even thought that the variable but predictable motions of the planets had some influence over human lives. It was a cause for concern and even terror, then, to have something brand-new and of strange appearance suddenly shining in the sky at unpredictable intervals. It must mean the upsetting of the seasons; famine; drought; destruction; catastrophe of one sort or another.

The something that appeared in the sky was not a sharp point like the stars and planets but a fuzzy patch with a long smoky extension. The ancients saw in it a resemblance to a distraught woman fleeing, with her long hair streaming out behind. The Greek word for long hair is *kometes*. The Romans called the objects *stellae cometae* (hairy stars), and we call them simply comets.

The Greek philosopher Aristotle thought the heavens were perfect and could not change. Only the earth and the regions below the moon could show change and corruption. Comets, therefore, must be part of the earth's atmosphere and were not really heavenly objects. (Other ancient philosophers disagreed but, as in other things, Aristotle's ideas, whether right or wrong, usually won out.)

In 1588, however, the Danish astronomer Tycho Brahe proved that the comet of 1577 had been much farther off than the moon. In 1704 the English astronomer Edmund Halley, studying comets, noticed similarities in the paths of the comets of 1531, 1607, and 1682. He declared they were the same comet, and predicted it would return about 1758. (It did, in 1759, seventeen years after Halley's death, and twice since then, in 1835 and 1910.)

This comet, now called Halley's Comet, was the first of a number whose orbits have been calculated, so that comets have been reduced to the status of normal members of the solar family and are no portents of disaster at all. We now know comets to have a nucleus and an envelope (or 'coma') of gas. One or more very long 'tails' may be driven out by radiation from the sun.

Competitive Inhibition

ENZYMES bring about, or catalyse, a chemical reaction, making it proceed with far greater speed than would be the case in the absence of the enzyme. Each different reaction requires an enzyme of its own that works to bring about that reaction and no other. Each enzyme need be present in only tiny quantities to do its necessary work, yet if that small quantity is prevented from operating, some chemical reaction is stopped, usually with serious, sometimes fatal results.

A substance which will block an enzyme's action, or inhibit it (from Latin words meaning to hold in), will seriously disturb or kill an organism. If one molecule of the inhibiting substance will stop one molecule of the enzyme, then no more of the substance is needed than of the enzyme. A small quantity of the substance can kill, and such a substance is a poison (from a Latin word meaning to drink, since poisons are usually thought of as being dissolved in liquid).

One way of keeping an enzyme from working is to present it with a substance possessing a chemical structure very like some portion of itself or some chemical it normally works with, but one that is not identical. This merely similar compound competes with the natural compound for combination with the enzyme. Once it is part of the enzyme, however, it is sufficiently different from the natural compound to keep the enzyme from working. This is competitive inhibition.

Some poisons that work by competitive inhibition are life-savers for us, since not all cells are identical. Some cells are more dependent on a particular reaction than are others. Some cells have enzymes more sensitive to a particular competition, or membranes more permeable to competing molecules. It is possible to find compounds that competitively inhibit the enzymes of bacteria without seriously affecting animal cells. Antibiotics probably work in this way, and are therefore selective poisons.

Continent

THE EARLY peoples did not, naturally, have accurate notions of the major features of the earth's surface. If they lived near a shore they recognized the existence of land and sea, however.

People who had much to do with the sea (the early Mediterranean people, for instance) could not avoid noticing that there were two types of land. There were first little pieces of land surrounded by sea. One Latin word for sea is *salum,* from *sal* meaning salt. (After all, the remarkable thing about sea-water, which is undrinkable, as compared with drinkable river-water, lake-water and well-water, is that sea-water is salt. We ourselves sometimes call the sea the briny.) Anyway, a piece of land in the sea is *in salo,* from which comes the Latin *insula* and our isle.

Then there was another kind of land, land that went on and on with no sign of sea that the ancients knew of. It was continuous land, and so was a continent, from the Latin *continens* (continuous). In English we speak of a continent also as the mainland (main coming from the Latin *magnus,* meaning great), as opposed to island (isle-land). In the days of the Spanish control of the Americas pirates had their strongholds in the islands of the

West Indies and went sailing against the Spanish Main.

To the Greeks there seemed to be three continents separated by sea. The Mediterranean Sea, in fact, gets its name from the Latin *medius* (middle) and *terra* (land). It was a sea that lay in the middle with three bodies of land about it. The three continents (Europe, Asia and Africa) are actually connected by land. Africa and Asia are connected by only the small Sinai isthmus (from the Greek *isthmos,* meaning a narrow passage) so that may be ignored, but Europe and Asia are connected by land for a thousand miles or more, and it is only custom that makes us speak of Europe as a continent. Many geographers speak of Asia plus Europe as Eurasia, while Eurasia plus Africa is sometimes called the World Island. The three together are, after all, surrounded by sea, so they form a large island, and they contain about 85 per cent of the earth's population, so it is an island that is almost the equivalent of the world.

Continental Drift

A GERMAN GEOLOGIST, Alfred Wegener, was intrigued by the fact that the eastern coast of South America looked as though it would fit the western coast of Africa. Could they once have been a single land mass that split and drifted apart? Wegener had explored Greenland, and, after collecting various determinations of longitude made there, decided that Greenland had moved a mile away from Europe over the previous century.

In 1912 he advanced the theory that at one time the major land masses of earth had formed a single body, Pangaea (from Greek words meaning all earth), surrounded by a single ocean, Panthalassa (all ocean). Pangaea had broken into fragments, the modern continents, and these had very slowly drifted apart, like big chunks of granite floating on the hot, thick fluid of earth's deeper layers. This theory is called continental drift.

The theory was not taken seriously to begin with. The increased interest in the ocean floors after the Second World War, however, and the discovery of mountain ranges in the mid-Atlantic, marked by a central rift, seemed to indicate that the Atlantic Ocean might indeed be widening.

In 1968 there came a most dramatic piece of evidence in favour of continental drift. A small piece of fossil bone, located in Antarctica, was clearly part of an amphibian animal that could not have lived in an Antarctica as cold as it is today. What's more, even if Antarctica had been warmer, the amphibian could not have crossed the stretches of salt water that separate the continent from other land masses today. Instead one had to assume that about 120 million years ago, Antarctica had been joined to South America and Africa in a more temperate latitude; that it broke off and drifted away, carrying an animal load with it. As it drifted into the polar regions a load of ice gradually accumulated on it and many of its life-forms died, perhaps to remain as fossils under the ice.

Continuous Creation

ALBERT EINSTEIN worked out the first theory, in 1916, that took into account the universe as a whole. (This originated the science of cosmology.) Einstein began with the assumption that the universe was the same in average structure everywhere. Everywhere there was an even spreading of galaxies. This was called the cosmological principle.

In the 1930s some astronomers began to think the universe had begun as a small sphere of extremely dense matter which had exploded (the big bang theory). The explosion set off a series of changes that finally brought the universe to its present state. This meant that the average appearance of the universe changed drastically with time.

To three astronomers in England, Thomas Gold, Fred Hoyle and Hermann Bondi, this seemed doubtful. They wondered if it might not be that the universe, taken as a whole, was the same in average structure not only everywhere in space but at all times throughout eternity. This they called the perfect cosmological principle.

But how could this be so when it was known that the galaxies were constantly receding from each other, so that the universe was gradually growing larger and larger and emptier and emptier? The three astronomers suggested, in 1948, that matter was continually being created throughout the universe, so that by the time two galaxies had doubled the distance from each other enough matter had accumulated between them to form a new galaxy.

With old galaxies moving apart and new ones forming between, the average appearance of the universe would remain the same throughout time. This is called the continuous creation theory, and is one that results in a steady-state universe.

Recent astronomical observations, however (such as those of Ryle), seem to place the weight of the evidence on the side of the big bang theory and to show that matter is not evenly distributed throughout the universe.

Corona

ASSOCIATED WITH A TOTAL SOLAR ECLIPSE is a ring of pearly light that surrounds the dark circle of the eclipsed sun and stretches out as far as two or three times the sun's diameter.

Oddly enough, this is not mentioned by early observers of eclipses. Apparently the eclipse was so frightening a phenomenon, and those observing it were in so strong a panic lest the sun disappear for ever, that no one really stopped to notice details. Plutarch mentioned something about a ring of light around the eclipsed sun in the first century A.D. The German astronomer Johannes Kepler said that such a ring had been seen about the sun in an eclipse of 1567. Finally, in 1715, the English astronomer Edmund Halley published a careful description of a solar eclipse that included the ring of light.

Even in the nineteenth century there was some question as to whether the ring of light was part of the sun, or part of the disc of the moon, which was eclipsing the sun. By 1860, when photography was first used in connection with an eclipse, it was definitely decided that the light was the glowing of the thin outer atmosphere of the sun. Ordinarily this glow was blotted out by the intense light of the deeper layers of the sun, but when the main body of the luminary was concealed by the moon's disc the outer atmosphere sprang into beautiful, pearly light.

Because the outer atmosphere surrounded the dark disc of the eclipsed sun like a crown resting on the head of a king (as seen from above), it was called the corona, the Latin word for crown.

The corona could only be studied during the few minutes of a total eclipse at first, but then in 1930 the French astronomer Bernard Lyot designed a telescope which blocked the light of the sun and made the corona visible even when an eclipse was not taking place. He called this instrument the coronagraph.

Cortisone

THE ADRENAL glands (see HORMONE) are actually two glands in one, an outer gland enclosing an inner. The inner gland is the adrenal medulla, medulla being the Latin word for marrow or, generally, something which is inside, as marrow is inside bones. The outer gland is the adrenal cortex, cortex being the Latin word for bark or, generally, something which is outside, as bark is outside the tree-trunk.

The adrenal medulla produces the hormone adrenalin (see HORMONE), while the adrenal cortex produces a whole series of hormones of an entirely different type. Edward C. Kendall, a biochemist at the Mayo Foundation, first isolated these in the 1930s and, before their chemical structure was known, called them simply Compound A, Compound B and so on.

As it turned out, though, the hormones from the adrenal cortex are all steroids (see CHOLESTEROL) so that they are now called adrenocortical steroids. Chemists don't like long names any more than do other people, so this has been shortened to cortical steroids, and even further to corticoids.

Some of the corticoids are ketones (see KETONE), so that in naming them the -one suffix, reserved for ketones, can be used. Compound B, for instance, once its structure was worked out, was named corticosterone (the ketone steroid from the cortex, you see).

That name could be used as a starting-point. The structure of Compound E was like that of corticosterone, except for a hydroxyl group present at the carbon-17 position and a couple of hydrogen atoms missing at the carbon-11 position. (The carbon atoms of organic molecules are often numbered for convenience according to an agreed-upon system.) Compound E can therefore be called 11-dehydro-17-hydroxycorticosterone. But this is too long, especially since it was discovered in 1948 at the Mayo Clinic that Compound E could give amazing relief in some types of arthritis. Anticipating much use of the name, Kendall and his group shortened it by taking certain letters out of the long name and reducing it to just cortisone. Cortisone has in the event proved to be something of a disappointment as a specific against rheumatism.

Cosmic Rays

AN EARLY way of detecting short-wave radiations was to use a box containing two very light sheets of gold leaf attached to a rod at one end. When the rod is charged with electricity the two leaves repel one another, taking on a V-shape. In the presence of X-rays or gamma rays electrons are knocked out of the air molecules in the box, and the air can then carry electricity. The electric charge leaks off the gold leaf into the air and the two leaves come together. This instrument is an electroscope. (The suffix -scope comes from the Greek *skopein,* meaning to watch. The instrument uses electricity, you see, to watch for radiation.)

In the absence of X-rays, gamma rays and such, the gold leaves of the electroscope ought to remain apart indefinitely, but they don't. Slowly they sink together. Scientists assumed there were small quantities of radioactive substances everywhere in the soil (and there are), and that these were the source of perpetual radiation that accounted for the slow leak of the electroscope's charge.

If this were so, however, an electroscope that was miles high, with a thick blanket of air shielding it from the radiating soil, ought to retain its charge indefinitely. Electroscopes were taken up in balloons in 1911 and thereafter just to prove this, and settle the problem so that scientists could forget about it.

But the unexpected happened. Several kilometres in the air, the electroscope leaked more rapidly than it did on the ground. There was radiation being detected, but not radiation from the ground. The Austrian physicist V. F. Hess was the first to publicize this by way of a name. He called the radiation *Höhenstrahlung,* which is German for radiation of the heights.

After the First World War the American physicist Robert Andrews Millikan took the lead in such balloon experiments, and in 1925 he suggested the name cosmic rays, since they came not from earth but from somewhere in the outer cosmos (a Greek word meaning order, particularly good order, which expresses the Greek idea that the universe forms an ordered whole).

Cosmogony

MEN HAVE ALWAYS BEEN INTERESTED in the question of how things began. Until modern times they have always supposed that supernatural forces were involved; that some god or gods had created the universe.

By the eighteenth century it was clear the universe was larger than had been thought, and by then scientists were attempting to account for the beginning of things without reliance on the supernatural. The French scholar the Comte de Buffon suggested about 1750 that the planets, including earth, had originated from matter knocked out of the sun when it collided with some other huge heavenly body about 75 000 years before.

But then how did the sun originate? In 1798 the French astronomer Pierre de Laplace supposed the solar system, both sun and planets, had originated out of a vast, whirling cloud of dust and gas.

As the astronomical horizon receded astronomers began to wonder how stars originated, clusters of stars, the Galaxy, and eventually how the entire vast universe began.

In the 1920s it became clear that all the galaxies were rapidly receding from each other (see RED SHIFT), and this gave rise to the thought that mankind was witnessing the after-effects of a huge explosion. In 1927 the Belgian astronomer Georges Lemaître suggested that to begin with all the matter in the universe had been compressed tightly into one dense mass, which had exploded. It was the aftermath of the explosion that created the universe as we know it.

The term cosmogony is from Greek words meaning birth of the universe. It is sometimes applied to the study of the origin of part of the universe—a galaxy, a star cluster, even a planetary system. Lemaître's suggestion, however, was the origin of true cosmogony, the study of the birth of the entire universe.

Cosmology

THE ANCIENT GREEKS thought the universe began as a heap of matter in utter disorder, which they called chaos. The creation of the universe, they felt, was the creation of order (cosmos) out of this chaos. For this reason we use cosmos to mean the orderly universe we observe about us.

Over the centuries men learned more about the universe, and gained a more accurate notion of its enormous extent. In Greek times the distance to the moon was worked out; later, in the eighteenth century, the distance of the planets; in the nineteenth century, the distance of the stars; and in the twentieth century, the distance of the outer galaxies. Astronomers know now that the universe measures several thousand million light-years from end to end.

How can astronomers learn to understand the workings of a universe so vast? One way is to work out the laws of science from observations on earth and hope that they apply to all the universe. Sir Isaac Newton worked out the law of universal gravitation in 1683—universal because he thought it applied to the universe as a whole.

It certainly applied to the solar system, and in the nineteenth century the German-English astronomer Sir William Herschel studied pairs of stars that circled each other and found that the law of gravitation applied to them too. Newton's gravitation alone, however, did not quite solve all problems.

In 1916 the German-born physicist Albert Einstein advanced his general theory of relativity, which so interpreted gravity as to make it possible to use it to deduce the structure and behaviour of the universe as a whole. For this reason, we may consider Einstein to have originated the science of cosmology (from Greek words meaning study of the universe), since he made it possible to study the universe as a whole.

Cosmonaut

IN JULES VERNE'S *From the Earth to the Moon,* published shortly after the American Civil War, Americans were described as making the first flight to the moon. As time wore on it seemed more and more likely that Americans indeed would. The United States grew to be the richest and most technically advanced nation on earth, and Americans gained a reputation for inventiveness and daring. In the early 1950s they were making plans to hurl an artificial satellite into orbit about earth.

The United States was in no hurry, however, for there seemed no competition. The Soviet Union announced plans of its own, but the Russians were considered technologically backward, and few Americans paid attention. Yet the Russians were really interested in space flight. A Russian, Konstantin Tsiolkovsky, had published a thoughtful book on the subject as early as 1903.

The Soviet Union did not forget this, and since Tsiolkovsky had been born in 1857, they endeavoured to have their first artificial satellite launched in 1957, his centenary (and the International Geophysical Year). In this, they succeeded. On October 4, 1957, they launched the first Sputnik (a Russian word meaning satellite) and initiated the space age.

Americans were caught by surprise and began to labour mightily to catch up to and surpass the Soviets. Eventually they did so, but not before the Soviet Union scored other firsts. On April 12, 1961, they sent Yuri Gagarin into orbit about earth, so that a Soviet citizen was the first man to venture beyond the atmosphere in space flight.

A number of Americans and Russians have now ventured out into space (with Americans the first to reach the moon). Each nation has a different name for these brave men. The Americans call them astronauts (from Greek words meaning star voyagers) but the Soviets even more ambitiously call their men cosmonauts (universe voyagers).

CPT Conservation

THE BASIC RULES of physics are a series of conservation laws which state that some numerical property of a collection of matter cannot change as a result of any alterations taking place within that collection alone. The law of conservation of energy, for instance, states that in any part of the universe shielded from interaction with the rest the total quantity of energy can neither increase nor decrease under any circumstances. (See ENTROPY.)

Physicists have worked out a series of conservation laws for sub-atomic particles by observing what changes never seem to take place among them. For a while a certain property called parity seemed to be conserved. In 1956, however, two Chinese-American physicists, Tsung Dao Lee and Chen Ning Yang, showed that in some cases it was not conserved. This meant that there were conditions in which particles acted as though they were left-handed or right-handed.

When a conservation law proves insufficient physicists may find ways of broadening it so that a new and more general property *is* conserved. For instance, particles occur in twin forms with opposite electric charges, or opposite magnetic fields, or both. When physicists describe this mathematically they make use of something they called charge conjugation.

Charge conjugation seems to change along with parity. If an electron (with a negative electric charge) should happen to behave as though it were left-handed its twin, the positron (with a positive charge), would behave under the same circumstances as though it were right-handed.

There is thus conservation of charge conjugation and parity taken together, or CP conservation. For theoretical reasons, this is linked with time. Any change in CP is reversed if time were imagined to be moving backward. The triple combination is CPT conservation.

Crucible

THE LATIN word *crucibulum* originally referred to a light that was kept burning before a crucifix (the Latin *crux* means cross, genitive *crucis*). When the alchemists melted materials to red-hot liquids in pots they called the pots by that name because they seemed to hold glowing lights. The word comes down to us as crucible. A more entertaining possibility is the story (probably untrue) that alchemists put crosses on their melting pots to keep away devils that might interfere with their experiments. (Any modern chemist will testify that something of the sort is badly needed on occasion.)

Another common vessel used in chemistry is a cylindrical glass container, usually with a lip for pouring. This is called a beaker, which might seem to be derived from the fact that the lip looks like a bird's beak, but which actually comes from the Greek *bikos*, meaning a jar or cask. Similarly, a glass container which narrows at the top to a small opening is called a flask, from the Latin *flasca*, meaning a wine bottle.

Oddly enough, the most familiar chemical vessel to the average non-chemist is one that went out of fashion a century or more ago. This vessel is a flask with a long, narrowing neck that is bent over to one side, slanting downward. It is for collecting vapours. That is, a liquid heated in the flask will give off vapours which will travel down the neck away from the flame and cool to liquid. The drops of liquid formed will drip out the end of the neck and be collected. Such an instrument is called a retort, from the Latin *re-* (back) and *torquere* (to twist), because the neck of the vessel is twisted back, or in Latin *retortus*.

The vessel is inefficient because the neck gradually warms up so that after a while the vapours no longer turn to liquid. Retorts have been replaced by devices in which the tube through which the vapour passes is water-cooled. The very word retort is no longer used in the chemical laboratory for any vessel. Nevertheless, the alchemist's retort has an unshakable association with chemistry in the popular mind.

Cryobiology

IT HAS BEEN KNOWN since prehistoric times that food spoiled more rapidly on warm days than on cold days. It always helped, therefore, to keep food in a cool cellar, or cave, during the warm seasons.

Man learned to pack perishables in ice, where that was available. In the twentieth century gas and electric refrigerators became part of the scene, and the still colder deep-freezer followed. This revolutionized the food habits of many people, for it made it possible to store foods a long time. In general the lower the temperature, the longer food can be preserved.

Since the Second World War the methods used for the production of very low temperatures have made it possible to refrigerate far more deeply than ever before at a reasonable cost. Liquid nitrogen, for instance, keeps temperatures below $-195\,°C$ and has no adverse effect on food. At liquid nitrogen temperatures food, biological samples and other perishables can be kept almost permanently. The study of preservation at such low temperatures is called cryobiology. The prefix cryo- comes from a Greek word meaning freezing cold.

During the 1960s, indeed, it occurred to some individuals that people themselves might be so preserved. Suppose a person were dying of some incurable disease, or just old age. At the instant before death, he might be placed in refrigeration at liquid nitrogen temperatures. Then at some time in the future, when the disease could be cured or old age reversed, he might be revived and treated. Cases have been reported of rich men leaving their bodies for freezing, maintenance and later revival. When the sleeper awakes, indeed!

Of course, there is as yet no method known for reviving a person frozen at liquid nitrogen temperatures, and it is questionable whether such freeze-now-revive-later procedures are psychologically and sociologically feasible. Still, societies have been founded for the purpose of doing this and they have called the techniques involved cryonics. Of recent years cryosurgery has become ever more important—techniques for freezing the body to the greatest possible extent short of damage, in order to minimize operational shock. A favourite futuristic theme is the freezing of interstellar passengers to reduce drastically their speed of aging (after all, the nearest star is over four years away, at the speed of light).

Cryogenics

UNTIL A HUNDRED YEARS AGO there was no way of getting temperatures much lower than those that occurred in nature. About 1860 two English physicists, James Joule and William Thomson (later known as Lord Kelvin), expanded gas under conditions where no heat leaked into it from outside, and found that its temperature dropped. This is called the Joule-Thomson effect.

This effect is made use of in refrigerators and air-conditioning devices. There a gas, liquefied under pressure, is allowed to evaporate and cool down everything about it. The process is repeated over and over, the compression taking place outside the system and the evaporation inside, so that heat is pumped out of the system continually. Scientists used the effect to reach lower and lower temperatures.

Ice melts at $0°C$ or $273°K$ (that is, 273 degrees above absolute zero). In 1877 the Swiss physicist Raoul Pictet brought temperature down to $133°K$, lower than any that occurs naturally anywhere on earth. At that temperature liquid oxygen was produced.

In 1900 the Scottish chemist James Dewar managed to produce a temperature of only $33°K$ and liquefied hydrogen by placing that gas under pressure. Finally, in 1911, the Dutch physicist Heike Kamerlingh Onnes produced $4·2°K$ and liquefied the last remaining gas—helium.

By allowing liquid helium to evaporate a temperature of $1°K$ can be reached, and still lower temperatures, to within a millionth of a degree of absolute zero, can be reached by methods more subtle than the Joule-Thomson effect. (Absolute zero itself cannot be attained.)

The study of the properties of matter at liquid helium temperatures has yielded surprising and useful results. This study is called cryogenics, from Greek words meaning to produce freezing cold. Liquid helium temperatures are therefore called cryogenic temperatures.

Cryotron

THE TREND in electronics is in the direction of miniaturization—that is, in making a device ever smaller. The smaller a device the less material is required to construct it and the more portable it is. One great step in this direction was the replacement of the relatively bulky radio tube by the much smaller transistor.

The transistor does not represent the most compact possible device for the delicate control of electron flow, a control that makes electronic devices possible. Two small wires would be sufficient.

This results from the fact that some metals lose all electrical resistance at very low temperatures, becoming superconductive. This superconductivity can be wiped out, even at very low temperatures, if a magnetic field of sufficient intensity is applied.

Consider a tiny, straight wire of tantalum. At temperatures below $4 \cdot 2\,^\circ$K, tantalum is superconducting, and a current sent through that wire could continue to exist indefinitely. But suppose a spiral of niobium wire is wrapped around the first. Niobium is superconductive up to $9 \cdot 2\,^\circ$K and can withstand at any temperature a larger magnetic field than can tantalum before losing superconductivity. If current is sent through the niobium a magnetic field is set up, and can be made strong enough to wipe out the superconductivity in the tantalum, but not in the niobium itself.

In this way the electron flow in the first wire can be delicately controlled by changing its resistance by means of an electric current through the second wire—but of course only while temperatures are maintained at those of liquid helium, very near absolute zero, the only condition under which superconductivity can exist. Such temperatures are cryogenic (to produce freezing cold). The two wires represent cryogenic electronics and are therefore termed cryotrons, an abbreviated form of that phrase.

Cryptogamous

THE TYPICAL and most advanced land plants are the sper-matophytes (see ANGIOSPERM), but there are common land plants that are more primitive. Of these, the best known are the ferns, which, like the higher plants, have leaves, stems and roots. They do not, however, have flowers or seeds.

Ferns reproduce themselves instead by means of spores, which are like seeds in that they will develop into adult plants, but unlike seeds in that they lack within them the tiny leaves and other differentiated parts that the true seeds of the higher plants have. (Nevertheless, spore is another form of the word *sperma,* meaning seed (see ANGIOSPERM); both spore and sperm come from the Greek *speirein,* to sow.)

The leaves of ferns are quite different from those of higher plants. In ferns, individual leaf-like extensions called fronds (from a Latin word *frons,* meaning leaf) come off a central stalk, so that the whole leaf rather resembles a feather. In fact, the Greek word *pteris,* meaning fern, comes from their word *pteron,* meaning feather. And the plant group to which ferns belong is today called Pteridophyta. Since the Greek *phyton* means plant, ferns are feather-plants. The word fern itself has been traced back to the Sanskrit *parna,* which means feather, among other things.

A still more primitive group of land plants includes as their most familiar representatives the mosses. The Greek word for moss is *bryon,* so this group is called Bryophyta (moss-plants).

Like the ferns, the mosses lack flowers and true seeds and reproduce by means of spores. For this reason an older classifica-tion, introduced by Carl von Linnaeus in the 1700s, listed both groups under the heading of Cryptogamia, from the Greek *kryptos* (hidden) and *gamos* (marriage). It is by means of flowers, you see, that higher plants 'marry' (i.e., are pollinated) and produce seeds. To produce the equivalent of seeds (i.e., spores) without flowers was to have a hidden marriage—no visible evidence. In the past the hypothetical 'fern-seed' was thought to be invisible, and to produce invisibility in whoever possessed it. This is now an out-moded group name, but ferns and mosses are still spoken of as cryptogamous plants.

Cryptozoic Aeon

GEOLOGICAL HISTORY is usually divided into long periods of time on the basis of the kind of fossils that are characteristic of those periods. The oldest rocks in which fossils are to be found are of the Cambrian era (named after Cambria, the Latin name for the region now known as Wales—where such rocks were first studied). The Cambrian rocks are up to 600 million years old, and anything older, with no fossils in it, was at one time simply considered part of the Pre-Cambrian era.

In recent years, however, it has become more and more evident that there are clear traces of life in Pre-Cambrian rocks. There aren't obvious fossils, to be sure, but there are microfossils and organic chemicals that seem to have been produced by one-celled creatures. For that reason all the geological eras from the Cambrian to the present day are now grouped as a single *aeon,* the Phanerozoic æon (from Greek words meaning visible life, thus implying the existence of previous life not so easily visible).

The Pre-Cambrian era has now become the Cryptozoic æon (hidden life), divided into two sections: an earlier Archæozoic era (ancient life) and a later Proterozoic era (early life).

The division between the Cryptozoic æon and the Phanerozoic æon is extraordinarily sharp. At one moment in time, so to speak, there are no fossils at all above the microscopic level, and at the next there are elaborate organisms of a dozen different basic types. Such a sharp division in the geologic record is called an unconformity.

Unconformities usually imply some sharp change in conditions to which earth is exposed. Explanations for the unconformity between the two æons have ranged from the rapid production of oxygen after the development of photosynthesis, which made elaborate life possible, to earth's capture of the moon, which created huge tides that wiped out the earlier record.

Crystal

THE ATOMS in a solid may have no particular arrangement, and the solid is then said to be amorphous (see DIAMOND). More often, however, the atoms are arranged in a definite pattern, and this results in the solid itself taking on a certain symmetry. (The Greek prefix. *sym-* or *syn-* means with, and implies things behaving in co-operation and not in conflict. They are with and not against. The word *metron* means measure; thus symmetry implies that the various measurements of an object combine smoothly and do not clash.)

The Greeks saw the best example of this in the case of snowflakes or the formation of hoarfrost patterns. The Greek word for frost was *kryos,* so they called these patterns of snow and ice *krystallos.*

Another remarkable thing about ice was that it could be transparent, and since the Greeks knew very few other transparent objects, that property struck them forcibly. Consequently, when they found pieces of rock that had symmetrical shapes and were transparent they called them *krystallos* too, and considered them a form of ice.

By early modern times, however, it was realized that many solids could take on symmetrical shapes if allowed to solidify slowly out of a solution, or out of molten form. They were not forms of ice, and they did not have to be transparent. It was the symmetry that counted. The word (crystal, in English) was applied to all of them.

It is also applied to glass objects which are cut into symmetrical shapes, even though glass itself is amorphous and its symmetry is purely artificial and not a natural reflection of atom arrangement.

In fact, sometimes it is the transparency of an object and nothing more that forces the use of the word. We speak, for instance, of a fortune-teller's crystal ball. This is nothing more than a glass sphere with no crystallinity about it at all, either real or artificial. It is crystal only because it is transparent, and because the Greeks were once amazed at the phenomenon of transparency.

Cumulus

THE VISIBLE sign of water vapour in the air is the cloud. This is derived from the Anglo-Saxon *clud,* meaning any round mass, as, for instance, a stone. The word clod for a lump of hard earth comes from the same root. To primitive people, unaware of the nature of clouds, water vapour or air, it isn't so strange that a cloud might well look like a pebble in the sky.

An ordinary cloud is a suspension of tiny liquid droplets in air; it might also consist of tiny solid particles, in which case it is smoke. This comes from the Anglo-Saxon, but there is a Greek word *smychein* (to smoulder). The word smother is also related to both smoke and smoulder.

A cloud which happens to be at the earth's surface is fog or mist. The former is a Danish word meaning spray or driving snow, the latter an Anglo-Saxon word meaning darkness. These are picturesque derivations. Fog and mist do indeed obscure the sun and bring on a premature darkness, just as driving water (spray) or snow would. In some industrial areas, such as Los Angeles, smoke may mix with a persistent fog, and a combination word, smog (*smoke-fog*) has been invented to describe this. (See SMOG.)

The most beautiful cloud is the wool-pack, the typical cloud of fair summer weather. It results from a column of warm air rising and meeting colder air aloft, so that water vapour condenses into fine droplets that look like a mass of wool heaped up high. These are cumulus clouds, *cumulus* meaning heap in Latin.

If the air contains enough moisture, then as it rises (or comes in from the sea), it will form a cloud that will spread over the visible heavens and be thick enough to let so little sunlight through as to take on a gray and threatening appearance. These are nimbus clouds, *nimbus* being Latin for rain-cloud, and, of course, such clouds often result in rain.

Sometimes very high clouds form in the shape of closely spaced feathers or tufts of hair. These are cirrus clouds, *cirrus* meaning a curl of hair.

Cyanide

COLOURED INORGANIC compounds suitable for use in various types of ornamentation have always been valuable. One blue mineral much used for decoration was lapis lazuli. *Lapis* is Latin for stone, while *lazuli* is apparently a corruption of the Arabic *lazaward*, meaning sky blue (azure is another corruption); lapis lazuli is azure stone. Powdered lapis lazuli was called ultramarine, from the Latin *ultra* (beyond) and *marinus* (marine) because it was imported from beyond the sea.

When in 1704 two Berlin dyers accidentally discovered a new deep-blue compound of iron, a kind of substitute lapis lazuli, they kept its method of preparation secret, so that it could be called only Prussian blue from its colour and place of origin.

Naturally, the secret could not be kept for ever. Chemists got to work. In 1783 the Swedish chemist Karl Wilhelm Scheele obtained a weak acid from Prussian blue which he called prussic acid. In 1815 the French chemist Joseph Louis Gay-Lussac isolated a gas from another source which could be easily converted to prussic acid. He found that the key atom group in these compounds consists of a carbon atom and a nitrogen atom. Prussian blue contains six such combinations, so he called this key group the cyanide group, from the Greek *kyanos* (blue). In prussic acid the cyanide group was attached to a hydrogen atom, so that became hydrogen cyanide. The gas he had discovered contained two cyanide groups in its molecule, and this he called cyanogen. (The –gen suffix is generally used by chemists to mean give birth to or produce. See HYDROGEN. Cyanogen is a substance that produces cyanide.)

Because compounds like cyanogen, hydrogen cyanide and

113

potassium cyanide are deadly poisons, the cyan- stem has come to have an unpleasant sound, and yet it is used in many innocent words for the sake of its 'blue' significance. There is a harmless blue compound called cyanidin, and a cyanometer (the Greek *metron* means measure) does not measure degree of poison by any means, but only the intensity of the blue of the sky.

Cybernetics

THROUGH THE PRINCIPLE of feedback (see FEEDBACK), any process is guided by the difference between the actual state of affairs at the moment and the desired state of affairs. The rate of approaching the desired state decreases as the difference decreases. When the desired state is reached so that there is zero difference between actual state and desired state, the change stops.

In 1868 a French engineer, Léon Farcot, used this principle to invent an automatic control for a steam-operated ship's rudder. As the rudder approached the desired position it automatically tightened a steam valve which made the rudder move more slowly. By the time the desired position was reached the steam pressure was shut off and the rudder moved no more. If the rudder moved out of place the valve opened and steam pushed it back. Farcot called his device a servomechanism (slave machine) because it was as though a slave sat there constantly adjusting the position. Since then more and more machinery has been designed to adjust itself automatically. In 1946 an American engineer, D. S. Harder, coined the word automation to describe such adjustment.

Electronic devices made automation more delicate, and radio beams extended the effect over long distances. The German flying bomb of the Second World War was essentially a flying servomechanism, and this quickly escalated to the horror of intercontinental missiles with nuclear warheads. Space exploration would be impossible without automated devices.

In the 1940s the American mathematician Norbert Wiener worked out some of the fundamental mathematical relationships involved in the handling of feedback. He named this branch of study cybernetics, from the Greek word for helmsman. A helmsman, after all, controlled a rudder by constantly observing its position, and this is what automated cybernetic devices did too, but more tirelessly and precisely.

Cyclone

WHAT WE call weather is largely a matter of air movements.
Cold air blows down from the north and warm air from the south.
Moist air blows in from the ocean, forming clouds and bringing
rain. Dry air blows in from the interior and brings drought.
Weather is an Anglo-Saxon word that has been traced back to an
old word that may be related to the Slavic *vetra,* meaning wind.
This is a reasonable theory. English folk-idioms frequently make
use of synonyms, as follows: spick and span, nook and cranny,
hale and hearty (often using old words that haven't survived in any
other fashion). And one speaks of wind and weather.

Because of the importance of wind, there are a number of words
for it in all its forms, most (such as breeze, gust, blast, storm, etc.)
being of Anglo-Saxon derivation, stretching back to old Germanic
words. There are exceptions, though. A gentle wind is a zephyr,
from the Greek *zephyros* (west wind), since to the Greeks the west
wind brought neither snow from the north nor heat from the
Saharan south, and was thus a gentle wind.

In some ways, weather depends upon seasons. There are hot,
cold, wet, and dry seasons. There are times of the year when it is
particularly apt to be stormy and the word tempest attests to that
since it comes from the Latin *tempus* (time). In Latin *tempestas*
means both period of the year and storm.

The severest forms that wind can take involve circular motions.
These begin when cold air and warm air meet and commence
whirling because of the motion of the earth. (A point near the
equator moves more quickly than a point farther removed from the
equator.) In the Northern Hemisphere, then, the southern part of
two colliding masses is dragged east more quickly than the northern part, and a counter-clockwise whirl—opposite in direction to
the motions of the hands of a clock, from the Latin *contra,* meaning
against—is set up; a circular wind, called a cyclone, from the
Greek *kyklos* (circle). In the Southern Hemisphere the northern
portion moves faster, and a clockwise wind is set up, an anticyc-
lone, from the Latin *anti-* meaning against or opposite to. Cyclone
thus means both a violent tropical storm and an area of disturbed
air, a depression.

Cyclic-AMP

IN 1885 the Swiss chemist Albrecht Kossel isolated a compound
which eventually turned out to be a component of some of the most
important substances in living tissue. He isolated it from the pan-

creas, the second largest gland in the body, and so he named it adenine, from the Greek word for gland.

In nucleic acids adenine is found combined with a sugar called ribose, and the combination was named by using letters from both names: adenosine. In addition, a phosphate group (containing atoms of the element phosphorus) is usually attached. Sometimes two and even three are attached. The resulting compounds are adenosine monophosphate, adenosine diphosphate and adenosine triphosphate, the prefixes mono-, di- and tri- coming from Greek words for one, two and three, respectively. The names are often abbreviated as AMP, ADP and ATP respectively.

ATP in particular is of key importance. It is involved in almost every kind of reaction which yields the energy that can be used to manufacture large molecules out of small ones.

In 1960 the American biochemist Earl W. Sutherland discovered a variety of AMP in which the single phosphate group was attached to the adenosine portion of the molecule in two different places. Thanks to this the molecules possessed atoms linked together to form a closed ring, or cycle. For this reason Sutherland called it cyclic-AMP.

Cyclic-AMP was found to be widespread in tissue, and to have a pronounced effect on the activity of many different enzymes and cell processes. It casts new light on how hormones achieve their results—something which has been hitherto mysterious. It may be that different hormones affect the production or destruction of cyclic-AMP in different ways, and that this in turn affects the cell chemistry in some crucial fashion.

Cyclotron

IN ORDER to change one kind of atom into another, sub-atomic particles must be sped into an atom's nucleus with great force. When such atomic changes were first brought about in 1919 the natural particles emitted by radioactive elements (see ALPHA RAYS) were used.

These were insufficient, however, and instruments were devised to speed up sub-atomic particles in greater quantities to greater energies. These instruments were popularly called atom-smashers.

A particularly successful type of atom-smasher was invented by the American physicist Ernest O. Lawrence. He rigged up a device in 1930 which forced protons into a circular path between two magnets. The protons were driven into greater and greater speed by the magnetic field so that they spiralled outward until finally they skidded out of the instrument altogether at a terrific velocity.

The instrument was named a cyclotron. The prefix cyclo- is frequently used in scientific names, coming from the Greek word *kyklos,* meaning circle. In this case it referred to the circular path of the protons. The -tron suffix is a false analogy with the names of some of the sub-atomic particles—electron, neutron and so on.

The -tron suffix has become customary in naming new atom-smashers that have been developed since the cyclotron. For instance, in 1940 the American physicist D. W. Kerst designed an instrument to accelerate electrons to great speeds. Since a speeding electron is a beta particle (see ALPHA RAYS), the new instrument was named a betatron.

The energies of a speeding particle are measured in electron volts, which is abbreviated ev. At the University of California an atom-smasher has been built that will speed up particles till they have energies of billions of electron volts. A thousand million (10^9) electron volts is abbreviated bev, so the instrument is called a Bevatron. Since such energies are about those of cosmic rays, similar instruments are called cosmotrons.

Cystine

ABOUT TWENTY different amino acids have been found in proteins (see GLYCINE), their discoveries being spread over a century and a quarter. Yet the first to be discovered wasn't connected with proteins for nearly ninety years.

It happened that in 1810 an English physician and chemist William Hyde Wollaston was analysing a bladder stone that has been removed from a human patient. (These stones form occasionally in kidney and bladder from insoluble substances that precipitate out of urine. There are different varieties, and Wollaston, as it happened, was dealing with one of the rare types.) He found the stone to be made up mostly of a sulphur-containing organic compound which he named cystine, from the Greek *kystis* (bladder). It was not till 1899 that the same amino acid was located in horn. Horn contains a protein, keratin (from the Greek *keras,* meaning horn) which is of all proteins the richest in cystine, so that was only right.

Then a similar amino acid was discovered into which cystine could be easily converted. The second amino acid was named cysteine to emphasize the similarity, but the additional e makes little impression on the eye, and the names are entirely too similar for comfort.

Others of the protein amino-acids were also named after the object in which they were first found. For instance, one was isolated from cheese in 1849, and named tyrosine from the Greek *tyros* (cheese). Then, another was isolated from silk in 1865 and named serine, from the Latin *sericus* (silken), which in turn comes from Seres, the name of a people in Eastern Asia.

More indirectly, an amino acid was isolated from asparagus in 1806 and named asparagine. In 1832 asparagine was converted into a closely related compound that was a stronger acid and was therefore named aspartic acid. It was not until 1875, however, that aspartic acid was recognized as one of the amino acids in proteins, and later still that asparagine also was added to the list.

Cytochrome

BEGINNING IN 1885, certain cell components of unknown function were detected through their ability to absorb light of certain wave-lengths. This meant that, if those cell components were isolated they would prove to be coloured. In 1925, the Russian-British biochemist David Keilin named these substances cytochromes, from Greek words meaning cell colours.

Keilin distinguished three different cytochromes by differences in the fashion in which they absorbed light, and labelled them a, b, and c. Since then, each of these three has been found to consist of two or more varieties.

All the cytochromes are proteins possessing an iron-containing portion like the haem in the well-known protein haemoglobin. Like haemoglobin, the cytochromes can attach oxygen molecules to themselves.

There is a difference, though. The haemoglobin molecules serve merely as a chemical-transport system, carrying oxygen molecules from lungs to tissues. To do so they attach oxygen to themselves loosely, without changing the chemical nature of the iron atoms to which they are attached. In the case of the cytochromes the oxygen molecules are attached more tightly and the iron atoms change character, each losing an electron on combining with the oxygen (and regaining the electron when giving up the oxygen).

In the 1940s it came to be understood that the various cytochromes formed a chain. Oxygen atoms passed from one to the other, energy being liberated at each step in small quantities that could be usefully stored by the body, until finally each oxygen atom was combined with a pair of hydrogen atoms obtained from fragments of food molecules that had been absorbed into the body.

Every cell that makes use of oxygen contains cytochromes, and these form part of the structure of small cell components called mitochondria.

DDT

THERE ARE nearly a million different species of insects known, and this is far more than the total number of different species of all other animals. Only a few thousand species are harmful to man and his environment, but these include mosquitoes, flies, fleas, lice, wasps, hornets, weevils, cockroaches, carpet beetles and the various insect species that live on the plant life or harass the animal life or damage the manufactured objects that man is trying to preserve for his own use.

As a result, men have long tried to kill insects in every way possible. As late as the 1930s copper- and arsenic-containing poisons were used in sprays, but these gradually poisoned the soil. In 1935 a Swiss chemist, Paul Mueller, began a search for some organic chemical that would kill insects but not other forms of life, and that would be cheap, stable and odourless.

Certain chlorine-containing compounds showed promise. In September 1939 Mueller tried one such compound called dichlorodiphenyltrichloroethane, a compound which had been known since 1873—and it worked!

The name was soon simplified to the letters that started its first, fourth and seventh syllables, and it became DDT. In 1942 it was produced commercially in the United States, but was reserved for military use. In January 1944 it was used to kill body lice in Naples and prevent a typhus epidemic. It was similarly used in Japan in late 1945.

After the war DDT came to be used against insects everywhere. It was by far the best-known and most popular insecticide (from Latin words meaning insect-killer). As the years went by it was recognized as harmful in the long run to creatures other than insects and as *too* stable, for it lingered in the soil and in living tissues. By the 1970s Government after Government had begun to control or ban its use.

Delta

THE LARGE RIVER best known to the Greeks of the time of Herodotus was the Nile of Egypt. Herodotus spoke of it with admiration, and called Egypt the gift of the Nile. The reason for that was first that the Nile was an unfailing flow of water through a rainless desert, and, second, that once a year it overflowed its banks, leaving behind as it receded new and fertile soil. This soil is carried down from the highlands of east central Africa where the main sources of the Nile are located.

The turbulent floodwaters deposit loosened soil and silt along its lower course, but manage to carry some right down to the Mediterranean. The slower the current the less material can be carried by a river, and at the sea, where the current flow stops altogether, all that remains of the silt is dropped.

There are virtually no tides in the Mediterranean to carry the silt away, so year after year it collects at the mouth of the Nile, and the Nile must find its way over the silt to the slightly more distant sea, where it again deposits silt. In this way a whole area of flat fertile soil is built out into the sea, across which the river splits up into slow-moving branches.

To Herodotus, looking across the sea to Egypt, the mouth of the Nile looked like a triangle, with its apex pointing away from the sea.

Now, it is customary for us to name things after letters if they have the proper shape (U-curve, I-beam, T-square and so on). The Greeks did the same. The triangular area resembled the capital form of the fourth letter of the Greek alphabet (Δ), and so, Herodotus tells us, they called it by the name of the letter, delta. The word is now used for all such built-up mouths, even when there is nothing triangular about them. The Mississippi delta, for instance, is shaped nothing at all like the Greek delta.

Dendrochronology

IN STUDYING THE PAST men are interested in when events took place; whether one event took place after another, or before, or at the same time. For recent periods there are records to consult. For times before the invention of writing (which is only about five thousand years old at most, and not nearly as old as that in most parts of the world), something else must be found.

Can natural events do the writing? In some ways nature is so regular as to be useless. The sun rises and sets in a fixed pattern; the moon changes its phases; the seasons come and go. The rain, however, varies, with plenty of rain in some years and drought in others.

The American astronomer Andrew E. Douglass, working in the desert regions of the American south-west in the early decades of the twentieth century, considered tree growth. Each growing season produces a layer of light wood about the trunk. The thin, dark layers between are the tree rings. A rainy season means considerable growth and a wide space between two rings; a dry season means little growth and a narrow space. A twenty-year-old tree would demonstrate a twenty-year pattern of rainfall that would not be exactly duplicated in any other twenty-year period.

Douglass studied many trees and found the pattern overlapping; the pattern in the early years of one tree would be like that in the later years of an older tree. He used wood salvaged from ancient buildings, and carried a continuous pattern back over a thousand years. A piece of wood from some man-made structure could be dated by fitting the pattern of its rings somewhere on Douglass's continuous pattern. Thus, one could tell when a prehistoric structure was built.

This device for dating early events was named dendrochronology by Douglass, from Greek words meaning telling time by trees.

Deuterium

ISOTOPES ARE varieties of atoms of an element that differ among themselves in their weight. For instance, some oxygen atoms weigh 16 units, some 17 and some 18. They are distinguished by being called oxygen-16, oxygen-17 and oxygen-18.

In 1931, however, the American chemist Harold C. Urey, and co-workers, discovered an unusual isotope. It was a hydrogen isotope present in small quantities wherever ordinary hydrogen occurred. Ordinary hydrogen, the lightest of the atoms, has a single proton (unit weight) in its nucleus. It is hydrogen-1. The new isotope had a proton (unit weight) plus a neutron (also unit weight) in its nucleus and was hydrogen-2 or, more colloquially, heavy hydrogen. An atom of hydrogen-2 weighed twice as much as one of hydrogen-1. There was the unusual property. No other isotopes in the list of elements differed so in weight (on a per cent basis).

Because of this large weight-difference, hydrogen-1 and hydrogen-2 are, for isotopes, unusually distinct physically and chemically. They seemed to deserve special names. The British physicist Ernest Rutherford suggested haplogen for hydrogen-1 and diplogen for hydrogen-2, from the Greek words *haploos* (single) and *diploos* (double).

Urey, however, suggested deuterium for hydrogen-2, from the Greek word *deuteros* (second), and it was this name that was adopted. By analogy, hydrogen-1 is protium, from the Greek word *protos* (first). Then, when the British physicist M. Oliphant discovered a still heavier isotope, hydrogen-3, in 1934 it was automatically named *tritium,* from the Greek word *tritos* (third).

Since the nucleus of a protium atom is a proton, the nucleus of a deuterium atom (a one proton/one neutron combination) is called a deuteron, while the nucleus of a tritium atom (a one proton/two neutron combination) is called a triton.

Diagonal

IN ANY POLYGON two adjacent angles are connected by sides of the figure. If the polygon has more than three sides, then it is possible to connect two angles that are not adjacent by a straight line that cuts directly across the polygon. Such a line is a diagonal, from the Greek prefix *dia* (through) and *gonia* (angle); it passes through the angles.

The most familiar polygon is, of course, the square, and this is usually drawn so that two sides are horizontal and two vertical. In a square drawn this way, the diagonal, moving from one angle to the opposite, must run slantwise. Diagonal has therefore come to mean a slantwise direction, though the proper word for that is really oblique, which comes from the Latin *obliquus—ob* (before) and *liquus* (crooked). An oblique line, in other words, runs crookedly on before, slanting away from true.

A circle has no angles and can have no diagonal, and yet a straight line can be drawn across the circle from one side to the other. Such a line is called a chord, from the Greek *chorde* (the viscera of an animal). The intestines were used, of course, to make the strings of musical instruments, which underlines another meaning of the word (see OCTAVE), but strings of some sort were the first measuring instruments, and a lute's strings would do at a pinch.

A chord that passes through the circle's centre is the longest possible chord for a given circle. It measures the greatest thickness of the circle. This is the diameter, from the Greek *dia-* (through) and *metron* (a measure); it is a measure through the circle.

The sum of the lengths of the sides of a polygon is, similarly, the perimeter, from the Greek *peri* (around) and *metron* (a measure); it is a measure round the polygon. This could logically be used for the length around a circle, too, but in the latter case mathematicians, for some reason, use the Latin equivalent. The length of the circle's boundary is the circumference, from the Latin *circum* (around) and *ferre* (to carry). The measure is carried around the circle.

Diamond

OCCASIONALLY an element can exist in two or more different forms, depending on temperature, pressure or other environmental factors. As an example, we have oxygen and ozone (see BROMINE). Generally one such form can be changed into another in the laboratory and the Swedish chemist J. J. Berzelius in 1841 suggested these forms be termed allotropes of an element, from the Greek *allos* (other) and *trope* (change).

The most dramatic example involves the element carbon, which is most familiar to us as coal, some forms of which are almost pure carbon. The Latin word for coal was *carbo* (genitive, *carbonis*), hence the name of the element. The word coal was originally applied to any burning ember. If wood is heated without being allowed to burn, for instance, a black residue is left, which will burn slowly. This is charcoal—i.e., coal formed by charring.

Coal, charcoal and various forms of soot are allotropes of carbon in which the carbon atoms are arranged in no particular order. They form no regular pattern or shape. A substance such as coal is therefore said to be amorphous (from the Greek *a-* meaning no or not and *morphe* meaning shape).

Carbon atoms can arrange themselves into rings of six, joined in sheets that resemble an orderly array of bathroom tiles, each sheet held loosely to neighbouring sheets. Such sheets are easily split apart, and if some of this material is rubbed over paper pieces flake off and remain behind. This is the secret of the pencil, and this allotrope of carbon is called graphite, from the Greek *graphein* (to write).

If carbon is put under great temperature and pressure the atoms take up a very symmetrical arrangement that holds together unusually tightly. The result is a transparent and extremely hard allotrope of carbon called diamond. This word is a corruption of the word adamant, once used for the jewel. Adamant in turn comes from the Greek *a-* (not) and *daman* (to subdue). The jewel was so hard, you see, that it could not be subdued. Nothing else, that is, could scratch or make an impression upon it.

Digital Computer

THE FINGERS were the first device used by man for counting, and for simple computations in adding and subtracting. Digit (from the Latin word for finger) therefore means both a finger and any number below ten. The fingers are thus the first digital computer.

Any mechanical device that represents whole numbers the way that fingers do, and can be manipulated as numbers can, so as to give correct answers to numerical problems, is also a digital computer. The abacus, in which numbers are represented by pebbles in grooves, or discs on wires, is a convenient digital computer.

In 1642 the French mathematician Blaise Pascal invented a mechanical device of wheels and gears. Each wheel had ten positions, one for each digit from 0 and 9. When the first wheel reached 9 and passed to 0 it engaged the second wheel, which turned to 1. When the second wheel turned as far as 0 the third wheel was engaged, and so on. In this way numbers could be added and subtracted mechanically. In 1674 the German mathematician Gottfried von Leibnitz arranged wheels and gears so that multiplication and division were also made automatic.

In 1850 an American inventor, D. D. Parmalee, pushed marked keys to turn the wheels, and this was the cash register.

Such mechanical devices were improved by electronic techniques. An electronic computer was first built during the Second World War, according to the plans of the American engineer Vannevar Bush, in which electric currents replaced mechanical gears. Such electronic digital computers were rapidly and vastly improved. Transistors took the place of vacuum tubes and reduced computer size; sophisticated memories were added; languages were developed for direction. Now computers can far outstrip the speed of the human mind in any purely mechanical computation for which a computer can be given the necessary complete and detailed instructions.

Digitalis

AS EVERY child knows, fingers are handy when it comes to solving simple mathematical problems, such as adding two and four. This shows up in the words we use. The Latin word for a finger or toe is *digitus,* and we still call fingers and toes digits. But the word digit also refers to whole numbers—one, two, three and so on. Fingers and numbers are so closely related that the same word will do for both.

Digits come into medicine, too. There is a plant with drooping purple or white flowers that in English is called foxglove. In German the name is *Fingerhut,* which means thimble, because the flowers form little tubes that look like thimbles.

In 1541 a German botanist, Leonard Fuchs, decided to give the plant a Latin name (only Latin names are official for plants and animals). With *Fingerhut* in mind, he called the group of plants to which the foxglove belongs digitalis. In Latin this means 'of or pertaining to the finger', and certainly a thimble is something that pertains to a finger.

Now, in the days before modern medicine there were people who cooked up various plants according to secret recipes handed down from generation to generation, and used the results to treat sickness. One of the plants valued for this purpose was the foxglove. However, as medicine advanced, doctors laughed at this sort of superstitious nonsense.

Just the same, an English physician, William Withering, wrote an article about the medicinal uses of foxglove in 1785, and was not ashamed to admit that he grew interested in the plant as a result of information received from an old countrywoman whose family secret it was.

And in fact for nearly two centuries now chemicals obtained from the foxglove have been used to improve the wavering heart-beat, slowing it and making it more even and intense. These medicines are the digitalis glycosides.

Dimension

GEOMETRY BEGAN as the practical art of measuring land areas (the word comes from the Greek *ge,* meaning earth, and *metron,* meaning measure) and determining the volumes of containers. It was important to know how many measurements were needed to fix an area or a volume. This thought is expressed in the word dimension, which comes from the Latin *dimensio*— *di-* (apart) and *metiori* (to measure).

A rectangle, for instance, must be taken apart, so to speak, into two measures—its length and its width. The two measures multiplied give the area. A rectangle, therefore, is a two-dimensional figure.

The basic two-dimensional figure is the plane, which can be visualized as a continuous sheet without thickness, spreading out in all directions, perfectly level and smooth. (Plane comes from the Latin *planus,* meaning flat or level.) Any figure which can be drawn on a plane is two-dimensional.

The true mathematical plane does not exist in the real world, because nothing real can have zero thickness. An object which has thickness as well as length and width can exist. To measure the volume of a cube, for instance, one must measure all three: length, width and thickness. A cube is a three-dimensional figure.

The importance of reality is shown by the fact that any three-dimensional object is called a solid, from the Latin *solidus* (dense).

A solid, you see, is dense; it has weight and substance; it is not a mere abstraction.

If we add a fourth dimension we are away from commonplace reality again. Mathematicians can suppose figures with any number of dimensions, and even find it useful to do so. In Einstein's theories, for instance, time can be considered a fourth dimension, though it is not experienced by our senses in the same way as are the ordinary three dimensions of space. To talk of a universe which contains such a fourth dimension, scientists use the expression space-time.

Dinosaur

OF ALL THE extinct forms of life on earth, the most dramatic are the gigantic reptiles that lived in the Mesozoic era (see FOSSIL). The most dramatic of modern-day reptiles are the snakes, whose characteristic method of movement gives the name to the whole group of animals, since reptile comes from the Latin *repere,* meaning to creep. The word serpent, incidentally, comes from the Latin *serpere,* also meaning to creep. The reptiles include all cold-blooded animals with scales and bones, and, while some do creep, others swim, leap, and even 'fly' for short distances.

The extinct giant reptiles are usually lumped together in the popular mind as dinosaurs (though, from a scientific standpoint, it should be realized that some dinosaurs were as small as hens, while some giant reptiles were not dinosaurs), from the Greek words *deinos* (terrible) and *sauros* (lizard)—and terrible lizards some were, indeed.

The most terrible dinosaur was *Tyrannosaurus rex,* which grew to about fourteen metres in length and stood as high as a giraffe on its tremendous hind-legs, with a skull over a metre long and a cavernous jaw equipped with tremendous teeth. It was the largest meat-eater ever to inhabit dry land, and its name is very descriptive.

To the Greeks, it is believed, a *tyrannos* was any individual who gained one-man rule of a city by rising from the common people (like a modern dictator), rather than by inheriting the rule as a member of a kingly line stretching back into antiquity. A tyrannos was not necessarily a worse ruler than a king, but in 510 B.C. the Athenians expelled the tyrant Hippias and established a democracy. After that they spoke ill of tyrants in general, and tyrant took on its modern meaning. Actually, though, master would be a better translation of the original *tyrannos.*

Since *rex* is the Latin word for king, *Tyrannosaurus rex* means

King Master-Lizard, which is an excellent name, for if ever there was a king of beasts, this was it.

Dipole Moment

MANY SUB-ATOMIC PARTICLES carry an electric charge, either positive or negative. Atoms are made up of two kinds of such particles: protons with a positive charge, and electrons with negative charge. The number of each kind in a complete atom is equal. What's more, both kinds of particles are distributed evenly about the centre of the atom. The average positions of the positive charges and of the negative charges appear to be at the centre of the atom. The effects of each cancel out, and the complete atom behaves as though it has no charge.

When two similar atoms join, as two chlorine atoms do in forming the chlorine molecule, they share electrons equally. The two types of charge are still distributed evenly about the centre of the molecule. When two unlike atoms cling together, however, and share electrons, one usually has a stronger hold on the electrons that does the other. The average position of the negative charge shifts away from the centre of the molecule and towards the atom with the stronger hold.

In the latter case the average position of the negative charge is in a different place from that of the positive charge. There is a separation of the two charges. We can say there is a positive pole in one place and a negative pole in another. (The terms are an analogy to the north and south poles in magnets.) Because there are two poles, such a molecule is a dipole.

In an electric field, such a dipole moves so as to have the line connecting the poles parallel to the direction of the field. The readiness with which it moves depends on the size of the charges and the amount by which they are separated. This readiness is the dipole moment, where moment is a form of the word movement.

Dirigible

IN 1782 two French brothers, Joseph Michel and Jacques Etienne Montgolfier, lit a fire under a large, light bag with an opening underneath, and allowed the hot air to fill it. The hot air, being lighter than cool air, lifted the bag, and this was the first balloon, a word which arises from the same source as ball, the -oon ending implying largeness. A balloon is just a kind of large ball, after all.

It was not until 1852, however, that another French inventor,

Henri Giffard, was able to mount a steam-engine in the gondola under a cigar-shaped balloon, and exert enough power in this way to guide the balloon against the wind. This was a dirigible balloon, from the Latin *dirigere* (to guide). The phrase was shortened simply to dirigible, which was then applied to powered balloons in general.

The man most responsible for making such dirigibles practical was Count Ferdinand von Zeppelin, a German inventor, who first built a large aluminium framework within which a bag might be expanded, thus making giant dirigibles possible. In his honour, such dirigibles are often called zeppelins. In fact, the most successful dirigible ever flown was the *Graf Zeppelin*, which made numerous crossings of the Atlantic and circumnavigated the world. It was named after the Count, since *Graf* is German for count.

Still another common name for the dirigible is airship, since it floats and moves through air, as an ordinary ship floats and moves through water.

In 1903, for the first time, a machine was driven through the air without a bag of hydrogen to hold it up. It was heavier than air, and that was the remarkable thing about it. What did hold it up was the moving air lifting upward under the large plane areas of the wings (plane being derived from the Latin *planus,* meaning flat or level). The device was therefore named an airplane or, using the Greek *aer* (air), aeroplane. (A seaplane, however, is not one that moves through the sea, but is an aeroplane equipped to land on the sea.)

Domains, Magnetic

ATOMS ARE MADE UP of electrically charged particles, and any electric charge always has an associated magnetic field. All matter, therefore, has the potentiality of displaying magnetic effects, and yet most types of matter scarcely do. Iron, steel and related materials are much more easily affected by magnetic forces than are other substances commonly found in nature. The particular type of magnetism displayed by iron is therefore called ferromagnetism, the prefix coming from the Latin word for iron.

An explanation of iron's peculiar relationship to magnetism was advanced in 1907 by the French physicist Pierre Weiss. He suggested that in general the tiny atomic magnets in matter were oriented in all directions so that the effects were neutralized. In iron, on the other hand, there were microscopic regions over which many millions of atoms orient themselves in such a way that all the tiny magnetic fields are in the same direction. Each tiny region is therefore a much stronger magnet than a similar region of any other substance would be. Such a region of strong magnetic field is called a magnetic domain.

In ordinary iron the magnetic domains are oriented in all ways, so that the iron is not particularly magnetic. The domains are much easier to orient than individual atoms are, however. Earth's magnetic field produces such an orientation in a ferric oxide called lodestone, and this can produce orientation in metallic iron or steel.

If a ferromagnetic substance is ground into particles smaller than the individual domains making it up each particle will consist of a single domain. If these are suspended in liquid plastic they can easily be aligned by the influence of a magnet. If the plastic is then allowed to solidify a particularly strong magnet results; one, moreover, that can be easily machined into any desired shape.

Duodenum

AS SOON AS you put food or drink into your mouth it has entered the alimentary canal (from the Latin word *alimentum,* meaning nourishment). Once you swallow the food enters a 25-centimetre tube that leads to the stomach. The tube is the gullet, from the Latin *gula* (throat), or oesophagus, a Greek word coming, possibly, from *oisein* (to carry) and *phagein* (to eat). It is a tube, in other words, that carries what you eat.

Leaving the stomach, the food enters a long coiled tube called the intestines, from the Latin *intestinus* (inner). The last part of the intestine is wider in diameter than the first part, so that we have a long small intestine and a comparatively short large intestine. (The terms small and large refer to diameter rather than to length.)

Anatomists have divided the small intestine into three parts. The first 28 centimetres of its length is the duodenum. This comes from the Latin *duodeni* (twelve each), because the early anatomists measured by using their hands as convenient rules, and the duodenum is twelve finger-widths in length. The Germans, who are much less patient with classical derivations than we are, call the duodenum *Zwölffingerdarm,* which is straight German for twelve-finger-intestine.

The next 2·4 metres of intestine is the jejunum, from the Latin *jejunus* (empty), because the Roman medical writer Celsus thought it retained no food but merely sent it on. The remainder of the small intestine is the ileum, a Latin word that comes perhaps from the Greek *eilein* (to roll up), because that is what must be done to the small intestine if it is to be made to fit in the abdomen.

The intestines are also referred to as the bowels. This comes from an Old French word *boel* which is in turn derived from the Latin *botellus* (a little sausage) and the intestines do indeed look like a string of sausages.

Dynamite

THE SWEDISH inventor Alfred Nobel was practically brought up in the explosives business. His father produced nitroglycerin for commercial use, and this, of course, is a dangerous occupation. In fact, an accidental explosion of nitroglycerin killed Nobel's younger brother.

Nobel therefore began to seek ways in which to make nitroglycerin safer to handle. In 1862 he found the answer in a kind of earth called kieselguhr. This name is German, from *Kiesel*, meaning flint (a very common rock) and *Guhr,* meaning an earthy deposit.

Kieselguhr is made up of microscopic pieces of silicon dioxide (of which flint is one form) that once made up the skeletons of tiny one-celled plants called diatoms. These plants are so called because in some common varieties each cell is divided into two nearly separate parts. The Greek *dia-* means through and *temnein* means to cut; the cell is almost cut through, in other words. Kieselguhr is often called diatomaceous earth, or diatomite.

Because kieselguhr is composed of individual skeletons with microscopic spaces within and between the pieces, it is porous and will absorb liquids. For instance, it will absorb as much as three times its own weight of nitroglycerin. The mixture of kieselguhr and nitroglycerin can be moulded into sticks which can then be handled with practically no danger of accidental explosion. When the sticks are detonated, however, in the appropriate manner, as for instance by an electric spark (set off from a distance), a powerful explosion results. This safe nitroglycerin was patented by Nobel in 1862 under the name dynamite, from the Greek *dynamis* (power).

Nobel made a large fortune from dynamite and other inventions in the explosives field, and when he died he left over £1 680 000 for the establishment of the Nobel Foundation. This awards each year the Nobel Prizes in six classifications: physics, chemistry, medicine and physiology, literature, peace, and (since 1969) economic science. In 1957 element 102 was originally produced at the Nobel Institute and named nobelium in its honour and in that of Alfred Nobel.

Dynamo

THE FIRST method for producing an electric current involved batteries (see VOLT), but batteries are not practical as sources of current on a large and continuous scale. The key to something better came when the Danish physicist Hans Christian Oersted discovered in 1820 that a wire through which a current was flowing could attract a compass needle, and that therefore electricity and magnetism were somehow related.

The English chemist Michael Faraday showed the reverse to be also true. In 1831 he discovered that a current was set up in a copper disc rotating between the poles of a magnet. Moving electricity resulted in magnetism, and moving magnetism resulted in electricity.

It was then only necessary to think up a way of rotating a coil of wire between the poles of a magnet (it doesn't matter whether the magnet moves and the coil is stationary or vice versa) and then bleeding off the electricity as it was formed. The coil can be made to turn continuously by means of a turbine run by water or steam (see ENGINE). In this way mechanical energy is converted into electrical energy in sufficient quantities to light cities and run huge factories.

Such a device is called a generator, from the Latin *gignere* (to produce); it certainly produces electricity. An earlier name, however, was dynamo-electric machine. The Greek *dynamis* means power, so the longer name means a machine producing electricity from ordinary non-electric power, so to speak. That name was shortened to dynamo.

This name had an unfortunate consequence for the city of Constantinople. When it was first suggested that the city be electrified it was explained to the Sultan of Turkey that it would be necessary to install dynamos. The Sultan, who was not a man of advanced education, knew only that dynamo sounded distressingly like dynamite, and he knew what dynamite was. So he vetoed the project, and Constantinople had to wait several additional years for its electricity.

Echinodermata

IT WAS a medieval theory that for every animal on land there was an equivalent animal in the sea. Some of that theory remains in the names given various sea animals. For instance, certain species of seals are called sea lions, sea bears and sea leopards, while a particularly large seal with a bulging snout is the sea elephant. The manatee, a sea mammal not related to the seals, is usually called the sea cow, while a porpoise (a small member of the whale family) was sometimes called a sea hog.

The fish also contribute examples, the most famous being the sea horse, a small fish with a head that resembles a surrealist version of a horse. There is also a fish with a spiny skin, capable of blowing itself up into a sphere when threatened. It is then both difficult and dangerous to grab. It is called the globe fish in consequence, or the sea hedgehog.

The term sea hedgehog also applies to another creature, a much more primitive one. It is an invertebrate animal (see PHYLUM) in the shape of a sphere, flattened on one side. It rests on the sea bottom on that flat side (in the centre of which is the mouth). The rest of the body is thickly covered with spines. The Latin word for hedgehog is *ericius,* from which we get urchin (usually applied these days to little boys, who can indeed have a prickly nature and be hard to get close to without damage) so the spiny sea creature is called a sea urchin. The Greek *echinos* means hedgehog, and the class of organisms to which the sea urchin belongs is Echinoidea.

The sea urchins, along with other creatures with spiny skins and additional features in common, belong to the phylum Echinodermata, from the Greek *derma,* genitive *dermatos* (skin), hence the spiny-skins. The best-known examples of the phylum are the various starfish (these are shaped like stars, but are not fish) which belong to the class Asteroidea, from the Greek *aster* (star). Plants as well as animals may have their equivalent in the sea, I should mention, since there are members of this phylum which owing to their appearance are commonly called sea cucumbers and sea lilies.

Echolocation

IN 1793 the Italian biologist Lazzaro Spallanzani grew interested in the manner in which bats could flit about in the dark, avoiding obstacles. Could they see in the dark? He blinded some bats and found that they could still fly without difficulty, and without blundering into obstacles. However, when he plugged their ears so that they could not hear they were helpless, and blundered into obstacles even though their eyes were open and working. Spallanzani had no explanation for this, and could only record the observation.

Studies in the twentieth century showed that bats constantly emit very high-pitched squeaks. Some of the sounds they make are ultrasonic (from Latin words meaning beyond sound). The sounds were of such high frequency and short wave-length, and therefore so highly pitched, that the human ear could not detect them.

Sound is reflected from objects, producing an echo. The size of the object required to produce an echo depends on the wave-length of the sound. Ordinary sound has such long wave-lengths that it takes large objects like walls to produce an echo. Ultrasonic sound has such short wave-lengths that small objects, even twigs or insects, could produce some echo.

The bat, emitting its high squeaks, listens for echoes with its large, sensitive ears. From the direction of the echo and from the time it takes the sound to reach the obstacle and return, it can tell the direction and distance, and even the nature of the obstacle. It can avoid a twig and snatch at an insect.

It is possible that dolphins also use such a system of echolocation, using somewhat deeper sounds and detecting larger objects, such as fish. The guacharo, a cave-dwelling bird of Venezuela, may also use echolocation.

Ecology

GENERALLY mankind has not concerned itself with other forms of life except as they suited its purpose. Useful plants or animals were preserved, cultivated or herded. Other animals were killed, sometimes for sport, even when they were not dangerous. Still others were ignored.

Men have come to realize, however, that organisms (even ourselves) do not live in isolation. Each depends on others. A species may even depend for its well-being on another species that preys upon it. Rabbits are better off because stoats exist, for instance.

In some areas the stoats were killed off by gamekeepers because they preyed also on game birds. The rabbits, freed of the menace, multiplied and outran their food supply. Many of them starved, and in the end there were fewer and weaker rabbits than before. Since stoats usually catch and kill the older and weaker rabbits, they serve to keep the rabbits younger and stronger than would be the case otherwise.

Man's interference often upsets the balance of nature and has created deserts. It has led to the extinction of harmless creatures preyed on by animals carelessly introduced into their areas by man. The rabbit was introduced into Australia to provide sport, and has become a major pest. Several of the native marsupials are dying out because of the competition of these more efficient mammalian counterparts. It can apply to plants too: the imported prickly pear ran riot in Australia till a parasite was found that would control it.

In fact, as man's numbers grow and his technological ability to change the environment increases, he is more and more rapidly and extensively changing the balance of nature. With this in mind scientists are beginning to study the interrelationships of life-forms with each other and with the environment, in order to learn best how to stop and reverse man's damaging effect.

The prefix eco- is from a Greek word meaning house (economy is house management) and it can be applied to the environment generally, for that is the house of life, so to speak. Consequently, ecology is the name given to the newly stressed study of the interrelationships of life-forms among themselves, and with the environment.

Ecosphere

CONSIDERING THE VASTNESS of the universe, it seems possible that earth might not be unique, and that there are other worlds with life—even intelligent life.

In the 1930s the solar system was thought to have been formed through the close approach of two stars, with mutual gravitational pull tearing out matter from each to form the planets. If this were so planetary systems would be rare, since stars are spread so widely apart that they would almost never approach each other in this fashion.

In the 1940s the German astronomer Carl von Weizsäcker advanced a theory of planetary formation from an original dust-cloud that made it seem that almost any star would be accompanied by planets. And since then some near-by stars have been shown to have planets.

Furthermore, investigation into the origins of life on earth have shown that, given earth-like conditions, the formation of life is almost inevitable. Consequently, any planet with the proper mass, temperature and chemistry ought to develop life (see EXOBIOLOGY).

What, then, are the requirements for the existence of earth-like planets? A particular star ought to have some region in its neighbourhood where a planet would get just enough radiation to maintain the conditions necessary for life—temperatures at which water would be a liquid, for instance. This life-suitable region, distributed in a spherical shell all about the star, is the ecosphere (where eco- is from the Greek word for house and therefore expresses a place man can inhabit).

The American astronomer Stephen Dole, considering the ecospheres of different types of stars and other information as well, has estimated there may be as many as 640 000 000 planets in our galaxy alone which might be suitable for earth-like life.

Elastomer

COMPLICATED MOLECULES can in some cases be easily broken down to other molecules much simpler in structure. It was as though the more complex molecule is built up of chains of one or more much simpler molecules. In 1830 the Swedish chemist Jöns Berzelius suggested that such a chain molecule be called a polymer, from Greek words meaning many parts. The single parts of which it was a chain would then be called monomers (one part).

Thus large starch molecules can be broken up and shown to consist of chains of small glucose molecules. Protein molecules consist of chains of twenty different, but related, amino acids. Rubber molecules consist of chains of a hydrocarbon called isoprene (a made-up name of no meaning).

Chemists found that simple compounds often hooked together in chains, even with very little encouragement. The simple compounds polymerized. Generally this resulted in a gummy, useless substance that chemists tried to avoid.

In the twentieth century, however, ways were discovered for deliberately creating polymers that might be useful because they were strong, stable, and could be moulded. These came to be called plastics, because something that is plastic can be moulded.

With the coming of the motor-car and its rubber tyres there was a demand for artificial rubber-like polymers that might be cheaper or more available than natural rubber. By the 1940s synthetic rubber was in production.

Rubber has the ability to deform its shape and then spring back to its original. This is called elasticity (from the Greek word meaning springiness). Synthetic rubbers were therefore elastic polymers, and this phrase was shortened to elastomers.

Electrocardiogram

THE REGULAR RHYTHM of the heartbeat is maintained by a periodic electrical change that travels along the heart muscle. If it were possible to follow this change one might detect abnormalities in the heart action impossible to catch by merely listening to the heartbeat.

It is impractical to place electrodes directly on the heart for some routine examination, but the tissues conduct electricity, and electrodes on the skin would do the trick if a device delicate enough to detect very tiny changes could be developed.

Such a device was first perfected by a Dutch physiologist, Willem Einthoven, in 1903. He made use of a very fine fibre of quartz that was silvered to allow it to conduct a current. Even tiny electrical charges caused noticeable deflections of the fibre. These movements could guide a pen which would mark out an irregular line on a slowly unrolling length of graph-paper. The result is an electrocardiogram, or ECG, from Greek words meaning written record of heart electricity.

The heart is not the only organ that works by rhythmic changes in electrical conditions. The organs most intimately connected with electric pulses are the nerves, and it would therefore not be in the least surprising to find that the brain was a source of varying electrical changes. In 1875 an English physiologist, Richard Caton, applied electrodes directly to the living brain of a dog and could just barely detect tiny currents.

Such an experiment on humans seems impracticable, but in 1924 an Austrian psychiatrist, Hans Berger, placed electrodes against the human scalp and by using a very delicate detecting device found he could just record electrical changes. He published his results in 1929, and improvements since then have made it easy to obtain and study electroencephalograms or EEGs (written records of brain electricity).

Electroluminescence

LIGHT IS ALWAYS PRODUCED by hot objects. The sun, an electric-bulb filament, burning gases from wood, coal, oil or a candle, all give off light because they are at temperatures of over 600°C. Such high-temperature light is called incandescence (from Latin words meaning to become hot enough to glow).

Nevertheless, light can also be formed in the absence of high temperature. Chemical reactions within the tissues of a firefly, for instance, produce small quantities of light at ordinary temperatures. This is sometimes called cold light, but the more formal name is bioluminescence (from the Greek for life and Latin, to become light).

Some substances can give off energy in the form of light after they have absorbed energy in some other form. For instance, the compound zinc sulphide when properly prepared will absorb energy from an electric field and then give it off again in the form of light. It will glow while remaining at room temperature, and this is called electroluminescence. This was first studied by the French physicist Georges Destriau in 1936.

In ordinary light bulbs, which work by incandescence, very little of the energy used is radiated in the form of visible light. Most is in the form of invisible infra-red radiation, which can be felt as heat. An electroluminescent powder, prepared over a sheet of glass, can be made to glow softly, producing only visible light and very little infra-red. Much less electrical energy is consumed to produce a given amount of light. The electroluminescent panel is much more expensive to prepare than is a light bulb, however, and it will deteriorate rather easily under improper conditions of temperature and humidity.

Substances like zinc sulphide that will exhibit electroluminescence are sometimes called electroluminors.

Electrolysis

ONCE VOLTA invented the battery (see VOLT), chemists had electric currents to play with. If such a current were passed through certain liquids chemical changes took place. Generally, there was a separation of the molecules of the substances in solution into smaller pieces. From dissolved copper sulphate, copper could be deposited. Hydrochloric acid in solution would break down into chlorine and hydrogen; water would split up into oxygen and hydrogen; and so on.

Because the molecules seemed to loosen and break apart, the process was called electrolysis. Since the Greek word *lysis* means a loosing, electrolysis is a loosing by electricity.

Distilled water will not carry electricity, but there are substances such as sulphuric acid or sodium chloride which when added to water allow an electric current to pass and electrolysis to proceed, and these are called electrolytes. A substance such as sugar, which when dissolved does not allow the electric current to pass, is a *non-electrolyte*.

In order for an electric current to pass through a liquid two metal rods—one connected to the positive pole of a battery, and one to the negative pole—must be dipped into the liquid. These two rods are electrodes, the -ode suffix coming from the Greek *hodos*, meaning route (exodus is the way out). The electrodes form the route of the electric current.

The electrode connected to the positive pole of the battery is the positive electrode; the other the negative electrode. The British physicist Michael Faraday first suggested (in 1834) that the positive electrode be called the anode, the negative the cathode, from the Greek prefixes *ana-*, meaning up, and *kata-*, meaning down. At that time it was believed the electric current travelled from the battery's positive pole to its negative pole, just as a current of water travels from a hilltop to a valley; hence the choice of prefixes. (As a matter of fact, just the reverse is now known to be true; electricity, or at least electrons (see ELECTRON), travels from negative pole to positive.)

Electron

THE ANCIENT Greeks noticed, back in 600 B.C. or thereabouts, that if pieces of amber were rubbed with cloth the amber became capable of attracting small feathers, light bits of wool and so on. Amber is a glassy, yellowish-brown substance that is the fossilized resin of long-extinct pine trees that once grew on the shores of the Baltic Sea. The Greeks (as well as other ancient peoples) used it for its ornamental value, and the Greek name for amber was *elektron*.

Other substances besides amber gained this attracting force when rubbed, but amber was the classic example. Therefore when William Gilbert (the court physician of Queen Elizabeth I) studied the attracting force he suggested it be termed electricity. Eventually people came to recognize the existence of an electric fluid which might stay put, as in amber, or might flow, as in a metal wire.

By the 1870s scientists grew to think more and more that just as matter was composed of tiny particles (atoms), so the electric fluid must also be composed of tiny particles. The Irish physicist G. Johnstone Stoney suggested in 1891 that the quantity of electricity present in each of these particles be called an electron. This was accepted, and soon the name was applied to the particle itself.

In 1932 the American physicst C. D. Anderson discovered a particle just the size of the electron, but possessing an opposite kind of electricity. Whereas the electron carried negative electricity, the new particle carried positive electricity, and was therefore named positron. (Actually, the r in positron is due to a false analogy with the r in electron. There is no r in positive, so that a more logical name for the new particle would have been positon.) For a while there was a move afoot to change the name of the ordinary negative-electricity electron to negatron, but the name never became popular.

Electrophoresis

GIANT PROTEIN MOLECULES come in many varieties, and

protein extracts from tissues usually contain a large number of very similar molecules which nevertheless can behave quite differently within the body. Attempts to separate and isolate the different components of these complex mixtures failed when ordinary chemical methods were used.

However, each protein molecule has negative and positive electric charges distributed across itself, and as a result will be attracted in one direction or another by an electric field. The direction and intensity of the attraction will depend on the pattern of charges on the molecule, and this is different for each molecule, even when two of them are otherwise very similar. This was first pointed out in 1899, by the English biologist Sir William Hardy.

If an electric current is passed through a solution of proteins, some molecules will travel toward one electrode, some toward the other, each at its own characteristic speed. As a result the protein molecules separate, and the number and identity of the components making up the mixture can be worked out. This procedure is electrophoresis, from Greek words meaning borne by electricity.

This way of separating proteins did not become practical until methods were worked out for detecting fine degrees of difference from point to point in the solution, as the nature of the mixture changed with the gradual separation of different molecules. In 1937 the Swedish chemist Arne Tiselius devised a set of tubes that could be put together at specially ground joints into a rectangular U. By the use of certain lenses he could follow differences in the protein mixture by changes in their ability to bend light rays, and by taking the U tube apart he could trap one component or another in the different sections.

Electrophoresis has been used to detect tiny changes in blood chemistry in the course of various diseases.

Ellipse

IF YOU were to hold up a circle (from the Latin *circulus,* meaning little ring) of cardboard to the light so that it cast a shadow on a smooth white surface the shadow would be circular if the disc were held just between light and wall, and just parallel to the line of the wall. If you were to tip the disc slightly the shadow would no longer be exactly circular, but would flatten somewhat into an oval. The more you tipped the disc, the flatter the oval.

This flattened circle is one of three related geometrical figures (see PARABOLA) studied by Apollonius about 250 B.C. He had a mathematical expression for each, and of the three the value of the expression was least for this flattened circle. Since the expression in this case was deficient compared to the others, and the Greek word for deficiency is *elleipsis*, the curve is called an ellipse.

There are two points within an ellipse, the foci (singular, focus). Imagine a large ellipse built out of mirrors facing inward. If a candle were placed at one focus of this ellipse the rays of light travelling in all directions out from the candle would strike the mirrors at all points, and at every point the rays would be reflected in such a way that they would all converge upon the other focus.

The German-born astronomer Kyepler discovered that the planets revolve round the sun in ellipses, and one of the foci is itself the sun.

Other curves also have foci, and lenses converge light to a focus. Different foci have different properties, but the interest always lies in what happens to light rays emerging in all directions from a focus. Light rays also emerge in all directions from a fireplace, and as a matter of fact *focus* in Latin means fireplace.

Halfway between the foci of an ellipse is its actual centre. The flatter the ellipse the farther apart are the foci, and the farther each focus is from the centre. Hence the flatter the ellipse, the greater is its eccentricity, from the Greek *ek* (out of) and *kentron* (centre). A circle, on the other hand, has its foci coinciding with its centre, so that it has an eccentricity of zero.

Energy

WE ALL use the Anglo-Saxon word work to mean any kind of continued and purposeful activity, but to the physicist the word has a more precise meaning. To him work involves the movement of a body against a resisting force, and only that.

The idea of work involves two things: (1) the amount of push or pull required to force an object into motion against resistance (this push or pull is, appropriately, called force, from the Latin word *fortis*, meaning strong); and (2) the distance through which the object is moved.

The Greek word for force is *dynamis*, so physicists measure the amount of force by means of a unit called a dyne. For instance, two objects each weighing 39 kilogrammes and separated by a distance of 10 centimetres attract one another with a gravitational force of one dyne.

If a force of one dyne moves a body through a distance of one centimetre the amount of work done is one dyne centimetre. The dyne centimetre is also called the erg, from the Greek word *ergon* meaning work. For example, a man weighing 68000 grammes lifting himself 244 centimetres against the resistance of gravity by climbing a flight of stairs does 68 100×244 or over 16 600 000 ergs of work. (As you see, one erg is a very small amount of work.)

An object with the capacity to perform work (pent-up steam, a suspended rock, your own muscles, a taut bowstring, an atomic bomb) is said to contain energy. It has work (ergon) in (Greek prefix, *en-*) it. Energy can not only be turned into work, but work can be changed into energy.

The British physicist James Joule proved the latter by actual experiment in 1843. He showed that a fixed amount of work was always converted into a fixed amount of heat (a form of energy).

Since the erg, as I have said, is an inconveniently small unit, the amount of 10 000 000 ergs was declared equivalent to one joule in Joule's honour. For everyday measurements the joule is more convenient. Thus, the work involved in climbing a flight of stairs is 1·66 joules.

Engine

ORIGINALLY the word engine simply meant an ingenious device. In fact the word engine is only a corruption of the word ingenious, which comes from the Latin *in-* (in) and *gignere* to produce). Clever ideas, you see, are produced in the mind of an ingenious man. After James Watt invented a practical steam-engine the word was applied more and more to those devices particularly which took power from some non-living source and turned it into work by means of the to-and-fro motion of a cylinder. Nowadays the internal combustion engines that power our cars, buses, lorries and aircraft are more important than the old steam-engine.

The older, more general meaning of the word still persists in cotton gin, a machine used to strip the cotton fibres from the seeds. Gin is only a slangy contraction of engine.

Devices which turn power into work by turning rather than by piston motion are called turbines, from the Latin *turbo* (a top, or other spinning object). Water-wheels, which are turned by running water, and which produce power as a result, are more properly water turbines. There are also steam turbines, where a jet of steam is the driving force, and gas turbines, where burning petrol or other fuel is the driving force.

A power-producing device which is turned by electricity is a motor, so called from the Latin *movere* (to move), because, after all, it is the part of the device that moves. My electric typewriter stays put while I work it, but if I lift the top I can look in and see a furiously moving motor.

In popular speech, however, the word motor is also applied to any vehicle that moves a person from here to there by use of mechanical power, so that we speak of a motor-car or a motor-boat, even though such objects are powered by engines rather than by motors.

Entropy

ENERGY CAN BE CONVERTED into work, and the law of conservation of energy states that the quantity of energy in the universe must stay for ever the same. Can one, then, convert energy into work endlessly? Since energy is never destroyed, can it be converted into work over and over again?

In 1824 a French physicist, Nicolas Carnot, showed that in order to produce work heat energy had to be unevenly distributed through a system. There had to be a greater than average concentration in one part and a smaller than average concentration in another. The amount of work that could be obtained depended on the difference in concentration. While work was produced the difference in concentration evened out. When the energy was spread uniformly no more work could be obtained, even though all the energy was still there.

In 1850 a German physicist, Rudolf Clausius, made this general and applied it to all forms of energy—not just to heat. In the universe as a whole, he pointed out, there are differences in energy concentration. Gradually, over the aeons, the differences are evening out, so that the amount of work it will be possible to obtain will grow less and less for ever, until all the energy is evened out and no more work is possible. This is the second law of thermodynamics, the conservation of energy being the first law of thermodynamics.

Clausius worked out a particular relationship of heat and temperature which, he showed, always increased in value as the differences in energy concentration evened out. He called this relationship entropy for some reason. (It comes from Greek words meaning to turn out, which seem unconnected with the case.) The second law of thermodynamics states that the entropy of the universe is always increasing.

With the discovery of quasars and other mysterious energy sources in the universe, though, astronomers are now wondering if the second law really holds everywhere, and under all conditions.

Enzyme

IN THE EARLY 1800s the notion was beginning to arise that the body produced certain substances that brought about particular chemical changes useful to the body. For instance, stomach juices contained something that digested and liquified meat. (The word digest comes from the Latin *digerere,* meaning to dissolve. The supine is *digestum.*)

At first the digestive action was attributed to the hydrochloric acid contained in the juices, but in 1835 the German physiologist Theodor Schwann reported that stomach juice contained something other than hydrochloric acid, and that this also had a digestive action. He called the new substance pepsin, from the Greek *pepsis,* meaning both cooking and digestion.

This news was greeted sceptically at first, but other such substances were found in saliva and in intestinal juices. This group of substances came to be known as ferments because their action seemed similar to that involved in the conversion of sugar and starch to alcohol (see FERMENT). In fact, by 1839 several people, including Schwann himself, proved yeast cells to be small living things, so that alcoholic fermentation was also associated with life.

For a while some scientists thought there were two kinds of ferments. Those in digestive juices, found outside the living cell, were unorganized ferments and were chemicals, no more mysterious than hydrochloric acid, which also digested meat. The ferments in yeast which brought about alcoholic fermentation, however, were organized ferments, and they involved a life force, since those ferments were only found within cells. In 1878 the German physiologist Wilhelm Kühne applied the name enzyme to the unorganized ferments, from the Greek *en-* (in) and *zyme* (yeast), to show that they were similar in behaviour to organized ferments in the yeast.

However, in 1897 the German chemist Eduard Buchner ground up yeast cells and filtered off the juice. The juice could still bring about fermentation, proving that neither intact cells nor life force were necessary. All ferments are basically alike, whether inside or outside a cell, and all are now called enzymes.

Equator

POSITIONS ON the surface of the earth can be described as being so many degrees east or west of the prime meridian (see MERIDIAN). This is the longitude, so called because the meridians by which it is measured are imaginary lines running north and south on the surface of the earth, or up and down on the conventional maps, so that they may be said to run 'longways'.

The prime meridian is 0 degrees, and measurements proceed both ways, so that there is a 10-degree east longitude and a 10-degree west longitude, for instance. The numbers increase until the series of meridians meet again directly opposite the prime meridian at 180 degrees.

In order to measure distance on the earth's surface north and south, another reference line, perpendicular to the prime meridian, must be used. The most natural such line is the equator, which circles the earth in an east-west direction and is equidistant from the poles. A series of lines parallel to the equator can be run north and south all the way to the poles, and these are called, naturally, parallels. (The meridians are not parallel, but converge at the poles.)

The parallels define the latitude of a point, since they run east and west or 'sideways' on the conventional maps, and the Latin word for side is *latus*. Latitudes are counted both north and south, so that the North Pole is 90 degrees north latitude and the South Pole 90 degrees south latitude.

Any imaginary line on the earth which cuts the world into two equal parts is a great circle, because it is the greatest circle that can be so drawn. All meridians are great circles, but the equator is the only parallel that is a great circle, so its name (from the Latin *aequator,* meaning one who equalizes) is appropriate. When the noonday sun is over the equator (see EQUINOX) day and night are equal in length, and it is that equalization that probably gave the line its name originally.

Equinox

THE TILTING of the earth's axis causes the noonday sun to seem to climb higher and higher in the sky for six months, then sink lower and lower for the next six months (see SOLSTICE). At its most northerly high point the sun is over the Tropic of Cancer at noon, and the days are then longest (and the nights shortest) in the Northern Hemisphere and vice versa in the Southern Hemisphere. At its most southerly high point, the sun is over the Tropic of Capricorn at noon and the days are then shortest (and the nights longest) in the Northern Hemisphere, and vice versa in the Southern Hemisphere.

As the noonday sun passes from its Tropic of Capricorn point (which it reaches on December 21) to its Tropic of Cancer point (which it reaches on June 21) it must pass midway the equator. It is over the equator on March 21, and at that time day and night are equal in length everywhere on earth. This is the vernal equinox. Equinox comes from the Latin *aequus* (equal) and *nox* (night). It is the time of equal nights. Vernal is from the Latin *vernalis,* meaning spring, since March 21 ushers in the spring of the year.

The noonday sun passes over the equator once again on its way south from the Tropic of Cancer back to the Tropic of Capricorn. September 22, which ushers in autumn, is the date of the autumnal equinox.

The earth bulges somewhat about the equator and the pull of sun and moon upon this bulge causes the earth's axis to twist about so that the North and South Poles trace a complete circle every 26000 years. From the earth's surface it seems as though it is the vault of the skies itself that slowly twists. The result is that every year the sun, when it crosses the equator at the time of the equinox, is seen in a slightly different spot in the sky, a little to the east of where it made the same crossing the year before. The point of crossing precedes (is earlier, is more eastward than) the last similar point of crossing, so that the 26000-year circular movement of earth's axis is called the precession of the equinoxes.

Erosion

ONE OF THE WAYS in which the face of the earth is being continuously changed is through the action of water, falling from the skies as rain and rolling across the land to the sea, dragging soil and pieces of rock with it. The amount of water involved in this process amounts to about 32000 cubic kilometres each year, so that a fair quantity of soil can be dragged along. In fact, enough soil is brought down to build flat areas far out into the ocean (see DELTA). In the process the river and its tributary streams cut out a bed for themselves in the rock and soil of the land. The process by which the energy of the flowing water cuts away a bed is called erosion, from the Latin *e-* (away, off, out) and *rodere* (to gnaw). It is a gnawing away at the soil, in other words.

The amount of erosion depends upon the gradient of the river (from the Latin *gradus* meaning step); in other words, upon the steepness of the steps taken by the river from highland to lowland. The steeper the gradient the faster the flow, and the greater the erosion. The softness or hardness of the rock across which the river flows also counts. Sometimes a river will fall over a cliff or resistant rock that it has scarcely touched into an eroded section of soft rock largely eaten away. There results a waterfall (derivation obvious) or cascade (from the Italian *cascare,* meaning

145

to fall, which in turn comes from the Latin *cadere*).

Over relatively flat land the river flows slowly, erodes little, and takes up an irregular winding course (called meandering, from the Meander river of Asia Minor, which does just that). On the other hand, under proper circumstances a river like the Colorado can cut deeply into the land and form a canyon. This is a Spanish word (Spanish-speaking people settled the Colorado territory before English-speaking people did), which comes originally from the Latin *canna* (a reed). We speak of the sugar cane, for instance, which is actually a reed. The gorge formed by the river is a long, hollow thing like a reed, and therefore canyon. The Colorado river forms the Grand Canyon, grand meaning large in many European languages.

Escape Velocity

WHEN AN OBJECT is thrown into the air the pull of gravity slows it more and more until it comes to a momentary halt, then begins to drop toward earth again. If it is thrown with greater force it moves upward more speedily, and it takes longer for the pull of gravity to slow it to a halt. The object moves higher before it begins to drop. The more forcefully it is thrown upward the higher its topmost point.

As it happens, the pull of gravity weakens as an object moves farther and farther from earth's centre. An object hurtled upward with such force that it attains a height of many kilometres finds the pull of gravity in the upper reaches of its flight to be significantly weaker. It is less effectively slowed up there, and reaches a greater height than would otherwise be expected.

Suppose a body were hurtled upward with such force that by the time it has lost half its upward velocity it was in a region of space where earth's gravitational pull was only half what it was at the surface. By the time its upward velocity was a quarter of the original the gravitational pull would also be only a quarter of what it was to begin with. Under these conditions, the object would move upward more and more slowly, but weakening gravitational pull would never bring it to a complete halt. It would never return to earth, but escape permanently into space.

The minimum initial velocity which will accomplish this is the escape velocity. On earth it is about 11·2 kilometres per second. On more massive planets, like Jupiter, the escape velocity is higher, and on less massive ones, like Mercury, it is lower. The matter of escape velocity, known since the time of Sir Isaac Newton in the 1680s, has become of particular importance since

1959, when the Soviet Union hurled the first object into space at greater than escape velocity and sent it past the moon on a trip eternally away from earth.

Ether

BEFORE MODERN times gases were not really understood. The very word gas, which is now used to represent any matter in an air-like state, was first invented about 1600 by J. B. van Helmont, a Flemish chemist, who got it from the word *chaos,* the Greek term for the mysterious unformed material out of which the universe was made.

This almost superstitious awe of gases, which could be neither seen nor touched, but which existed, was also shown in the way volatile liquids (those, that is, that turned easily into vapour —vapour being the Latin word for steam) were named. They were referred to as spirits.

One such spirit, known since the 1200s, resulted from the action of sulphuric acid on alcohol (itself a spirit—and alcoholic beverages are still called spirits today). The new spirit was the most volatile liquid that had yet been discovered. If left standing it vanished with extraordinary quickness. The chemist Frobenius named it *spiritus aethereus* in 1730.

The *aethereus* was a reference to the Greek *aither,* their term for an imaginary and incorruptible substance supposed to fill the heavens. The volatile liquid Frobenius studied seemed so eager to leave our crude, imperfect earth and depart for the heavenly realms above that it could be nothing but a 'spirit of the aither' anxious to return home.

Eventually, however, the poetry of this thought was lost under the stress of practical usage (and better understanding of gases and vapours) and the name was shortened to a simple ether.

Nowadays the term ether applies to a whole class of organic compounds with a structure similar to that of the original ether. The original ether was found to contain a pair of two-carbon groupings as part of its molecule. Such groupings were named ethyl groups, from *ether* and the Greek word *hyle,* meaning matter. The full name of the original ether is now diethyl ether, the prefix *di-* coming from the Greek word *dyo,* meaning two.

Eukaryote

IN 1831 the Scottish botanist Robert Brown detected a noticeable oval region within the plant cells he was studying, and because it seemed to him to be located within the cell, like the kernel within a nut, he called it the nucleus of the cell, from a Latin word meaning little nut.

Virtually all cells have nuclei, and although they represent a small portion of the total volume of the cell, they are a vital portion without which the cell could not survive. The nucleus contains the chromosomes carrying the genes that control the formation of specific proteins. The genes determine the nature of the cell machinery, and govern the inheritance of that machinery in the course of cell-division, and also in the course of reproduction of entire organisms.

The separation of the reproductive and hereditary machinery of the cell into a special section at its centre seems to represent a move in the direction of security and efficiency. All cells that possess a nucleus are called eukaryotes. This is from Greek words meaning well-nucleated.

The cells of our own tissues and of the tissues of all multicellular plants and animals are eukaryotes. The larger unicellular plants and animals are also eukaryotes.

Still, there must have been a time when the primitive cells of aeons past had not yet developed the efficient isolation of the nucleus; when the reproductive and hereditary machinery was as yet scattered throughout the body of the cell. As a matter of fact, remnants of that early kind of life still exist. Bacteria, for instance, do not possess distinct nuclei, but have nuclear material distributed throughout the cell. They are prekaryotes (before the nucleus). Blue-green algae are examples of prekaryotes that possess chlorophyll.

Eutrophication

LIFE ON EARTH depends on a balance among different species. One species serves as food for others. The wastes of one species are the fertilizing substance of others.

The activities of mankind can serve to upset this balance. For instance, man's chemical fertilizers and his detergents contain nitrates and phosphates which he gets out of the minerals of the soil. In the soil these would slowly be washed into the freshwater lakes and into the ocean. Man, when he is through with these materials, quickly dumps them into earth's waters.

In the ocean, there has been enough water to survive the shock, but in the confined waters of a lake the sudden influx of nitrates and phosphates supplies a huge bonanza for various bacteria. These multiply tremendously as a result of an increased supply of minerals needed for their tissues. In the process the multiplying bacteria consume the oxygen dissolved in water at an abnormally high rate. The oxygen content of the water consequently drops and the animal life present begin to suffocate and die. They decay, which means a further growth in bacterial activity and a further reduction in oxygen supply. Even the bacteria finally suffocate and die.

The algae, one-celled plants, don't need oxygen, and they grow brilliantly. With no decay bacteria to break them down when they die, they form a green scum on the water, while undecayed sewage collects in the lake and causes the body of water to stink and silt up.

This process, whereby the addition of fertilizing substances causes a wild initial growth of some forms of life, followed by the death of all, or nearly all, is called eutrophication, from Greek words meaning good nourishment.

Evolution

THE FRENCH naturalist Georges Cuvier founded the science of comparative anatomy. Anatomy itself is the science that deals with the study of the physical structure of an organism. This structure can be properly studied only if the organism is carefully cut up so that its interior can be looked at. From the Greek *ana* (up) and *temnein* (to cut) comes the word anatomy (to cut up).

By comparative anatomy is meant the study that compares the anatomy of one creature with another to show relationships among them. Cuvier even compared the anatomy of existing creatures with those of extinct creatures, as revealed by fossil remains (see FOSSIL). In this way he showed that there were series of extinct creatures, existing through time, each a little different from the one before. He decided that there must be periodic catastrophes that wiped out life on earth, and that after each new and somewhat different life-forms were created.

Others, however, suggested that life was a continuous thing, but that individual forms might with time slowly change into new and different species. Life would then be like a scroll, rolling out and revealing new and still newer life-forms until out of very simple beginnings all the complex variety of modern life came into existence, while some forms of life, equally complex, might first have been formed and then have passed out of existence. This theory is termed evolution, from the Latin *e-* (out) and *volvere* (to roll); a rolling out (Latin *evolutio*), in other words. The most celebrated of the early evolutionists was Jean de Lamarck, who worked out evolutionary theory but contended that animals inherited directly characteristics that their parents had acquired.

Charles Darwin was not the first to think of evolution by any means, but he collected so much evidence in its favour and in 1859 published such an excellent book about it (*The Origin of Species,* which sold out its entire first printing on the first day of publication) that he might as well be considered the inventor of the theory.

It was only a generation later that biologists first learned of the method of change (see MUTATION) by which differences could be brought about between parent and offspring so that evolution might take place.

Exchange Forces

IN THE EARLY 1930s it was discovered that the atomic nucleus consisted of positively charged protons and uncharged neutrons. Positive electric charges repel each other with enormous force when crowded together in the tiny space of the nucleus, however. What held the nucleus in place? In 1932 the German physicist Werner Heisenberg suggested that the nucleus held together because neutrons and protons were exchanging electric charge very rapidly.

Such an exchange might act to keep the particles together even against the repulsion of the electric charges—and yet not allow them to get too close together. The situation would resemble two boys playing catch. In order to throw the ball back and forth, the boys must not be too close or the game will be no fun; nor too far apart, or they will not be able to reach each other. As long as they play catch, therefore, they must remain at a certain distance from each other, and anyone watching from a distance who could not see the ball might be puzzled at seeing the boys move back and forth and roundabout yet always maintain a certain distance.

Heisenberg called this an exchange force, because it seemed to be a force of attraction resulting from exchanges of charge. As it turned out, though, the theory did not allow a strong enough attraction to account for the nucleus. In 1935 the Japanese physicist Hideki Yukawa introduced the meson. When that particle, with mass, was exchanged, rather than electric charge alone, the exchange force became strong enough.

It is possible that other forces which seem to extend across empty space can be accounted for by the continual exchange of particles. Electromagnetic forces may result from the emission and absorption of photons; gravitational forces from the emission and absorption of gravitons; the weak nuclear force by the emission and absorption of W-bosons (which see).

Exobiology

IN THE DAYS when earth seemed to man to be the only world in existence, earth was naturally assumed to be the only abode of life. When it came to be known that the moon and the planets were worlds, and there were other worlds that couldn't be seen without a telescope (like the satellites of the other planets), and that there might even be, and very likely were, planets circling other stars, the question arose as to whether there was life on those other worlds.

The first impulse was to believe there was. Might not there be intelligent creatures on the moon? As late as the 1830s a series of hoax articles in the *New York Sun,* telling about the discovery of intelligent beings on the moon, was believed by millions. In 1877 the Italian astronomer Giovanni Schiaparelli noted markings on the planet Mars that came to be thought of as canals (although he had called them simply *canalli*—channels). Many thought these might have been formed by intelligent creatures.

As the twentieth century advanced, however, such notions faded. The worlds of the solar system, except for earth itself (and just possibly Mars), were found to lack the environment necessary for life as we know it. They were too hot or too cold, or lacked water or oxygen.

Nevertheless, astronomers wondered whether some life-forms (quite different from our own) might not adapt themselves to the conditions on other planets, even though we might find them unsuitable for ourselves. They also wondered if planets of other suns, with earth-like conditions, might not exist (see ECOSPHERE). The study of life elsewhere than on earth was given the name exobiology (study of life outside) by the American biologist Joshua Lederberg.

It is, however, a science without a subject, for there is as yet no actual evidence for the existence of life anywhere outside earth. No trace of life has been found on the moon. An alternative name for the science is xenobiology (study of stranger life), so the science with no subject has two names.

Fallout, Radioactive

WHEN THE FIRST NUCLEAR BOMBS were exploded in 1945 the attention of the public was caught by the enormous effects of the blast and heat. The bombs were compared in their destructive effect to so many thousands of tons, or kilotons, of dynamite.

Eventually the comparison was to be to millions of tons, or mega-tons.

It became apparent, though, that there were effects of the nuclear bomb which no dynamite explosion, however great, could duplicate, effects that were more dangerous than either blast or heat.

The nuclear bomb explosion is produced by the sudden break-up or fission of vast numbers of uranium atoms. The breakup not only liberates energy (which produces blast and heat) but also produces uranium atom fragments, most of which are intensely radioactive. These fission products are carried upward in the mush-room cloud and are blown by the wind for varying distances, during which time they gradually fall out of the air to come to rest on the surface of earth. This represents radioactive fallout.

If a nuclear bomb is small and is exploded at ground-level the fission products are attached to soil particles and settle out quickly within a hundred miles of the blast. Large bombs exploded in the open air, however, blast fission products high into the strato-sphere, where they may drift all around earth, gradually settling out over a space of years. In this way nuclear bombs can poison and pollute the entire atmosphere, ocean and soil of earth. Radioactive elements can collect in the body, strontium in the bones, iodine in the thyroid.

It is chiefly for this reason that all-out nuclear warfare may destroy everybody and produce victory for no one. It is because of the insidious poisoning by fallout, even without war, that in 1963 the United States, the Soviet Union and Great Britain (but not China or France) agreed to test no more nuclear bombs by explosion in the open air.

Feedback

WE ARE SO USED to certain intricate abilities of our bodies that we take them for granted.

Suppose a pencil is on the table and you reach for it. Your hand goes to it unerringly, grasps it and brings it back. Yet the act is a complicated one. As your hand moves out to the pencil its motion must slow down as the pencil is approached, so that by the time the fingers touch the the pencil the hand is no longer moving. The fingers must begin to close before the pencil is touched, so that by the time you do touch it only the smallest possible movement of fingers is left and you can begin withdrawing at once.

To do all this you must look at your hand and the pencil and make continuous corrections. If it looks as though the hand is slowing up too much it must be speeded up; if it is going too quickly it must be slowed; if it is veering off to one side or another, it must be brought back to the desired path. All this is done so automatically, delicately, and quickly that you are not conscious of doing it at all, and seem to make but one smooth and simple motion.

But suppose you look at the pencil, memorize its position, then close your eyes and reach. The chances are you will have to grope a bit. Some people with brain damage are unable to make the necessary corrections properly, and make wild motions when they try to pick up the pencil, overshooting and undershooting the mark.

Under normal conditions the position of the moving hand and the pencil is fed back by the eyes to the brain centre controlling the hand motion. It is this feedback that makes things work so well.

The principle of feedback operates in the mechanical world too (see CYBERNETICS). A thermostat works properly because it can constantly sense the actual temperature of the system whose temperature it is regulating.

Ferment

ONE CHEMICAL discovery that was definitely prehistoric was the fact that fruit juice if allowed to stand changed in nature. The taste changed and the effects on the human being who drank it were odd and sometimes striking. The change was accompanied by the formation of bubbles in the juice, and this was the most noticeable outward sign of the change. (As we know today, the sugar in the juice was being converted to alcohol and to the gas carbon dioxide.)

Similarly, a dough made of mashed grain moistened with water also begins to change if allowed to stand. (The starch is converted to alcohol and carbon dioxide.) Bubbles form and are trapped in the sticky dough, causing the whole mass to be raised up. If baked in the risen position a light, fluffy loaf of bread results (with the alcohol driven off by the heat), rather than a hard, compact pancake of baked flour. (The latter is just as nourishing but not as pleasant to eat.) A piece of the rising dough, if saved and added to a fresh batch, would hasten the change in the new batch.

The lump of rising dough is called leaven, from the Latin *levare* (to raise). Bread made with it is leavened bread; without it, is unleavened bread.

This changing of fruit juice and moistened grain mixtures is called fermentation, from the Latin *fermentare* (to cause to rise), which in turn comes from *fervere* (to boil), because bubbles appeared in the process as they did in ordinary boiling.

Exactly what is was that brought about the changes was not known until the invention of the microscope made it possible to see the tiny yeast cells that were responsible. The whatever-it-was, however, received names, which were all derived from the notion of boiling.

In Latin it was called *fermentum,* from which comes our word ferment. In Sanskrit the word for 'it boils' is *yasati,* from which, perhaps, comes our word yeast.

Fermi

AN ATOM is about 10^{-10} metres in diameter. That means that 10^{10} (or ten thousand million) atoms placed side by side will stretch across a metre.

Within the atom is the nucleus, which contains almost all the mass of an atom, but which is far tinier still. It takes about 100 000 nuclei placed side by side to stretch across a single atom. The atomic nucleus is about 10^{-15} metres in diameter. It would take 10^{15} nuclei (a thousand billion of them) placed side by side to stretch across a metre.

The unit 10^{-15} metres is therefore a convenient one for measuring diameters of sub-atomic particles.

It also comes up in another respect. Sub-atomic particles freed in the course of energetic nuclear reactions move at almost the speed of light (3×10^8 metres per second). The most unstable particles only last for about 10^{-23} seconds before breaking down. In 10^{-23} seconds, moving at nearly the speed of light, particles move only a little more than 10^{-15} metres, so nuclear physicists must frequently use that distance in their work.

Enrico Fermi was an Italian physicist who first studied the action of uranium under neutron bombardment; studies that eventually led to the discovery of uranium fission. He emigrated to the United States in 1938, and during the Second World War he was one of the leaders in the development of the nuclear bomb. He died of cancer in 1954, when he was only fifty-three. In his honour, the unit 10^{-15} metres was named the fermi, so that sub-atomic particles are said to be so many fermis wide, and very unstable particles are said to travel so many fermis before breaking down.

In 1962 the unit 10^{-15} metres was officially renamed the femtometre.

Ferromagnetism

EVERY ATOM contains electrically charged particles, and each gives rise to a tiny magnetic field. The charged particles in some atoms are so arranged that their magnetic fields tend to cancel out, and the atom as a whole shows no magnetic effect. In other atoms the magnetic fields do not balance and the atom as a whole acts like a tiny magnet.

If a collection of such atoms is placed with their magnetic fields lined up in the same direction the total effect can be quite strong. In most substances such lining up is not possible at ordinary temperatures. In a few substances, however, there are small regions called magnetic domains (see DOMAINS, MAGNETIC) within which all the atoms are lined up. Usually, these magnetic domains are oriented in all directions, but they can be lined up more easily than individual atoms can be. With the domains lined up, a strong magnet is produced.

Natural magnets, with domains lined up by earth's magnetic field, are found, and these can be used to form still stronger magnets. Strong magnetic effects are found most commonly in iron and in cobalt and nickel, metals closely related to iron. This strong magnetic effect is called ferromagnetism, therefore, from the Latin word for iron.

Ferromagnetism vanishes at high temperatures where individual atoms vibrate so strongly as to lose their alignment. On the other hand, certain substances, not ordinarily ferromagnetic, can become so at low temperatures. Nickel is no longer ferromagnetic at a temperature over 356 °C, while the metal dysprosium becomes ferromagnetic below –188 °C. The French physicist Pierre Curie first discovered this relationship of ferromagnetism and temperature in 1895. The temperature below which ferromagnetism exists is therefore called the Curie temperature of a substance.

Field-ion Microscope

IN THE SEVENTEENTH CENTURY the microscope was invented and mankind was able to enter the world of the invisibly small. During the next three centuries the microscope was improved and objects the size of tiny bacteria could be seen with ever greater clarity. However, objects that were smaller than a single wave-length of visible light could not be seen clearly, no matter how perfectly an ordinary microscope was designed.

In the 1930s microscopes making use of electrons rather than light were designed. The electron is associated with a wave form that has a length equal to that of X-rays, and much shorter than that of visible light waves. X-rays, however, cannot be focused easily because of their great energy, while electrons, which carry an electric charge, can easily be focused by magnetic fields. Such electron microscopes made objects visible that were far too small to study in ordinary microscopes.

While electron microscopes could focus on single giant molecules, single atoms remained beyond the horizon.

In 1936 the German physicist Erwin Müller began work on a new principle whereby a very fine needle tip was made to emit electrons, or perhaps positively charged ions, in a vacuum. These would travel in a straight line to a fluorescent screen which would be lit at the points of impact.

In 1955 Müller built the first field-ion microscope (in which ions were stripped off a needle tip by an electric field). The pattern that appeared on the fluorescent screen was the pattern of atom arrangement on the needle tip, magnified about five million times. In effect individual atoms could be seen as dots in the pattern. So far, this works only for a few metals that have particularly high melting points, but even so, valuable information on atom arrangements, in perfect and imperfect crystal forms, has been obtained.

Fission

UNTIL 1939 the only nuclear reactions (see NUCLEUS) known were those involving rather minor changes in the nucleus—a re-arrangement of particles or loss of one to four of them. At most a nucleus lost only about 1 to 1½ per cent of its mass in the form of small particles. It got so that physicists accepted this sort of thing and expected no more.

In 1934 the Italian physicist Enrico Fermi bombarded uranium with neutrons and found he got confusing results. Supposing that the uranium nucleus had been changed only slightly, he tried to explain his results on that basis and just got entangled. Others who repeated the experiments had no better luck.

In 1938 the German physicists Otto Hahn and Fritz Strassmann finally decided that they just had to believe their own chemical testing. They had barium mixed in with the bombarded uranium, and although the barium atom was so much smaller than the uranium atom, they could not see how it could have come there. Later that year a German physicist (in exile), Lise Meitner, suggested that when the neutron hit the uranium nucleus it split that nucleus into two nearly equal parts, one of the parts being barium.

It was unheard of; her theory created a sensation. Physicists all over the world (and in America especially) began checking and she was right! The uranium nucleus did break in two! This new nuclear reaction was called fission from the Latin *fissio* meaning splitting. The nucleus, after all, was not merely chipped; it was split in two.

Fission gives off several times as much energy as ordinary nuclear reactions, and, besides, liberates neutrons that can cause neighbouring atoms to undergo fission. Each atom-split causes more atom-splits which, given certain conditions, proceed like links in a chain. This is called a chain reaction.

This chain reaction made it possible to develop a nuclear reaction that would be self-sustaining (i.e., would continue of its own accord, so to speak, once started; just as wood burns of its own accord once started). As it happened, the first such self-sustaining fission was achieved under the leadership of Fermi, who started it all.

Fossil

THE HISTORY of the earth during the millions of years of its existence has been deduced from a study of the rock formations of its crust, and in particular from the remains of once-living organisms which have been dug out of the ground. These remains (called fossils, from the Latin *fossilis,* from *fodere,* meaning to dig) were observed long before modern times, but for many centuries were considered to be ordinary rocks that accidentally resembled living things, or perhaps the remnants of animals drowned in Noah's flood.

In 1791, however, an English land-surveyor, William Smith, showed that different rock layers contained different types of fossil, and that a given layer could be traced across broken ground by following the location of its characteristic fossils.

The French anatomist Georges Cuvier studied these fossil remains in about 1796 and began to show that as much sense could be made out of them as if they represented living animals. Furthermore, he showed that some fossils represent animals completely different from any existing today.

Because of the importance of fossils, earth's history was divided into broad stretches of time according to the nature of the life then existing. At first the three major divisions, counting backward, were the Cainozoic, the Mesozoic and the Palaeozoic eras. The suffix -zoic comes from the Greek *zoon* (animal). The various prefixes are from the Greek *kainos* (new), *mesos* (middle) and *palaios* (old). So there was the era of 'new animals', with mammals dominant, in the last 60 million years; of 'middle animals', with reptiles dominant, in the 100 million years before that; and of 'old animals', with fish and land invertebrates dominant, in the 300 million years before that.

Still earlier eras are sometimes described: the Proterozoic, with traces of primitive marine invertebrates dominant (from the Greek *proteros,* meaning earlier), and the Archaeozoic, with only one-celled animals existing (from the Greek *archaios,* meaning ancient).

Fraction

EVERY CHILD begins his study of mathematics with a consideration of the whole numbers: 1, 2, 3, and so on. So did mankind as a group. Slowly, and in steps, men manipulated digits and came up against unexpected problems that taught them about numbers other than digits. For instance, there are two twos in four, three twos in six and so on. In other words, $^4/_2$ (four divided by two) is 2, and $^6/_2$ (six divided by two) is 3.

But how many twos are there in five? More than two, certainly, but less than three. In this way men were forced to think of numbers lying between digits. The quantity $^5/_2$ (five divided by two) can only be two and a half; that is, one unit plus one unit plus half a unit. Half a unit can be expressed logically as $^1/_2$ (one divided by two).

Because a number like ½ represents a unit broken into two equal parts, such numbers are termed fractions, from the Latin *fractus* (broken).

By using fractions, one can locate numbers at various points between digits. The midpoint between 1 and 0 is ½, of course. The midpoint between ½ and 0 is ¼, while the midpoint between ½ and 1 is ¾. You can also locate ⅓, ⅔, and so on. Between 5 and 6 you can do the same by having 5½, 5¼, 5¾, 5⅓, 5⅔, and so on.

Such numbers, which seem to fill up all the space between the digits, are obtained, you see, by comparing two digits. The figure two-thirds is obtained by comparing 2 and 3. The digit 2 is two-thirds as large as 3, hence ⅔ means two-thirds. Such numbers seem to involve a mental comparison of digits, thought, and judgment. The Latin word for accounting or reckoning is *ratio.* Two numbers put side by side and compared form a ratio.

Any number which can be expressed as the ratio of two digits is a rational number. Such numbers include digits as well as ordinary fractions, since 5, for instance, can be written as the ratio $^5/_1$ or $^{10}/_2$.

Friction

ISAAC NEWTON'S First Law of Motion states that a moving object will continue moving in a straight line for ever unless acted on by some external force that will serve to slow it, speed it or change its direction. It took a long time to discover this truth, since in actual fact there always exist external forces which are easy to overlook.

For instance, slide a ball along ice. It will move a long way in a straight line, but it will slow constantly and finally come to a halt. If Newton's Law is correct why should that happen? Nothing seemed to touch it or affect it in any way.

Actually, all the time the ball is moving it is rubbing against the ice. There are tiny irregularities in the ball's lower surface, and tiny irregularities in the ice. These catch at one another as the two surfaces rub together and this absorbs some of the ball's energy of motion, so that it slows. If the ball were sliding along wood, which is less smooth than ice, it would stop sooner. If it were sliding along brick it would stop still sooner.

This slowing force resulting from the rubbing of uneven surfaces is called friction, from the Latin *fricare* (to rub).

Even if motion does not involve one solid rubbing against another, there are forces slowing motion. A ship moving through water has to force molecules of water apart, and must overcome the attraction between the water molecules. This has the same effect as friction. Just as one solid may be rougher—hence display more friction than another (as brick is rougher than ice)—so one liquid may be thicker and harder to move through than another (as molasses is thicker than water). This thickness is referred to as viscosity, from the Latin *viscum,* a word for a kind of thick and sticky birdlime.

Even gases have viscosity, and a great deal of aeroplane design involves itself with methods for reducing air resistance (called drag for obvious reasons) so that as little power as possible is wasted.

Galaxy

THE SUN IS one of a gigantic cluster of perhaps 100 000 million stars, arranged in lens form. The diameter of this lens is about three times its thickness. We are near the outer, thin end of the lens.

The stars near our sun are seen as individual points of light, and, looking through the thickness of the lens, we see to the emptiness beyond. However, if we look through the long diameter of the lens, the stars dim with distance and are too numerous to be seen through. They form a soft luminous band that encircles the sky.

The Greeks thought of it as a belt of milk in the sky, and called it *galaxias* from their word *gala*, meaning milk, and today we call the entire cluster of stars of which we are part the Galaxy.

The Latins called it *via lactea*, meaning road of milk, and we call it just that ourselves—the Milky Way.

In the early days of telescopic observations a number of cloudy objects were detected among the stars, particularly by the French astronomer Charles Messier, by Sir William Herschel (see URANIUM) and his son, Sir John Herschel. These objects were called nebulae (singular, nebula), which meant clouds in Latin and came from the Greek word *nephele,* meaning cloud. Many of the nebulae turned out to be patches of dust within the Galaxy, seen as black clouds because they blotted out sections of the Milky Way behind them, or as illuminated clouds, lit by stars within them. A number, however, like the Andromeda Nebula, turned out to lie far outside the Galaxy, and, indeed, to be collections of stars as large as the Galaxy, appearing small only because of their tremendous distances.

These outside nebulae are called extragalactic nebulae (*extra* being a Latin prefix meaning outside) to distinguish them from the relatively minor dust-patches inside the Galaxy. There are thousands of millions of such extragalactic nebulae in existence. Sometimes they are referred to, poetically, as island universes, but more and more the term galaxy is being applied to all of them, so that we can speak of galaxies in the plural. In fact, because the galaxies themselves exist in clusters, the phrase galaxy of galaxies is now being used.

Gallium

THE DISCOVERY of an element affords an excellent chance for its discoverer to combine science and patriotism by naming the element after his native country. The most recent case was that of element 95, which was first prepared in 1944 by a group of American chemists under the direction of Glenn T. Seaborg. The new element was named americium. (Local patriotism may also be invoked: in 1950 a team headed by Dr. Seaborg produced californium.)

Earlier, in 1939, the French chemist Marguerite Perey detected element 87, and eventually (in 1946) named it francium.

Still earlier, in 1898, Pierre and Marie Curie, in their researches on a uranium ore, had located small amounts of element 84. (Later that same year they were to make their famous discovery of radium—see RADIOACTIVITY—but that was the second element they discovered, element 84 being the first.) Madame Curie was a Pole by birth, her maiden name being Marie Sklodowska, so the element was named polonium.

And still earlier, in 1886, the German chemist Clemens Alexander Winkler discovered element 32, and named it germanium. (Of course, to a German the name of his country is Deutschland, but Winkler used the Latin name Germania.)

The most interesting case is the earliest, and it involves a certain point of ethics.

Back in 1875, the French chemist Lecoq de Boisbaudran discovered element 31 and named it gallium. The usual explanation of the name is that it is derived from Gallia, the Latin name of his native country. However, it has been pointed out that Lecoq (meaning the rooster in French) is in Latin *gallus*. There is therefore the strong suspicion that de Boisbaudran successfully defied ethics by naming an element after himself.

Game Theory

A GAME, in its usual sense, is some artificial activity designed to give pleasure. Usually, though, man's competitive instinct is such that the pleasure arises from pitting the individual's luck or skill against that of others and, of course, winning.

The game may involve pure chance, as in tossing coins; pure skill, as in chess; a mixture, as in bridge; physical prowess, as in most athletic games. Even patience games are not without an adversary. The chance fall of the cards (that is, the randomness inherent in the universe) is in this case the adversary.

Games with known and limited rules, dealing with fixed numbers of pieces, limited areas and times of play, lend themselves to the mathematical analysis of optimum strategy, the steps that will ensure the greatest chance of winning. To achieve the best result one must assume the adversary is also using optimum strategy (but he has a different hand, different abilities, or differs perhaps only in that he moves second while you move first).

What applies to an ordinary game may also apply to the serious

161

aspects of life. Business is a game played between competing producers and the consumer; war is a game between nations; even scientific research is a game between scientists and the universe.

The Hungarian-born mathematician John von Neumann applied mathematical analysis to the development of schemes of optimum strategy in these more serious games, based on the principles developed by dealing with games as simple as matching pennies. This originated the subtle mathematical treatment called game theory. He collaborated with the economist Oskar Morgenstern in writing a book, *The Theory of Games and Economic Behavior,* in 1944, and this, together with the development of computers, has brought game theory into prominence since the Second World War.

Gamma Globulin

SCIENTISTS OFTEN have to deal with substances that belong to the same family or group but are different in little ways. In naming them it is often convenient at first to give them numbers or letters. To vary the monotony Greek letters are sometimes used. The first three letters of the Greek alphabet are α, β and γ. The names of these, spelled out in ordinary Roman letters, are alpha, beta and gamma. (From the first two we get the word alphabet.)

An example of the use of these Greek letters involves the proteins of blood. The first protein to be obtained from blood came from the red corpuscles and was called globulin (see HAEMOGLOBIN). This protein differed from egg albumin, the protein of egg white (see PROTEIN) and the one best known, by having a larger molecule and being less soluble in water.

Following the lead of Felix Hoppe-Seyler, a German physiologist, it became customary to divide the simpler proteins into two groups: the smaller, more soluble albumins, and the larger, less soluble globulins. (Oddly enough, the protein of the red corpuscles, the original globulin, turned out to be a poor representative of the group, is now called globin, and is not considered to be a true globulin.)

The liquid part of blood (the plasma) contains proteins of both the albumin and globulin variety dissolved in it. These proteins can be separated by subjecting the plasma to the influence of an electric field. The various proteins in it move under the influence but at different rates, so that they gradually separate. The albumins move more quickly than the globulins and stick together. The globulins break up into three groups. And now we have the example of Greek-letter-naming I was going to give you.

162

The fastest-moving globulins were named alpha globulins, the next beta globulins and the slowest gamma globulins.

The gamma globulins turned out to be the headline-makers, since they include chemicals that govern the body's resistance to many diseases.

Gemini, Project

ONCE THE ONE-MAN SPACE FLIGHTS of Project Mercury had been successfully completed in 1963 the National Aeronautics and Space Administration (NASA) went on to the next step. The attempt was now to be made to send up two men at a time. With two men in a larger capsule, it would be possible to arrange to manoeuvre the vessel to change its orbit, to attach to a second capsule that had been sent into orbit at another time, and so on.

The two-man launchings planned by NASA were referred to as Project Gemini. Since Gemini is the name of one of the constellations of the zodiac, one might think that there was some astrological significance to this, but there wasn't.

The Latin word *geminus* means twin, and twin brothers (or sisters) would be *gemini*. In the Greek myths the most famous twins are Castor and Pollux (who were brothers of Helen of Troy). In the skies there are two neighbouring first-magnitude stars with no other bright stars very close. They looked like twins, so one was named Castor and the other Pollux. They retain those names to this day. Naturally, the dim stars in the neighbourhood were thought of as representing the images of two young men, their arms linked, and the constellation was called Gemini.

Project Gemini, however, refers not to the constellation but to the two men in the capsule who would be circling in the heavens like the Heavenly Twins, as the constellation Gemini is sometimes called in English.

Project Gemini proved successful. On March 16, 1966, one of the Gemini vessels joined an orbiting vessel while both were moving through space, the first docking in space. In 1965 an American astronaut left his vessel for a space walk, as a Soviet cosmonaut had done three months before.

Gene

ONE OF THE first ways in which biologists tried to find out something about the inside of the cell was to subject it to the action of various dyes. The different substances within the cell reacted differently, as was to be expected, so that some objects would show up coloured against a colourless background (or vice versa). Within the nucleus of the cell, for instance, there were certain small areas that would bind dye tightly and show up coloured. In 1879 the German anatomist Walter Fleming, therefore, called these bits chromatin, from the Greek *chroma* (colour).

If this method of staining is used on cells in all stages of cell division it turns out that at one stage during the process the chromatin collects itself into small thread-like bodies. These threads of chromatin were called chromosomes, or coloured bodies, since the Greek *soma* means body. Furthermore, since these threads of chromatin play such a noticeable part in cell division, that process was given the name mitosis, from the Greek *mitos* (thread).

It became quite apparent that the chromosomes somehow directed the chemistry of the body and that children, because they inherited half their chromosomes from their mother and half from their father, had some characteristics resembling those of the family of each parent. Since human cells contain forty-six chromosomes while each individual is made up of thousands of inherited traits, biologists have assumed that each chromosome is made up of hundreds of smaller units, each of which controls one characteristic. These smaller units they call genes, from the Greek suffix *genes* (giving birth to)—see OXYGEN, HYDROGEN, NITROGEN —which in turn arises from the Greek *gignesthai* (to be born or be produced).

From the word gene comes the name for the science that makes a study of how traits are inherited—genetics. The study of how to encourage the inheritance of valuable traits (a study that is still in its infancy) is eugenics, the Greek prefix *eu-* meaning well.

Genetic Code

IN 1902 biologists decided that the chromosomes, small structures in the cell nucleus, controlled the inherited characteristics of the cell. Cell chemistry, and therefore its characteristics, are controlled by the many enzyme proteins present in the cell, each of which hastens a particular chemical reaction. Therefore the chromosomes must somehow direct the production of particular enzyme molecules.

The chromosome consists of protein and nucleic acid and it was taken for granted that it was the protein portion that was important. The chromosome protein might consist of models of the proteins making up the cellular enzymes. Each cell might manufacture enzymes according to the nature of its chromosome protein.

In 1944, however, the American biochemist Oswald T. Avery demonstrated that it was the nucleic acid that governed the synthesis of enzyme molecules. Biochemists were now faced with a riddle. How could the nucleic acid molecule, which was utterly different from protein, serve as a model for enzymes?

Investigation into nucleic acid structure moved into high gear. It turned out that nucleic acids were every bit as large and complicated in molecular form as were proteins, but nucleic acid molecules were built up of units called nucleotides, while proteins were built up of units called amino acids.

It further turned out that groups of three neighbouring nucleotides represented a particular amino acid. A given nucleic acid, made up of a chain of such triplets, could serve as a model for the construction of a protein with a corresponding string of amino acids. But which group of nucleotide triplets corresponded to which amino acid? This correspondence was called the genetic code (from a Greek word meaning birth, because you were born with a set of controlling chromosomes) and it was worked out in the 1960s.

Geochemistry

CHEMISTRY, the study of the composition of matter and the transformations it could be made to undergo, is a very old science that stretches back into prehistory. When men learned to ferment grape juice, to make brick and glass, to smelt metals, they were being chemists.

With the passing of the centuries men learned more and more about chemistry. Chemists studied various minerals, learned to form new gases, identified new elements and evolved the atomic theory. It was not until the nineteenth century that the concept was advanced of the chemical structure of earth as a whole and the transformations of the planetary chemistry in the course of earth's evolution. The Swiss chemist Christian Schönbein coined the word geochemistry in 1838 for this branch of the science, the prefix geo- coming from the Greek word for earth.

To gain real knowledge of the chemistry of earth as a whole there had to be a wide study of minerals from every region of its crust. Methods had to be devised for studying atomic arrangements within the minerals, and for developing theories as to which arrangements were most probable. Finally, methods were developed for determining something about the chemistry of earth's deep interior.

It was not until the 1920s that planet-wide crust-sampling, the use of X-ray techniques and the study of earthquake waves travelling through earth's interior supplied the necessities. In the 1910s and 1920s the science reached adulthood in Norway, particularly as a result of the work of the Swiss-born Norwegian chemist Victor Goldschmidt.

Since then astronomical studies have brought in more and more information about the chemistry of the universe as a whole to give rise to the still newer subject called cosmochemistry (which means, obviously, chemistry of the cosmos).

Geoid

EVEN THE ANCIENT GREEKS knew that earth was a sphere but it was not until the time of Sir Isaac Newton in the 1680s that it was possible to show that it could not be a perfect sphere. Newton showed that, because earth rotated, it had to bulge outward in the equatorial regions. Careful measurements in the eighteenth century showed that Newton's theory was correct.

The diameter of earth, measured from one point on the equator to an opposite point, is nearly forty-two kilometres greater than the diameter from pole to pole. Earth is an oblate spheroid, where spheroid comes from Greek words meaning sphere-shaped. (If the diameter from pole to pole were the greater the earth would be a prolate spheroid. Both oblate and prolate come from Latin words meaning to bring forward, as though either the equator or the poles were reaching out.)

Of course, earth's surface is actually quite irregular, but even the highest mountain top is less than nine kilometres above sea-level, and even the deepest ocean bottom is only eleven kilometres below sea-level. What's more, mountainous regions are composed of fairly light rock, while the sea bottom is made up of dense rock, so that gravitational pull doesn't change as much as one would expect from the height alone.

In recent decades, geologists measured the actual pull of gravity at different places on earth and calculated what the shape of earth would be like if the actual surface were lowered or raised to make the pull of gravity equal everywhere on the surface. This equal-gravity shape is the geoid (earth-shaped). Under ideal conditions the geoid would coincide with a perfect oblate spheroid.

Actually, it turns out that there are small bumps and flattenings. In some places the geoid is as much as fifty metres farther from earth's centre than it ought to be if the planet were a perfect oblate spheroid, and in some places fifty metres closer. This is because of uneven densities.

Gerontology

THROUGH MOST OF HUMAN HISTORY life-expectancy was short. Until the middle of the nineteenth century the average life expectancy, even in the most prosperous and advanced regions, was about thirty-five; and the general level of nutrition was such that men were usually distinctly old at fifty. In the 1860s, however, the germ theory of disease was put forward by the French chemist Louis Pasteur, and methods were developed to combat a number of serious ailments that had till then killed millions of people each year.

This was followed by better notions of hygiene, by the discovery of vitamins in the early decades of the twentieth century, and by the discovery of antibiotics still later. Nowadays the general life expectancy is seventy or more in the advanced regions of the world and has climbed rapidly in even the under-developed regions.

To be sure, even in the times when life was most insecure and short there were always a few who lived on into extreme old age, but these were few indeed. In modern times the increase in life-expectancy has meant a rapid rise in the numbers of the aged, and this has brought special problems to medicine and sociology. Considerable research is now being conducted into the manner in which cells and tissues age, and this is called gerontology (from Greek words meaning the study of old men). The medical treatment of aged people generally is geriatrics (from Greek words meaning medicine of old age).

Despite all the advances that have brought about increases in average life-expectancy, the maximum life-expectancy stays where it was, at not much over a hundred years. But even if old men must still die, it is the aim of specialists in gerontology and geriatrics to make the last period of life as free of pain and discomfort as possible.

Gibberellin

JAPANESE RICE FARMERS had long been aware that every once in a while some rice plants would suddenly greatly elongate their stalks. They would grow so tall that their stalks would buckle and they would die. The Japanese farmers called this *bakanae*, which can be translated into English, rather charmingly, as foolish seedling disease.

In 1926 a Japanese botanist, E. Kurosawa, was able to show that the seedlings that were so foolish as to overgrow and die were actually infested by a fungus, which he felt must be liberating a compound that accelerated growth. The fungus in question bore the scientific name of *Gibberella fujikuroi*. (The first part of the name is from a Latin word meaning humpbacked, from the appearance of its cells. The second is from the name of the Japanese scientist who first described this particular species.) The substance which caused the growth was therefore named gibberellin.

In 1938 two closely related compounds were isolated from extracts of the fungus by three Japanese chemists. They were called gibberellin A and gibberellin B. (Partly because of the break in communications between Japan and the rest of the world during the Second World War, it was not until the 1950s that any work in the field was done outside Japan.)

It turns out that gibberellin is produced in higher plants too, not just in fungi—something first established in 1956. Gibberellin is, indeed, a group of plant hormones, rather similar in effect to the auxins, but with more complicated molecules. It is merely the fact that the fungus produces too great a quantity of the hormone that makes it dangerous to the plants.

Gibberellins, in appropriate quantities, can be used to hasten plant growth to a non-dangerous extent, and thus increase crop yields.

Gigantopithecus

THROUGH THE 1920s and 1930s archaeologists uncovered fragments of fossil bones in a cave about fifty kilometres southwest of Peking, China. These belonged to a primitive type of man with a brain distinctly smaller than that of modern man, but larger than that of any living ape. (All man-like species with brains larger than apes and with characteristics, such as upright posture, resembling those of modern man, are referred to as hominids, from a Latin word for man).

Among the places in China where the search for hominid remains was intense were the native pharmacies. These sold powdered dragon bones (actually fossils) because they were thought to have medicinal properties. In 1935 a Dutch archaeologist, Gustav von Koenigswald, came across a large tooth in such a pharmacy, and in the next few years discovered three more. They looked very much like human teeth, but were unusually large. For a man to have teeth so large he would have to be three metres tall or so.

Speculation arose concerning the possibility of a prehistoric race of giant men. After all, the legends of many people speak of giants; even the Bible does.

The Second World War put an end to further investigation, but between 1957 and 1968 four jawbones were discovered into which teeth would fit that would be large enough to match von Koenigswald's find. The jawbones, however, were clearly ape-like in nature. Apparently there once existed a giant species of primate that was not man-like, but gorilla-like—the largest primate who ever existed. It was about three metres tall, and its diet was very much like that of man, so that it developed similar teeth.

It was now named Gigantopithecus, from Greek words meaning giant ape. It probably was not the source of the legends about giants, though, for it has been extinct for at least a million years.

Global Village

ALTHOUGH the future of mankind seems dark in many ways, there are bright aspects. Proper education might teach the world's people how to live together in peace, and how to take those measures needed to preserve the environment.

For education to reach all the world's millions some technological advance is needed, and a possible solution was pointed out in 1945 by the scientist and science-fiction writer Arthur C. Clarke. He was the first to explain the uses of communications satellites, and the manner in which as few as three such satellites at a height of 35 000 kilometres above earth's surface might provide world communication at the speed of light.

Since then the dream has come to birth. In 1965 *Early Bird,* the first commercial communications satellite, was launched. In 1971 the more sophisticated *Intelsat IV* was launched with a capacity for 6000 voice circuits and 12 TV channels. In the future the possible use of laser light in combination with communications satellites may make many millions of channels available for the world's population.

It seems within the realm of possibility that every person on earth can be placed in potential touch with every other. Closed TV circuits might exist in such numbers that all business conferences can be held through images, with no one having to move physically from his place. Documents and books could be facsimiled and their images transported at once. Every person on earth could be placed easily in touch with any source of information and with any event of cultural importance. There would be no backwaters, no yokels.

The entire planet earth might then have the character of a village in which everyone knows everyone, and where all information is totally available. Those who look forward to this kind of future speak, therefore, of the global village.

Glucose

IN THE EARLY days of chemistry, when modern instrumentation was absent, the tongue had to substitute as one means of distinguishing chemicals. In this way the class of sour substances (see ACID) was identified.

In the same way sweet substances were noted. The Greek word for sweet is *glykys* and so a particular variety of sugar (there are a number of chemical varieties) is now named glucose. (The -ose suffix is now commonly used by chemists for sugars and related compounds.) As it happens, glucose is not as sweet as ordinary table sugar.

Glucose is found in small concentration in blood, and is the immediate source of the body's energy, so that it is sometimes called blood sugar. In 1857 the French physiologist Claude Bernard discovered in liver a starchy substance which the body could convert to glucose at need. It was called glycogen, the Greek suffix *-genes* meaning producing. Glycogen was the producer of sweetness.

Other chemical compounds which tasted sweet received similar names. An organic liquid which is actually sweeter than table sugar to the taste, but fairly poisonous and used mainly as an antifreeze, is called glycol. (The -ol ending is used by chemists for alcohol and related compounds.) A similar compound, slightly more complicated, which makes up part of the molecules of oil and fat and is quite harmless (in fact, it is used in confectionery), is as sweet as table sugar and is called glycerol.

Glucose, glycol and glycerol are all chemically similar in that they contain within their molecule atom-groups made up of an oxygen atom and a hydrogen atom, such a combination being called a hydroxyl group. Nevertheless, molecules without the hydroxyl group, such as the amino acid glycine, can be sweet, and glycine is so named for that reason (see GLYCINE). (The -ine ending is mostly reserved by chemists for nitrogen-containing organic compounds.)

Even the element beryllium has the alternate name of glucinum because some of its compounds are supposed to taste sweet.

Glycine

THE THICKENING, sticky material in dough is gluten, and from it comes our word glue for anything thick and sticky. Important sources of glue are the hides, hoofs and bones of animals. The substance in these that is responsible for the glueiness of the final product is a fibrous protein named collagen, from the Greek *kolla* (glue) and *-gen* (the suffix meaning producing). Collagen, in other words, produces glue.

If pure collagen is made to undergo prolonged heating the protein molecule breaks up somewhat to form gelatin. The word comes from the Latin *gelare* (to freeze), since a warm liquid solution of gelatin, if allowed to cool, freezes into a jelly. (The word jelly also comes from *gelare,* by the way.)

In 1820 the French chemist H. Braconnet investigated gelatin chemically. Earlier it had been shown that cellulose (see PROTO-PLASM) on treatment with acid decomposed into simpler molecules that were a kind of sugar. Would gelatin, which came from collagen, a fibrous constituent of animals, do the same?

Gelatin did indeed break down into smaller fragments on treatment with acid, and at least one of these fragments when purified was sweet to the taste. Braconnet was sure it was a sugar, and called it, forthrightly, sugar of gelatin. It wasn't till 1838 that sugar of gelatin was found to contain nitrogen, which ordinary sugars do not. It was renamed glycine, from the Greek *glykys* (sweet). An alternate name, not much used, is glycocoll, meaning sweet glue.

Braconnet had done more than he knew. Glycine was the simplest amino acid (see AMMONIA); not the first discovered, but the first to be shown to form part of a protein molecule, a milestone in biochemistry.

And it also turns out, derivationally speaking, that though the story of glycine starts out with glue, the coincidence of the first two letters has no significance.

Grammar

GRAMMAR IS the science that deals with the correct (or at least accepted) use of language. It is derived from the Greek *gramma* (letter), the verbal form of which is *graphein* (to write). In ancient and medieval times, grammar was one of the most important branches of learning (see LIBERAL ARTS), but in modern times it has been crowded by the advance of the physical sciences.

Yet in one respect it resembles some of these physical sciences. The words with which the science of grammar deals are as carefully classified as are the animals with which zoology deals, or the elements with which chemistry deals.

For instance, there are nouns—words that name persons or things—John, rat, table. The word noun comes from the Latin *nomen,* meaning name.

A pronoun is also a name, but one for an object already mentioned by a specific name. The pronoun is not specific: I, which, who. One of the meanings of the Latin *pro-* is substituted for, so a pronoun is a word substituted for a noun.

A verb denotes an action (to smash), a state (to be) or a happening (to age). These are the words which can least be spared, perhaps. Certainly, the derivation would seem to make them most important, since the Latin *verbum* means simply word.

An adverb explains the nature of a verb (walk 'slowly'; insist 'stubbornly'). The Latin prefix *ad-* means to; so an adverb is added to a verb to explain it.

Similarly, an adjective describes or defines a noun ('white' hat; 'fast' train). It is derived from the Latin *ad-* (to) and *jacere* (to throw). The noun and adjective are thrown together, so to speak.

One more example is the interjection, an exclamation that intrudes on ordinary speech ('Ouch,' 'Oh'), from the Latin *inter* (between) and *jacere* (to throw). It is a word thrown between ordinary words or sentences.

Granite

THE ROCK which makes up most of the dry land on earth is granite. This is in turn made up of a mixture of three different kinds of rock: mica, feldspar and quartz.

Mica can be easily split into thin transparent sheets, and has therefore been used to form windows in ovens where it is necessary to have something transparent that won't burn or melt. The thin sheets are glossy and shiny, so that the name may come from the Latin *micare* (to shine).

Feldspar is the most common rock about us. It makes up perhaps 60 per cent of the dry land, and the name shows that. *Feld* is the German word for field, while spar is from an Anglo-Saxon word for any rock that is not metal-containing. (A rock that is metal-containing is an ore. This word is much more familiar to us because ores are rarer, more useful and more valuable than spars, so that the ores are more spoken of.) Feldspar is thus field rock, the common rock of the fields.

Quartz is from a German word, *Quarz,* which is of uncertain origin. It is also very common, and, when broken into small fragments by the action of wind and water, it is sand (a word of Anglo-Saxon origin).

Granite is an igneous rock (see IGNEOUS), and when it forms by cooling the three components, mica, feldspar and quartz, form crystals large enough to be seen separately. Granite, therefore, does not present a smooth and homogeneous appearance but is obviously made up of mixed grains of different substances. The Latin word for grain is *granum,* and so granite is the grained rock, so to speak.

An even more common igneous rock is basalt, which underlies the granite of the continents and makes up most of the earth's crust under the oceans. It is darker and heavier than granite. According to Pliny (Gaius Plinius Secundus) the Roman naturalist, the word originated in Ethiopia, where it was applied to a dark variety of marble. The meaning later spread to include any dark or blackish rock, particularly the one we call basalt today.

Gravitational Lens

ACCORDING TO ALBERT EINSTEIN'S theory of relativity, light travelling through a gravitational field moves in a curved path. If a ray of light from a distant star passed very close to the surface of the sun on its way towards earth the light would curve slightly towards the sun's centre, and would seem to change its position compared to other stars farther from the sun.

Ordinarily we can't see stars close to the sun, but during a total eclipse we can. In 1919 astronomers observed a total eclipse to see if stars near the sun had changed their positions. They had, and the results were dramatic support for Einstein's theory, promulgated three years earlier.

Light also curves in passing from air to glass. When light passes through a lens it will bend and meet at a point on the other side. Light-rays gathered from a large area and concentrated in this fashion make the objects emitting or reflecting the light-rays seem larger and brighter.

If one astronomical object is located directly behind another as viewed from earth the light from the farther object in approaching us will skim the surface of the nearer object on all sides. All the light from the farther object will bend slightly inward towards the centre of the nearer object. The rays of light converge considerably by the time they reach us, and the farther object will look larger and brighter than it really is. This effect is that of a gravitational lens.

Some astronomers wonder if the quasars, which seem so unusually bright, are affected by a gravitational-lens effect. Probably not, but in 1988 it is expected that one particular star will move in front of a more distant one, and on that occasion the idea of the gravitational lens will be tested.

Graviton

THE FOUR FORCE FIELDS are (1) the strong nuclear interaction, (2) the electromagnetic interaction, (3) the weak nuclear interaction; and (4) the gravitational interaction. All make themselves felt across a vacuum, possibly through the continual exchange of particles. The electromagnetic interaction works through the constant interchange of photons, and the strong nuclear interaction through the constant interchange of pions.

Both photons and pions are known and have been studied. Are there exchange particles for the other two interactions also? In particular, what about the gravitational interaction? It is by far the weakest of the four interactions, and if there is a particle associated with it, it must carry very little energy and be extremely hard to detect. Physicists have suggested that such a particle must exist. Using the -on ending common for sub-atomic particles since the discovery of the electron, they have named it the graviton.

All particles can be detected in the form of waves as well as of particles. Very energetic particles are much easier to detect in particle form, while very unenergetic particles are much easier to detect in wave form. Thus, while all masses are constantly giving off and absorbing gravitons, these might be detected most easily as gravitational waves.

Since 1957 the American physicist Joseph Weber has been trying to detect gravitational waves. He makes use of a pair of aluminium cylinders, 153 centimetres long and 66 centimetres wide, suspended by a wire in a vacuum chamber. A gravitational wave moving over it would distort the cylinders very slightly. Such a wave, coming from far out in space, would sweep over the entire planet earth. Weber therefore places his two cylinders hundreds of kilometres apart. When both distort in precisely the same way at precisely the same time he feels that a gravitational wave has passed by. In 1969 he reported a number of such events.

Gravity

ISAAC NEWTON'S First Law of Motion states that every object in motion will move in a straight line unless forced to change

174

direction by some external push or pull. If you were whirling in a circle at a great speed, held only by the handle of an elastic rope in your clenched fist, and were to relax your hold at any time, you would fly off instantly out of the circle in which you had previously been moving. It is the constant pull of the rope that forces you to change the direction of your motion constantly.

The force tending to move you out of the circle is centrifugal force, from the Latin *centrum* (centre) and *fugere* (to flee). It is the force that causes you to flee the centre. The force pulling you towards the centre (the pull of the elastic rope) is a centripetal force, from the Latin *petere* (to move towards). It is the force that causes you to move towards the centre. The balance of the two keeps you in the circle.

The rope keeps you from flying away because it is held strongly together by molecular cohesion (see CAPILLARITY). When centripetal force is the result of such cohesion it works only if there is a continuous material connection between yourself and the centre.

All matter, however, attracts all other matter, even without physical contact. If the lump of matter is large enough the attraction is considerable. For instance, an object released in space 1600 kilometres from earth is attracted by the earth, and moves towards it, although there is a vacuum between. The moon, 384400 kilometres away, is also falling, but it has a motion of its own, and centrifugal force balances its falling motion, so that it stays in a closed orbit about the earth, never falling altogether and never escaping altogether.

This pull without physical contact is what gives us the sensation of weight or heaviness. The Latin word for heavy is *gravis,* and so the attraction of one body for another is gravity, and that is the centripetal force that holds the solar system together, and everything on earth in place.

Green Revolution

THE FIRST POPULATION EXPLOSION we know of in human history came with the development of agriculture some ten thousand years ago. Once plants were deliberately cultivated, a given tract of ground could support more people than could be supported by simply gathering or catching what food happened to be available.

Ever since then world population has managed to expand because more and more land has been given over to agriculture, because fertilizers have increased crop yields, because pesticides have killed off insect pests that devour crops, and because the use of power machinery makes it possible to do the work of farming more efficiently.

Another method for improving the yield of the land is to concentrate on those strains of particular plants that grow more rapidly than others or that are more resistant to cold or to disease.

The last method was the course followed by the American agricultural scientist Norman Borlaug when he was sent to Mexico in 1944 to study methods for improving wheat-production.

He developed new strains of dwarf wheat by crossing a Japanese variety with native Mexican wheat. The new strains were high-yield and resistant to disease, so that by 1960 Mexico's wheat production had increased tenfold and it had become an exporting nation.

Borlaug went on to develop varieties of wheat and rice that would flourish in the Middle East. The result was called the Green Revolution, for land turned greener, with denser, healthier growths of plant life. The immediate danger of famine receded.

Borlaug himself has pointed out, however, that the increase in food supply would serve no purpose if the human population continued to increase wildly; for then the danger of famine would return—and pose a worse threat than ever.

Greenhouse Effect

IT IS POSSIBLE to grow plants in large glass buildings, even in cold weather. This is because glass is transparent to the short waves that make up visible light but not to the longer infra-red waves.

When sunlight strikes the glass buildings that house the plants it passes through. Its energy is absorbed within and is re-emitted in the form of infra-red. Because the infra-red cannot get out easily, it piles up, raising the temperature within the structure. Such a structure is a greenhouse because the plants within are green, even when the land is barren outside. The phenomenon whereby energy enters in one form and cannot escape in another is called the greenhouse effect.

In earth's atmosphere oxygen and nitrogen are transparent both to visible light and to infra-red. Carbon dioxide and water vapour, however, transmit only visible light. This means that sunlight can reach earth's surface easily during the day, but at night, when earth is re-radiating infra-red, it has difficulty escaping. Earth's surface is warmer than it would be, therefore, if there were no carbon dioxide or water vapour in the air.

The planet Venus has a very thick atmosphere that is mostly

carbon dioxide. The greenhouse effect there is enormous, and Venus's surface temperature is nearly 400°C. This was determined from radio-wave emission in 1956, and was confirmed when the Venus probe *Mariner 2* passed near Venus in 1962.

The greenhouse effect may be of increasing importance on earth. The burning of vast quantities of coal and oil since 1900, and especially since 1940, is slowly increasing the carbon dioxide content of the air. And even a small increase may raise earth's temperature to the point where the ice-caps will melt and raise the oceans' level so that millions of hectares of the continental lowlands will gradually be flooded.

Guyot

UNTIL MODERN TIMES virtually nothing was known about the sea-bed. It was covered by miles of sea-water, and most people assumed that there was just a flat, featureless plane under all that water. In the 1870s, however, the vessel *Challenger* was the first to study the ocean bottom carefully. By dropping lines overboard, the scientists on board found that the Atlantic Ocean was shallower in its mid-region than on either side.

Detailed studies could not be made, however, until the technique of listening to sound echoes was developed in the 1920s. Once that was introduced geologists began to make out the outlines of a vast mountain chain, larger and higher than anything on dry land, running the length of the Atlantic Ocean and into the Pacific and Indian oceans. This chain, the Mid-Oceanic Ridge, virtually encircles earth, and its highest peaks emerge through the ocean surface to form islands such as the Azores.

In 1942 the American geologist Harry H. Hess, a naval officer, had echo devices installed on his ships, and while he crossed and recrossed the Pacific Ocean on war business he also studied variations in ocean depth. In this way he discovered isolated mountains studding the sea-bottom, mountains whose tops were flat, as though they had been worn away by wave action, but which were 600 metres below the surface where wave action is nil. They may have been mountains which were originally above the surface till the sea-bottom subsided or the sea-level rose.

These submerged mountains, discovered by the hundreds, are called seamounts or tablemounts, but their most dramatic name was given to them by Hess. He called them guyots (pronounced gee-oze´, with a hard g) in honour of an old professor of his at Princeton, a Swiss-born geographer, Arnold H. Guyot, who founded modern physical geography.

Hadron

THERE ARE FOUR TYPES of force fields known to man, each of which makes attraction (and sometimes repulsion) felt across a distance. The first to be studied was the gravitational interaction, which is by far the weakest of the four but is associated with such huge conglomerations of matter, like the sun and earth, that the total effect is very large.

A second is the electromagnetic, which is associated with the electrically charged particles that make up the atom, the electron and the proton chiefly. Atoms and molecules hang together because of electromagnetic interactions. All the pushes and pulls we associate with living beings and with mechanical devices (except for gravitational effects) are electromagnetic.

It was not until the middle 1930s that any other force fields were studied. These arose out of the necessity of explaining the atomic nucleus, which held together despite the fact that it contained protons and neutrons only. The protons were all positively charged and should have repelled each other strongly. Yet they stayed together. It seemed there must be a nuclear interaction that held them together (with the help of the uncharged protons) against this repulsion.

In order to account for what went on within the nucleus, two kinds of interaction were required. There was the strong nuclear interaction, which was 130 times as strong as the electromagnetic, and the weak nuclear interaction, which was far weaker than the electromagnetic but was stronger than the gravitational.

Only the more massive sub-atomic particles, such as the proton and neutron, were capable of responding to strong nuclear interactions. These massive particles (including some mesons and the very massive hyperons) were grouped together as hadrons, from a Greek word meaning strong.

Haemoglobin

ANTON VAN LEEUWENHOEK, the Dutchman who first used a microscope systematically (see MICROBE), was the first to see the small objects in blood which contain the red colouring matter. He called them corpuscles, from the Latin *corpusculum* (little body). Today they are generally called red corpuscles to distinguish them from other little bodies in blood which do not contain colouring matter and are the white corpuscles.

Sometimes these are referred to as red cells and white cells, but in the case of the former this is a misnomer. The white corpuscles are actually cells, but the red corpuscles are not. The red corpuscles of mammals lack a nucleus (see PROTOPLASM), which all true cells must have. And yet the name most used by scientists keeps this misnomer, since a red cell is most often called an erythrocyte from the Greek *erythros* (red) and *kytos* (a hollow space or cell). Similarly, the white corpuscle, with greater justification, is a leucocyte, from the Greek *leukos* (white).

The red corpuscles were also called globules by van Leeuwenhoek, from the Latin *globulus* (little ball). This too is a misnomer. Van Leeuwenhoek's early microscopes could not show him the exact shape, but we know now that the red corpuscles are not globes or spheres, but are disc-shaped, with a depression on each flat side. In fact, the red corpuscles are sometimes called red discs for this reason.

Nevertheless, in 1805 or thereabouts, when the Swedish chemist Jöns Berzelius obtained a colourless protein from the red corpuscles, he named it globulin because he got it from the globules. Inside the corpuscles the globulin was associated with the colouring matter, then called haematin (see PORPHYRIN). The combined molecule was named haematoglobulin. This in time (even scientists are lazy) got shortened to haemoglobin by omitting four letters, and that today is the name for the red protein of blood.

Haemophilia

IN THE LAST half-century two royal families were afflicted with a disease marked by excessive bleeding. The son of Tsar Nicholas II of Russia was a 'bleeder'. A small scratch was enough to put him in danger of bleeding to death. This was also true of certain sons of the Spanish king, Alfonso XIII. This received so much attention in the newspapers that the condition became known as the royal disease. Its proper name is haemophilia, from the Greek words *haima* (blood) and *philia* (love), since the patient apparently 'loved to bleed'.

But haemophilia hit royalty only in modern times. The Spanish haemophiliacs could be traced back to Queen Victoria through her youngest daughter, Beatrice, while the Russian haemophiliacs could be traced back to the same Queen through her second daughter, Alice. It seems fairly certain that the royal disease originated with her, at least as far as royal families are concerned.

People suffer from haemophilia when they are born without one of the numerous substances in blood that contribute to clotting. The disease is tied up with the sex of a person in such a way that while men can have the disease, they can't pass it on to their children; whereas women may not show it and yet be able to pass it on. Queen Victoria gave birth to at least two daughters who could transmit this failure of blood-clotting mechanism and who did, without themselves showing it, of course.

Actually many perfectly non-royal people have the disease, or other similar bleeding diseases brought on by lack of clotting factors in the blood. One odd variety was first described in 1952. It occurred in a five-year-old boy whose last name was Christmas. The missing factor was therefore named Christmas factor, and this particular variety of bleeding condition was, with peculiar insensitivity, named Christmas disease.

Half-life

IN THE 1890s it was discovered that certain atoms constantly give off particles from their nuclei and change into different types of atoms. They are said to undergo radioactive breakdown.

If every radioactive atom of a particular kind had a fixed life —that is, if each existed a particular length of time before breaking down—then a group of identical atoms would exist unchanged for a while and then suddenly break down all at once, loosing vast energies.

That doesn't happen. Instead a large collection of identical radioactive atoms are continually giving off small quantities of energy, as though a few atoms were breaking down during each small interval of time. Some might break down today, some tomorrow, some millions of years from now. The time when a particular atom breaks down would seem to be entirely a matter of chance. There is no use, therefore, in speaking of the lifetime of a radioactive atom. That can have any value.

In any large collection of such atoms, though, there is a particular probability that within a certain length of time a certain fraction of the total number will have broken down. We can tell exactly when one-tenth the total number will have broken down, even if we can't tell which tenth. (Insurance companies can predict how many English males will die in the next year, though they couldn't tell which particular ones would die.)

It proved particularly convenient to select a period of time in which exactly half the radioactive atoms in any conglomeration would break down. This period of time is called the half-life. Thus the half-life of uranium 238 is 4500 million years, while those of its sister varieties, uranium 234 and 235, are only 248 000 and 713 million years respectively. Some half-lives are much shorter. The half-life of radium 226 is 1620 years, and that of polonium 212 is 0·000 000 3 seconds.

Hallucinogens

THE BRAIN, like every other part of the body, performs its functions through certain chemical reactions. These are produced by stimuli brought to the brain through the senses. It is possible to change the brain chemistry by taking into the body substances that interfere with these chemical reactions. In that case the body will respond to stimuli that don't relate to the outside world. Objects seem to be sensed that are not really there; while other objects which are really there may be ignored. The results are hallucinations, from a Latin word meaning to wander in the mind.

Certain plants contain chemicals which can produce hallucinations. The peyote cactus and a mushroom called *Amanita muscaria* contain such chemicals. Sometimes these plants are eaten in primitive religious celebrations because the hallucinations are thought to be glimpses of another world (or an escape from this one). Another substance that produces hallucinations is hashish, one form of which is marijuana.

In 1943 a Swiss chemist, Albert Hofmann, was studying an organic compound called lysergic acid diethylamide and accidentally got a few tiny crystals of it on his fingers. He happened to touch his fingers to his lips and was soon overcome by odd hallucinations. It took him a full day to regain normality. He began careful studies, and found the chemical could always produce hallucinations in very small doses. The name was soon reduced to an abbreviation of the three words. Since the German word for acid is *Saure* and Hofmann spoke German, the abbreviation was LSD.

Since many young people foolishly began to play games with their minds by taking LSD and other such substances, hallucination-producing drugs became important to study. They are now lumped together under the general name hallucinogens (producers of hallucination).

Helium

IN 1868 a total eclipse was visible in India, and for the first time the sun's atmosphere (best observed during eclipses) could be studied by the new technique of spectroscopic analysis (see SPEC-TRUM). This had been developed only nine years earlier, and con-

181

sisted of passing the light radiated from a white-hot substance through a glass prism. The light is split up into lines of different colours, and each element forms its own characteristic pattern of coloured lines in fixed positions.

The French astronomer Pierre Janssen allowed the light of the solar atmosphere to pass through the prism during the Indian eclipse and noticed that among the familiar lines of earthly substances a yellow line was produced which he could not identify. The British astronomer Sir Norman Lockyer compared the position of this line with those of similar lines produced by various elements, and decided that this new line was produced by an element in the sun that was not present, or had not yet been discovered, on earth. He called it helium, from the Greek word for the sun, *helios*.

For decades that was how matters stood. Helium remained an oddly coloured line in sunlight and nothing more. Few chemists took it seriously.

In 1888 the American chemist William F. Hillebrand found that a uranium ore named uraninite, when treated with strong acid, gave off bubbles of gas. He studied this, and decided it was nitrogen. To be sure, some of the gas was nitrogen, but Hillebrand unfortunately ignored the fact that when it was heated some of its spectrum lines were not those of nitrogen.

The Scottish chemist Sir William Ramsay read of this experiment and was dissatisfied. He used another uranium mineral, cleveite, and, in 1895, repeated the experiment. He and Lockyer studied the spectral lines of the gas, and almost at once they realized what they had. Fully twenty-seven years after helium had been discovered in the sun, it was finally located on earth.

Hieroglyphic

LANGUAGES associated with a religion persist even after they have gone out of everyday use. Thus Latin has always been the language of the Roman Catholic Church and ancient Hebrew of the Judaic ritual, although both are usually dismissed as dead languages.

This sort of thing was observable in ancient times too. Greek tourists of the centuries before Christ, in an Egypt that was already ancient, found monuments with carved figures on them that ordinary Egyptians could no longer read. Only the priests preserved the language. The Greeks therefore called these figures *hieroglyphika* (English, hieroglyphics), from *hieros* (sacred) and *glyphein* (to carve); they were sacred carvings.

After about 600 B.C. the Egyptians as a whole used a much less elaborate writing for ordinary work, and this was called by the Greeks *demotika* (English, demotic) from *demos* (people). It was the writing of the common people.

The only other written language as early as (or possibly earlier than) the Egyptian was the Sumerian. The Sumerians lived in the Tigris-Euphrates area (modern Iraq), and in the absence of stone inscribed their writings on clay with slanted jabs of a stylus. The Babylonians, Assyrians and Persians all adopted this style of writing, and it did not die out until 300 B.C. Because the marks on clay were wedge-shaped (▼), the writing was called cuneiform from the Latin *cuneus* (wedge) and *forma* (form). It was writing in the form of a wedge.

These early writings were not alphabetical. Each different sign represented a separate word or idea. (Eventually they reached the point where they represented only syllables, and sometimes even letters, but this was a late development after the Phoenicians had pointed the way with a true alphabet.) Such idea signs are ideographs. Idea itself is a Greek word meaning form, shape or resemblance to reality, and *graphein* means to write. An ideograph is an idea writing. The best known example of a modern ideographic language is, of course, Chinese.

Holography

IN ORDINARY PHOTOGRAPHY a beam of ordinary light, reflected from an object, falls on a sensitized film. Wherever light falls the film darkens, forming a negative. From the negative a positive is formed, a flat, two-dimensional representation.

Suppose instead that a beam of light is split in two. One part strikes an object and is reflected with all the irregularities that this would impose on it. The second part is reflected from a mirror with no irregularities. The two parts meet at the photographic film, and the interference between the two light beams is recorded. The film that records this interference pattern seems to be blank, but if light is made to pass through it takes on the interference characteristics and produces a three-dimensional image, which can then be photographed in the ordinary manner from different angles.

The notion was first worked out by the Hungarian-British physicist Dennis Gabor in 1947, and he called it holography, from Greek words meaning whole writing. The whole of the image, not part, was recorded.

Gabor's idea could not be made practical with ordinary light, in which wave-lengths of all sizes moved in all directions. The interference produced by two such beams of light would then be too chaotic to give sharp images.

The introduction of the laser changed everything. Laser beams had uniform wave-lengths, all moving in the same direction. In 1965 Emmett N. Leith and Juris Upatnieks at the University of Michigan were able to use laser light to produce the first holograms. Since then the technique has been sharpened to the point where holography in colour has become possible, and where the photographed interference-fringes produced with laser light could then be viewed with ordinary light.

Hormone

THE LATIN WORD *glans* means acorn. People have always been impressed by the smallness of acorns (probably because of the contrast of the giant oak which is the final result), and small lumps of tissue in the body were therefore called glands by the early anatomists (see LYMPH).

Such lumps are to be found over the kidney. The Latin word for kidney is *ren* and the glands are the suprarenals or the adrenals. (The Latin *supra* means above and *ad* means on, so the two names are pretty much the same.)

In 1895 the biochemists George Oliver and Edward Albert Sharpey-Schafer found that there was something in the adrenal glands which caused a contraction of the arteries and a rise in blood-pressure. The actual substance that did this was isolated in 1901 by the Japanese biochemist Jokiche Takamine.

This was the first example of a chemical, formed in a gland and liberated in small quantities into the bloodstream, that could excite specific organs to specific activities. In 1902 the English physiologists William Bayliss and E. H. Starling suggested that such chemicals be known as hormones, from the Greek *horman*, meaning to set in motion.

Chemicals, and fluids generally, formed by small organs have become so important to physiologists that the word gland has come to mean any bodily organ that secretes a fluid, regardless of size. The liver is a gland though it weighs two kilogrammes. On the other hand, the lymph glands—perhaps the first to be so named—should no longer really be called glands since they secrete no fluid. They are better termed lymph nodes, from the Latin *nodus* (knot), since they resemble knot-like swellings in the string-like lymph vessels (see LYMPH).

As for the first hormone, the one from the suprarenals, it is called adrenalin from, as stated above, the Latin *ad* (on) and *ren* kidney; or the equivalent name epinephrine, from the Greek *epi* (on) and *nephros* (kidney).

Humour

THE LIVER JUICE, which may be called either bile or gall (see BILE), has also a Greek name, *chole,* which comes from their word *cholos* (bitter), and bile is bitter indeed. (We have a common simile, bitter as gall.)

The Greek word is used in English only in combination. For instance, the ancient Greeks thought there were four chief fluids in the body: blood, phlegm, yellow bile and black bile. (Actually, there is only one bile, but when freshly formed it is golden yellow, whereas after standing a while chemical changes take place that turn it a greenish black, and this is what may have given the Greeks their notion of two biles.)

If any one of these fluids was present in too great a quantity, thought the Greeks, the person suffered from immoderation in one way or another. If there was too much yellow bile, for instance, he had too great a tendency to anger, and was choleric (the Greek *chole* showing up, you see). If he had an overbalance of black bile he was too given to sadness, and was melancholic (the Greek *melas* means black, hence melancholy is black bile).

In later times these fluids were referred to as humours, from the Latin word *humere* (to be moist). The theory of humours is gone now, but the word still lingers for any pronounced type of temperament that is changed from the normal. A person may be in a bad humour or a good humour or a contrary humour, and so on.

In Elizabethan times there was quite a fad for the writing of plays about people who had some pronounced one-track personality. One might be of a boastful humour, one cowardly, one avaricious. Ben Jonson went to the deliberate extreme of a play that included only such characters, called *Every Man in his Humour.* Such plays were generally comedies, and the characters, each harping away at his particular personality trait, were laugh-provoking. For that reason the word humorous has come to mean funny.

Hurricane

MOST STORMS are cyclonic in character (see CYCLONE), and in general are mild enough. However, every once in a while, conditions are such as to make the whirling of a cyclone too rapid for comfort.

An all-too-familiar condition to inhabitants of the eastern and Gulf seaboards of the United States is a cyclone that begins over the Caribbean in late summer or early autumn, forms a gigantic whirlwind with winds of over 160 kilometres an hour, and starts moving north-westward. This is called a hurricane, from a Caribbean Indian word, Hurakan, which was the name of one of their evil spirits. Anyone who has lived through a hurricane (as I have) will testify that the evil-spirit theory has its points.

Hurricanes (at least by name) are confined to the Atlantic. There are, however, similar severe cyclonic storms in the west Pacific. These are typhoons. This comes from the Arabic *tufan*. Why Arabic? Well, the Arabs explored the south-eastern reaches of Asia before Europeans did, so that most of the population of Indonesia are Moslem and there are many even in the Philippines to this day. (Some tribes in the southern Philippines are called Moros, from the Spanish word meaning Moors because of their religion.) The Arabic word probably comes in turn from the Greek Typhon (or from some common ancestor), the name of an evil giant who fought with Zeus in a battle of storm and lightning. (Again the evil-spirit theory.)

Sometimes over land surfaces much smaller but more intense cyclones are set up, with winds so fierce (their speed has never been measured since they destroy all measuring devices) that they devastate whatever they pass over. Fortunately, their path is short and narrow. These are tornadoes. The Spanish word *tornado* means return (the wind moves and returns in a circular path over and over), but it has also been derived from the Spanish *tronada* (thunderstorm). Why Spanish? Well, the Spaniards colonized the south-western and middle areas of the United States before the English-speaking Americans arrived, and they experienced these storms (which breed best in the south central USA) before they did.

Hydrogen

THE BRITISH chemist Henry Cavendish was the first (in 1766) to study systematically a gas he obtained by treating iron filings with acid. Because the gas burned when heated he called it 'inflammable air from the metals'.

To the early chemists, still more remarkable than the mere fact that the gas burned was that after burning it left behind a liquid substance that proved to be pure water. Now, the chemists had not forgotten that for many centuries the early Greek notions of the structure of matter had prevailed, these notions being that all matter was made up of differing proportions of four fundamental elements—fire, air, water and earth. Although this was no longer accepted, even in the 1700s, some of the force of this intellectual tradition remained. Here was the case of a kind of air reacting with ordinary air when heated, becoming fire and turning to water. From one element via a second element to a third element.

The French chemist Antoine-Laurent Lavoisier emphasized this startling property of inflammable air some years after its discovery by giving it a name that reflected the transmutation. He called it *hydrogène* (converted in English to hydrogen) from the **Greek word** *hydor,* **meaning water, and a Greek suffix** *-genes,* meaning born or produced. Hydrogen, therefore, is something from which water is produced.

The Germans, who have less of a tendency for turning to Greek and Latin for their scientific words than do the French and British, **named the new 'air' in straight German. They too however paid** honour to the strange transmutation, and called it *Wasserstoff,* meaning water substance.

Nowadays, of course, hydrogen has gained a fearful new importance through another transmutation it undergoes; not into a substance incorrectly labelled an element by the Greeks, but into a substance, helium, that is truly an element. This new transmutation provides the energy for the destructive force of the hydrogen bomb (see THERMONUCLEAR REACTION).

Hydrophobia

ALL HUMAN beings know fear. Some fears (of lions or homicidal madmen when these are in your vicinity) are normal and even sensible. However, there are abnormal fears too, or morbid fears (from the Latin *morbidus,* sickly, the substantive form of which is *morbus,* meaning disease). Psychologists call such a morbid fear a phobia, from the Greek *phobos* (fear).

Phobias are sub-divided according to subject. A familiar example is claustrophobia, which is a morbid fear of enclosed places (from the Latin *claustrum,* meaning a closed place). Another is agoraphobia, the morbid fear of open spaces (from the Greek *agora,* meaning a market-place, which of course was the chief open space in a Greek city). Thus a claustrophobe may absolutely refuse to stay in a room with closed doors, while an agoraphobe may absolutely refuse to stay in a room with open doors, even when no conceivable danger is involved either way. There are dozens of specialized fears named in this way, even panphobia, which is a morbid fear of everything (from the Greek *pan,* meaning all) and phobophobia, which is a morbid fear of being afraid.

Less extreme cases (and more political than psychological) are extreme dislikes of things English or Russian, let us say. This is called Anglophobia or Russophobia, respectively.

But one phobia is actually a physical disease and not a state of mind. There is a virus infection that attacks the nervous system. A person with that disease cannot swallow. The attempt to swallow water or even just the sight or sound of water (which automatically makes him try to swallow) throws him into convulsions. The ancient Greeks considered these convulsions to result from a morbid fear of water and they called the disease hydrophobia from *hydor* (water).

This disease is generally transmitted through the bite of an infected animal, and in animals it is usually called by its Latin name of rabies from *rabere* (to rave), since the animal is in such distress and agony that it behaves as though in a senseless rage. We usually call a dog with rabies a mad dog.

Hydroponics

SINCE THE INVENTION of agriculture, about ten thousand years ago, the tilling of the soil has been the most important method by which human beings have assured themselves of a food-supply. Agriculture is from Latin words meaning to cultivate a field. A similar word from Greek roots would be geoponics (meaning to till the ground).

People always assumed that plant life was nourished by the soil, and indeed some soil was fertile and some barren; some yielded good harvests, unless poorly watered, and some did not, even when well watered. In the nineteenth century, however, it came to be realized that plants made use of small quantities of mineral substances in the soil, by dissolving those substances in water. The main nourishment came from the air, while the soil itself (aside from the small quantities of useful minerals) served merely to hold the plant.

It therefore seemed possible to grow plants in a water solution of those substances necessary to its nutrition. In that way the nutritive substances would be particularly available to the plant, and their proportions would be easily controlled. Plants ought then to grow larger and healthier despite the absence of solid soil. This would be hydroponics (to till the water).

Hydroponics requires a highly developed chemical industry for support, and as long as ordinary agriculture is possible it is unlikely that hydroponics will ever replace it. Hydroponics can merely be used as supplement. In one respect, though, hydroponics may soon come into its own. When men take off on long space voyages beyond the moon, it would be impractical to carry with them enough food to last for years. It would surely be better to grow food of one sort or another, and on board the spaceship the techniques of hydroponics would be most suitable. Astronauts will then be expert hydroponicists, or water farmers, as well.

Hyperon

ALTHOUGH THE ELECTRON and the proton have electric charges of identical size, the proton is 1836 times as massive as the electron. In 1930 the neutron was discovered. It has no electric charge, but is even slightly more massive than the proton.

For nearly twenty years the proton and neutron were the most massive single particles known, and it was suspected they might be the most massive possible. In 1947, however, two English physicists, George Rochester and Clifford Butler, detected a V-shaped track in a cloud chamber, and found that one of the branches of the track was that of a particle about 2200 times the mass of an electron, and therefore about a fifth again as massive as a proton or neutron.

Since the capital form for the Greek letter lambda is shaped like an upside-down V, Rochester and Butler, in honour of the shape of the track that revealed the new particle, called it a lambda particle.

In the years that followed other particles, even more massive, were detected and studied. All were unstable, enduring for not more than 10^{-10} seconds before breaking down to protons, neutrons and smaller particles.

Each of the different groups of massive particles was given a Greek letter name, following the lambda precedent. There was a group of sigma particles with a mass 2360 times that of an electron; a group of xi particles with a mass of about 2600 times that of an electron; and in 1964 an omega particle was discovered that was 3300 times as massive as an electron, nearly twice as massive as a proton.

All these massive particles were grouped together under the name of hyperons, from the Greek word *hyper,* meaning above or beyond. The masses of the hyperons were, you see, above or beyond those of the proton and the neutron.

Hypotenuse

A LINE drawn on a perfectly level surface is horizontal. That is, it points towards the horizon at either end, and not above or below it. Horizon itself comes from the Greek *horizo,* meaning bound, since the horizon bounds the visible earth in all directions.

A weight suspended over the surface would hang by a vertical line—one, that is, that points to the vertex, the topmost point of the sky. Vertex comes from the Latin *vertere* (to turn) and was originally used for the centre of a whirlpool about which all turned, and was then applied to that point on the scalp about which the hairs seemed to turn when each lay in its natural position. Since this latter point is at the top of the head, more or less, vertex came to mean the top of anything.

A vertical line is perpendicular to a horizontal line, from the Latin *per* (through) and *pendere* (to hang); the vertical line hangs down and through the horizontal one as in the familiar plus sign (+). The angles formed in the plus mark are right angles. Our right comes from the Latin *rectus,* which means several other things, including upright. The right angle is formed when one of the lines stands exactly upright.

Of course, if the plus sign is tilted to form the multiplication sign (×) so that neither line is upright, the two are still perpendicular and the angles are still right angles.

The two lines of an angle meet at a point called the vertex (see above) because that point can be drawn topmost. If a right angle is so drawn (∧) a third line drawn underneath will form a triangle (from the Latin *tres* meaning three, so that a triangle is a figure with three angles). A triangle which includes a right angle as one of the three is called specifically a right triangle.

The third line 'stretched under' a right angle to form the right triangle is called the hypotenuse, from the Greek *hypoteinousa* —*hypo* (under) and *teinein* (to stretch). What could be clearer?

Idiot

IT IS perhaps only human that there are a great many different words used to express mental deficiency, most of them slang; the vocabulary of insult is always great. Psychologists, however, have tried to make objective use of three of them to indicate various grades of mental deficiency.

A moron is only mildly deficient. He is capable of doing useful work under supervision. The term was adopted in 1910 by psychologists, and is derived from the Greek *moros* (stupid).

More seriously retarded is an imbecile, who cannot be trusted to do useful work even under supervision, but is capable of connected speech. Whereas moron has always applied to mental deficiency, imbecile referred originally to physical deficiency, since the word is derived from the Latin *in-* (not) and *baculum* (staff); that is, it refers to a person too weak to get along without a staff. In the modern meaning, it is the mind that cannot get along without help.

Most seriously retarded is the idiot, one who is not capable of connected speech or of guarding himself against the ordinary dangers of life. This word has the oddest history of the three. The ancient Greeks were the most political of people. Concerning oneself with public business was the pet hobby of everyone. The Greek word *idios* means private, idiopathy is a 'self-generating' disease, so a Greek who despite all this was odd enough to concern himself only with his private business rather than with public business was an *idiotes*. The Greek view concerning such a person is obvious, since *idiotes* and idiot are the same word.

Of the more colloquial words, fool comes from the Latin *follis,* meaning bellows, obviously implying that a fool is someone whose words, though many and loud, are so much empty air. The slang expression windbag is the exact equivalent. Stupid is from the Latin *stupere* (to be stupefied; to be rendered speechless). Here the implication is of someone without words. Apparently for one to be intelligent his words must be neither too few nor too many, and in my opinion that's not a bad way of putting it.

Igneous

THE INTERIOR of the earth is at a high temperature, high enough to melt iron in the central depths. This heat arises from radioactive decay of certain elements within the earth (chiefly uranium, thorium and one isotope of potassium).

The rocky crust of the earth is exposed to more heat with increasing depth, and also to more pressure from the weight of the rock above. Eventually heat and pressure combine to turn rock (which we know as hard and brittle) into a soft, doughy mass. Such rock is called magma, a Greek word meaning dough, which is akin to *massein* (to knead).

This magma may work its way up towards the surface, and as the temperature and pressure both decrease it slowly solidifies and becomes hard. Such rock may eventually be exposed on the surface of the earth, and the large crystals that make it up show that it has solidified slowly out of a liquid state. These rocks are called plutonic rocks, from the Greek god of the underworld, Pluto.

Occasionally pockets of magma occur quite close to the surface, perhaps because of local accumulation of radioactive materials, and this may break out as flows of molten rock (see VOLCANO), which quickly congeal and harden into small-crystal solids. These are volcanic rocks.

Together, plutonic and volcanic rocks are igneous rocks, since both are formed from the fires beneath the earth, and *ignis* is the Latin word for fire.

Another kind of rock is formed out of material worn away from solid land by the action of rain and rivers and carried down to the ocean, where it settles out and sits at the bottom. Such settled material is sediment (from the Latin *sedimentum,* which comes from *sedere,* meaning to sit). As more sediment accumulates the lower layers are compacted together by the pressure of the upper layers. The result is sedimentary rock, and appears on the surface in regions where shallow seas once existed and are now gone.

Imaginary Number

A NUMBER MULTIPLIED by itself yields a square number and the original number is itself the square root of that square number. Thus, the square of 5 is 5×5, or 25. This means that 25 is the square of 5 and 5 is the square root of 25. A square root doesn't have to be an integer; in fact, it usually isn't, but is an endless decimal instead. The square root of 2 is 1·414214 . . . and so on, endlessly.

Now consider numbers with signs before them, positive and negative numbers. Two positive numbers multiply to yield a positive product, and two negative numbers do so likewise. This means that $+5 \times +5 = +25$ and $-5 \times -5 = +25$. The square root of $+25$ is both $+5$ and -5.

But what about the square root of a negative number, such as -25? There isn't any in the ordinary system, since neither positive numbers nor negative numbers will do. You can write any negative number, $-x,$ as $+x \times -1$. The square root is then the square root of $+x$ times the square root of -1. The whole problem boils down to the square root of -1.

The square root of -1 was called an imaginary number, because it didn't exist in the ordinary number system. In 1777 the Swiss mathematician Leonhard Euler symbolized the square root of -1 as i (for imaginary).

Imaginary numbers are not, however, imaginary. Consider a horizontal line with a zero point located on it. Points to the right can represent positive numbers, points to the left negative ones. Another line, up and down through zero, can represent positive and negative imaginary numbers. The entire plane is defined by a combination of the two numbers called complex numbers. Complex numbers are indispensable in the mathematics of electrical engineering, for instance. And in the 1960s physicists talked of tachyons with masses expressed in imaginary numbers. Ordinary or real numbers are thus no more real than the imaginaries.

Incisor

THE HUMAN mouth contains thirty-two teeth of different shapes, each shape designed for a special purpose. The eight teeth in front (four upper and four lower) are broad, and taper into a chisel-like cutting edge. These are the incisors, from the Latin *in-* (into) and *caedere* (to cut); they cut into food. Mammals such as rats, squirrels and beavers have very prominent incisors which are constantly used for gnawing. These animals are examples of rodents, from the Latin *rodere* (to gnaw).

Next to the incisors are a total of four conical teeth, one on each side, upper and lower. They tear at food, rather than cut, and are particularly useful to meat-eating animals. They are prominent in dogs, for instance, and are sometimes called the dog teeth for that reason, though more often they are termed canines, which is much the same, since it comes from the Latin *caninus* (of the dog) which in turn comes from *canis* (dog).

The twelve rear teeth, three on each side, upper and lower, have flat, irregular surfaces, between which food can be ground, and which are sometimes called the grinders for that reason. More usually, since just such a grinding action takes place in a mill (where grain is ground to flour between two millstones) these back teeth are called *molars,* from the Latin *mola* (millstone).

Between the molars and the canines are eight teeth (two on each side, upper and lower) that possess two conical points, and which are therefore called bicuspids, from the Latin *bi-* (two) and *cuspis* (point). They are also called premolars, because they come before the molars, the Latin *pre-* meaning before.

The Latin word for tooth is *dens* (genitive, *dentis*), hence dentist for a man who specializes in the treatment of teeth. The Greek word for tooth, however, is *odous* (genitive, *odontis*) and that gives the name to several specialized varieties of dentists. For instance, an exodontist is a dentist who specializes in tooth-extraction. The Greek *ex-* means from; he takes teeth from a jaw.

Again, an orthodontist is one who specializes in straightening teeth, from the Greek *orthos* (straight).

Indigo

BEFORE 1856 the only dyes available to mankind for use on textiles were those that could be found in nature, and of those only three were really good. That is, only three were brilliant in colour, would not fade in air and sunlight, would attach firmly to textiles, and would not wash out. Consequently, the three were immensely valuable products for centuries (until the chemist learned to improve on them with hundreds of synthetic dyes, and threw the natural dyes out of business).

One of the natural dyes was the red-purple substance obtained from a small Mediterranean shellfish. Its production was the specialty of the ancient Phoenician city of Tyre; even today the dye is called Tyrian purple. It was so expensive that its use was restricted to royalty.

A second dye (orange-red) was obtained from the root of a plant called madder, and is called alizarin, from the Arabic *al asarah,* meaning the juice. To the Arabs it was so valuable that there was no need, apparently, to specify which juice.

The third dye (blue-purple) was obtained from an Indian plant which in Sanskrit was called *nili*. The Arabs called it *al nil,* which the Spaniards further changed to *anil.* The name anil has led to certain important chemical names (see ANILINE), but it is not what English-speaking people call the plant.

To the Romans the plant was simply *indicum* (Indian), and that has come down to us, again through Spanish, as indigo. The word was transferred from the plant to the dye, and from that to the colour. Indigo also gave rise to a number of chemical names, including that of an element.

In 1863 two German physicists, Ferdinand Reich and Hieronymus Richter, discovered a new element which when heated showed a bright spectral line (see SPECTRUM) that was indigo in colour. They therefore named the new element indium.

Inertia

THE ANCIENT Greeks, more than any people who ever lived, had a profound appreciation of talent, both physical and mental. A person without the ability to do something well, without the possession of some art, was not a complete man. Later civilizations have inherited a little of this feeling, and it still shows up in our word inert, which comes from the Latin *in-* (not) and *ars* (art). A person who has no art merely vegetates; he lacks an essential spark of life. And so the word is applied to anything without life, anything that is sluggish, heavy, unresponsive, resistant to change, and so on. (Of course, art also has an evil meaning, when it is used in the sense of wile rather than talent. A person without wile is artless and that kind of no art is considered a virtue.)

In 1687 Isaac Newton presented the world with three simple Laws of Motion on which all modern mechanics is based. The first law is this: 'Every body persists in a state of rest or of uniform motion in a straight line unless compelled by external force to change that state.'

This means that a brick resting on a board would rest there through all eternity unless it were pushed or pulled and made to move. Left to itself, it would never move. This seems to stress the inertness of matter; it raises inertness to the state of a natural law. For this reason Newton's First Law is called the principle of inertia.

Of course, the law also states that if the brick were hurtling through space it would continue moving for ever, unless stopped. That doesn't seem very inert of it, but actually it is. Remember that inert implies a resistance to change. You could look at it this way: that the brick having started moving at last is now too 'lazy' to stop moving, or even to change its direction of motion, unless made to do so by some external force (see FRICTION).

That, after all, is still inertia, but in another guise.

Infinite

THE ORDINARY objects that surround us have a beginning and an end. A piece of paper, a ruler, a locomotive, the British Isles—all have some end on all sides such that if you go past that end, you reach a place where the piece of paper or the ruler or the locomotive or the British Isles does not exist. The universe itself, according to Einstein's theories, has an end. The Latin word *finis* means end (we see it on cinema screens sometimes, for that reason) so objects that come to an end are finite.

Despite the finiteness of the universe, not everything comes to an end. If you were to start counting, and continue counting until you came to an end, you would find that there was no end. After every number, no matter how large, it is always possible to add another number. The total number of numbers, in other words, is not finite. It is (using the Latin prefix *in-* which means not) infinite.

The quality of being endless is infinity. You may speak, for instance, of the infinity of numbers or the infinity of ideas. Infinity is not, however, itself a number. You must not imagine infinity to be the last number, since there is no last number. Mathematicians find it useful to use the symbol (∞) to represent a condition of endlessness, but the symbol is usually read by everyone (even mathematicians themselves, when in a hurry) as infinity, which, strictly speaking, is wrong.

It is also possible to imagine something that is infinitely small. In the number system, for instance, one is the first number and as close as you can get to zero without actually being zero. However, if we consider fractions, one-half is smaller than one, one-quarter still smaller, and one-eighth smaller still. In fact, no matter how close to zero the fraction gets, one can get still closer without actually reaching zero by increasing the denominator without increasing the numerator of the fraction. An infinitely small thing is said to be infinitesimal.

Influenza

THE HEAVENLY bodies were thought in ancient times to govern the fates of men, and the study of the exact manner in which they did this was known as astrology, from the Greek *astron* (star) and *logos* (word). An astrologer talks about the stars, in other words. Usually the suffix -logy implies a branch of respectable science, but the false notions of astrology have so discredited the word that the science of the stars is today known as astronomy instead. The suffix -nomy comes from the Greek *nemein,* meaning to arrange. Thus an astronomer was originally one who studied the arrangement of the stars, but the word has come to mean one who studies everything about the stars.

The mysterious power residing in the stars was supposed by the astrologers to flow down to earth and into men, governing them. The Latin word for to flow in is *influere,* so the stars were said to have influence over men.

In the days before the real cause of disease was understood, it was natural to assume that sickness, like everything else, was caused by the influence of the stars, so one of the most common diseases was called just that—influenza (which is Italian for influence).

Another disease which reveals pre-scientific notions of cause is malaria, which is now known to be caused by a one-celled animal that infests our red blood cells and is spread from person to person by mosquitoes. In Italy, during the decline of the Roman Empire

196

and during the worse times of the barbarian invasions that followed, fields were abandoned and allowed to revert to swampland along the coast. Mosquitoes flourished among the swamps, and malaria spread.

The Italians noticed the connection between malaria and the swamps, but overlooked the mosquito. To them the mosquito was only a nuisance; the real fault lay with the bad air that hovered over the swamps. (It was undoubtedly musty with decaying vegetation, and smelled.) The Italian words for bad air are *mala aria,* and hence the name of the disease.

Infra-red

LIGHT BEHAVES as waves in at least some ways. The waves of a particular ray of light will cover a fixed distance with each undulation. This distance is the wave-length of that light. Such wave-lengths are generally measured in Ångstrom units (abbreviated Å), one of which is equal to 10^{-10} (one in ten thousand million) metres. (It is named after the Swedish physicist Anders Jönas Ångstrom, one of the pioneers of spectroscopy; see SPECTRUM.)

Light with wave-lengths of about 7000 Å seems red to our eyes. A wave-length of 6500 Å strikes us as orange, 5900 Å as yellow, 5400 Å as green, 4800 Å as blue and 4200 Å as violet. The spectrum gives the entire range of visible colours from 7200 to 4000 Å.

But visible light is not the only form of radiation. A photographic plate placed beyond the violet end of the spectrum, where nothing is visible, is strongly fogged, so there is some sort of radiation there. This is sometimes called black light, but its more formal name is ultra-violet light, *ultra* being a Latin word meaning beyond. The wave-lengths of ultra-violet light range from 4000 Å down to 100 Å. (X-rays and gamma rays have still shorter wave-lengths.)

Similarly, there can be light beyond the red at the opposite end of the spectrum. The astronomer William Herschel (see URANIUM) noticed in 1800 that a thermometer placed beyond the red showed a rise in temperature. There are radiations there too. These are sometimes called heat rays because of this heating effect. The more formal name, however, is infra-red light, *infra* being a Latin word meaning below. Why below rather than above? Because red light contains less energy than any other form of visible light. In terms of energy content it is at the bottom of the scale. Infra-red, with still less energy, is below it.

Infra-red light has wave-lengths of from 7200 Å all the way up to

197

3 000 000 Å. Beyond it lie the radio-waves, with still longer wave-lengths.

Infra-red Giant

IN 1905 the Danish astronomer Ejnar Hertzsprung considered the fact that some red stars, like Antares and Betelgeuse, were very bright, and some, like Barnard's Star (which see), were very dim. Both were equally low in surface temperature, for all were no more than red-hot; warmer stars glow yellow, white, and even blue-white. The only way that a cool star that gives off only a dim red glow from each portion of its surface can nevertheless appear bright is for it to have a large surface.

Hertzsprung therefore suggested that there were red giants and red dwarfs among the stars. There seemed to be no red stars of intermediate size, and that began the line of reasoning that developed modern notions of the way in which stars evolved with time.

To begin with a large cloud of dust and gas slowly condenses, getting hotter and brighter as it does so, for a million years or more, until it finally reaches a stable situation in which it can remain for thousands of millions of years. In the process of condensing, when it glows red-hot, it will pass through a red-giant phase.

After it has consumed its chief nuclear fuel (hydrogen) through many millions of years of shining, a star will begin fusing helium and still more complicated atoms. When it does so it expands enormously, and in the process its surface cools and again it passes through a red-giant phase.

In condensing originally, a star passes through a phase when it radiates chiefly infra-red light. Again, when a dying star expands, it may expand so much that its surface radiates chiefly infra-red. In 1965 astronomers at Mt Wilson Observatory used a special telescope with a large plastic mirror to scan the sky for spots rich in infra-red. Within a couple of years they found thousands of objects that were barely shining, and those were the infra-red giants.

Insect

INSECTS ARE the great pioneers of the animal kingdom. They were the first animals to invade the land, the first to learn to fly, the first to develop complex social groups. The pioneering has paid off; there are more different kinds of insects in existence today than the total of all kinds of all other animals put together.

One of the distinguising characteristics of insects is the division of their bodies into three relatively thick regions—head, thorax and abdomen—which are connected by short and sometimes very thin stalks. (We have the expression wasp waist for an unusually small waistline.)

The insect outline is cut into, so to speak, between the main regions. The Latin word for to cut into is *insecare* (past participle, *insectum*), hence insect.

Many insects emerge from the egg in the form of worm-like creatures that bear no resemblance to the adult insect that laid the egg. Out of a butterfly's egg, for instance, comes a caterpillar. This immature form is a larva, which is the Latin word for ghost. The connection is rather far-fetched, but just as a ghost originates from a human being but is quite different in appearance, so a larva originates from an ordinary insect but is quite different in appearance.

Eventually the larva enters a quiet stage during which its body is reorganized into the adult form. It sometimes weaves a silken covering, a cocoon (from the French *cocon,* meaning a little shell) as protection in the meantime. In this quiet, reorganizing state the insect is a pupa, a Latin word meaning doll or puppet, since in that stage it shows no visible life. The pupa stage of some butterflies is golden in colour, so another term for pupa is chrysalis, from the Greek *chrysos* (gold).

An insect in the pupa stage or in the late larval stage is also sometimes called a nymph. The nymphs in Greek mythology were demi-goddesses who were ever young and ever beautiful. The Greeks gallantly applied the term to young girls on the point of marriage and adulthood, and we spoil it by applying it to insects on the point of adulthood.

Insulin

THERE ARE glands which, unlike the liver (see BILE), do not pour the chemicals they manufacture through a duct, but transfer them directly into the bloodstream. These are the ductless glands, or the endocrine glands. The latter name comes from the Greek *endon* (within) and *krinein* (to separate); the fluid is separated within the body and distributed directly into the blood. Such glands produce hormones (see HORMONE) and it is for this reason that doctors who specialize in the workings of hormones are called endocrinologists.

The pancreas is a gland with a duct, and a pancreatic juice travels through the duct into the intestine. In 1869, however, a German pathologist, Paul Langerhans, discovered little clumps of cells scattered throughout the pancreas that were different from the rest. In his honour these are called the islets of Langerhans (perhaps the most romantic-sounding name in the body).

In 1889 it was discovered that a dog from which the pancreas had been removed did not live more than a few weeks, and for those few weeks he showed symptoms similar to those of a human disease called diabetes mellitus. Diabetes is a Greek word meaning something which goes through. It is applied to those diseases which involve an over-production of urine; liquid just seems to go through the body without pausing. In diabetes mellitus sugar is not handled properly by the body; it accumulates in the blood, and spills over into the urine. The resulting sweetness of the urine is described by mellitus, which comes from the Latin *mel,* meaning honey.

In 1916 the British physiologist Sharpey-Schafer suggested that the islets of Langerhans were a group of ductless glands buried in an ordinary gland, and that they produced a hormone that controlled the manner in which the body handled its sugar-supply. Since the hormone was produced by 'islets', or 'islands', he suggested the name insulin, from the Latin *insula* (island). The hormone was indeed eventually isolated, and Schafer's name stuck. In 1955 the structure of the complicated insulin molecule was worked out, and in 1966 it was finally synthesized.

Interferon

ANTIBIOTICS have helped the medical profession control many bacterial infections. That is because a particular chemical can affect bacterial cells more seriously than it will affect host cells. With proper dosage bacteria can be stopped with no harm to the host. Viruses are not so easily handled. They do their work inside the cell they parasitize, using the cell's own chemical machinery. To kill one by outside chemicals means killing the other.

Nevertheless, virus diseases don't always kill. In most cases, the affected organism recovers; sometimes it is hardly affected. Thus the organism must have natural defences. The organism forms protein molecules capable of reacting with the virus and preventing it from harming the cells. These antibodies sometimes persist through life, so that anyone who recovers from measles, chickenpox, mumps and certain other virus diseases is nearly always immune to them thereafter.

In the 1930s there began to be signs that cells could fight off viruses even without producing antibodies. There was virus interference. The fact that viruses invaded some cells made it harder for them to invade other cells in the organism. In 1957 the English biochemist Alick Isaacs showed that the entry of viruses into cells stimulated the formation of tiny quantities of a small protein that left the cell, entered other unaffected cells, and guarded them against penetration. This limited the effects of the virus infection. The protein which interfered with the virus Isaacs called interferon.

Each species produces an interferon that works only on its own cells. You can't infect chickens, isolate their interferon, and hope to help humans with it. However, attempts are being made to learn how to stimulate the production of interferon in humans by the addition of some harmless substance that will then leave the body immune to particular viruses thereafter.

Iodine

IT IS common to have a liquid turn into a vapour when heated. It is less common to have a solid substance turn into a vapour directly without ever going through the liquid state. The best-known example these days is solid carbon dioxide, which has the appearance of a cloudy ice, but is much colder, and which doesn't turn to liquid when warmed but to gas. Because of this absence of liquid and wetness, the common name for solid carbon dioxide is dry ice.

A number of other substances behave the same way. A good example is iodine. We are most familiar with iodine when it is in solution in a mixture of water and alcohol. This is tincture of iodine, from the Latin *tinctura* (dyeing), from *tingere* (to dye), which is the source also of the common word tint. In pharmacy alcoholic solutions are commonly called tinctures because many dyes will dissolve in alcohol but not in water. Tincture of iodine is reddish brown in colour.

Iodine itself, however, is an element which is solid at room temperature and forms slate-grey crystals. If some is heated gently in a test tube it will not liquefy but will form a beautiful violet vapour which will solidify into grey crystals again in the upper portion of the test-tube where the glass is cooler. The iodine has been raised to a higher place in this manner.

The Latin word for high is *sublimis*, from *sub-*, meaning under, and *limen*, meaning the lintel (upper wooden crossbar) of a door. Something just under the top crossbar of a door is relatively high. So lofty ideas or thought or behaviour are said to be sublime, and when a solid turns directly into vapour it too is said to sublime.

The violet vapours of iodine were first observed in 1811 by the

French chemist Bernard Courtois. He was working with a water extract from burned seaweed and by adding sulphuric acid he isolated from it a black powder. When heated this formed a violet vapour turning into grey crystals and Courtois had unwittingly discovered a new element. Three years later Sir Humphry Davy named it iodine from the Greek *iodes* (like the violet).

Ion

IN ELECTROLYSIS parts of a molecule seem to travel to one electrode and parts to the other. The water molecule, for instance, breaks up into hydrogen and oxygen, the hydrogen appearing about the negative electrode, or cathode, and the oxygen appearing about the positive electrode or anode (see ELECTROLYSIS).

The travelling parts of the molecules were given the name ion in about 1834 by the British physicist Michael Faraday, from a Greek word *ienai* meaning to go. The Greek word *ion* means going. After all, they were going to one electrode or the other. The ions that travelled to the cathode were cations; those that travelled to the anode were anions (both words are pronounced in three syllables). Exactly what ions were, however, remained a mystery for about fifty years.

Then, in 1884, a 25-year-old Swedish physical chemist, Svante August Arrhenius, presented a dissertation at the University of Uppsala, with which he hoped to earn the degree of Doctor of Philosophy. He suggested that under the influence of the electric current molecules broke up to form atoms or groups of atoms that carried electric charges. Those with a negative charge were anions, attracted to the positive electrode, while those with a positive charge were cations, and were attracted to the negative electrode.

Since chemists had at that time never heard of atoms carrying an electric charge, his theory was considered quite ridiculous, and he got his degree with a minimum passing grade. However, in 1903 Arrhenius received the Nobel Prize in chemistry for that same dissertation.

Between 1884 and 1903, you see, the negatively charged electron had been definitely shown to exist, and to form a part of all atoms (see ELECTRON). It was recognized that an atom or group of atoms might lose one or more electrons to become positively charged, or gain one or more to become negatively charged. In the former case, a cation resulted; in the latter, an anion.

Ion Drive

A ROCKET makes its way out of the atmosphere and through space by means of jet propulsion. A propellant burns, sending a jet of exhaust gas in one direction, so that the rest of the rocket is pushed in the opposite direction. The propellant is usually a chemical fuel that burns in some active substance like oxygen, so that this method of accelerating a rocket may be called a chemical drive.

A chemical drive can lift many tonnes of matter into space against the pull of earth's gravity. Chemical fuels are, however, rapidly consumed. Many tonnes of propellant are required to launch a rocket. Once out in space, where the gravitational pull of distant worlds is comparatively small and there is no air resistance, it takes only short bursts of fuel exhaust to correct orbits, but even they use up fuel quickly.

Is there any kind of propellant that would last longer? One suggestion involves positive ions, atoms that have lost one or more electrons, and which therefore carry a positive electric charge. These ions can be repelled by another positive charge and forced out the rear of a rocket engine, at an enormous speed.

The ions, moving backward, will push the rocket ship forward. They will do so only very slightly because the ions are so small, so such an ion drive would not work in the neighbourhood of earth's surface. Only far out in space where there is no necessity to overcome air resistance and no strong gravitational field will the ion drive manage to make its very slight push felt.

However, the ions produced by a large supply of matter will last for a very long time. There would be only a tiny push, but it would be kept up for years, and that tiny push–push–push would eventually bring the rocket to speeds close to that of light. It may be that only by some sort of ion drive can we ever hope to reach the distant stars.

Irrational Number

ANY NUMBER which can be expressed as the ratio of two digits is a rational number (see FRACTION). To a non-mathematician, who doesn't think of rational as being derived from ratio, there is a natural tendency to think of such numbers as being reasonable or sensible ones, because the word rational means that in the ordinary vocabulary. It comes from the Latin *rationalis,* which in turn comes from the verb *reri,* meaning to think. So does the word ratio, which accounts for the confusion.

When fractions were first discovered it seemed logical to assume that any conceivable number could be written as a fraction. For instance, a number halfway between $1/2$ and $1/3$ is $5/12$. A number halfway between $5/12$ and $1/3$ is $9/24$. To express some numbers it might be necessary to use big digits, as, for instance, $28\ 067\ 048/57\ 134\ 097$, but that doesn't affect the principle of the matter.

But now what about square roots? If you restrict yourself to digits only, the square root of 4 is 2 and that of 9 is 3 (see SQUARE ROOT). But there is no digit that is the square root of 8, for instance. Yet what if you use fractions? If you multiply $14/5$ by $14/5$, you have $7 \cdot 84$. That makes $14/5$ almost the square root of 8, but not quite. Try a slightly larger fraction and work out $141/50 \times 141/50$. This comes to $7 \cdot 9524$, which is better. Multiply $707/250 \times 707/250$ and you have $7 \cdot 997\ 584$, which is still better. It certainly appears that if you keep adjusting fractions, you will eventually find one that will, when multiplied by itself, give you exactly 8.

But this is not so. You can get awfully close to 8 but you can never hit it exactly. There is no fraction which is the square root of 8 (or of the vast majority of other digits). The square root of 8 (and of most other digits) simply cannot be expressed by a ratio of two digits in any way. The square root of 8 (and most other digits) is an irrational number. To the ancient Greeks who discovered this (and to many a schoolboy since), this has seemed so odd that irrational numbers seemed irrational indeed, in the ordinary sense.

Isomer

UNTIL 1800 chemists were quite certain that every different compound was made up of a different combination of elements (regardless of what they thought the elements were); or if the same elements were found in different compounds, at least they would exist in different proportions. And this is generally true in the realm of inorganic chemicals.

Organic chemicals (see ORGANISM) are, however, made up for the most part of only a handful of elements. Carbon, hydrogen, oxygen and nitrogen occur most frequently. And since there are uncounted thousands of organic compounds, it is not surprising that occasionally two should be found with molecules made up of equal numbers of the same atoms. By 1830 enough such pairs had been found for the Swedish chemist Jöns Jakob Berzelius to think it wise to propose a name. He suggested that compounds of identical composition but different properties be called isomers, from the Greek *isos* (equal) and *meros* (part); they were made up of equal parts of the various elements.

For a number of years the reason for the existence of isomers was not quite understood. Then in 1874 the French chemist Joseph Le Bel and the Dutch chemist Jacobus van't Hoff independently suggested that the carbon atom attached itself to four other atoms arranged in specific positions about itself. Obviously, then, chemicals might have molecules made up of the same number of the same atoms, but in different arrangements.

As a result of this it sometimes happens that when a compound is discovered which differs from an old-established compound only in atom arrangement it is named by adding the prefix iso- to the older name, as a short version of 'isomer of'. For instance, in 1818 one of the amino acids (now known to be found in proteins) was first isolated, and was named leucine from the Greek *leukos* (white) because it showed up as white crystals. (This is a poor excuse for naming, by the way. The majority of organic chemicals are white.) When in 1905 another amino acid was discovered, differing from leucine only in a minor detail of atom arrangement, it was named isoleucine.

Isosceles

A CLOSED FIGURE made up of straight lines may contain any number of sides from three up, and the same number of angles, if the lines don't cross. The Greek word for angle is *gonia* and for much is *polys*. A figure containing a number of angles is therefore a polygon.

The particular variety of polygon is named from the Greek numbers. A five-angled figure is a pentagon (*pente* is five); a six-angled one is a hexagon (*hex* is six); an eight-angled one is an octagon (*okto* is eight) and so on. On this system a four-angled figure ought to be a tetragon and a three-angled one a trigon. These last two can, indeed, be found in the dictionary but they are not often used.

Instead the Latin equivalents are used. A four-angled figure is a quadrangle (*quattuor* being Latin for four) or, even more commonly, a quadrilateral (four sides, see PARALLEL). The three-angled figure is a triangle (*tres* being Latin for three).

A polygon with all sides equal is equilateral, from the Latin *aequus* (equal) and *latus* (side). The only equilateral polygon with a special name of its own is the equilateral quadrangle. This is a square (see PARALLEL) if all the angles are right angles, or a rhombus if they are not. (The word rhombus comes from the Greek name for a small block of wood of that shape which was placed on a string and whirled as a noise-maker during certain festivals. The Greek word *rhembein* means to whirl.)

A triangle with all sides equal is an equilateral triangle. One with only two sides equal is an isosceles triangle; the word isosceles comes from the Greek *isos* (equal) and *skelos* (leg). (A man's two legs, standing apart, make an isosceles triangle with the ground.)

Isotope

UNTIL THE middle 1800s there seemed no orderliness in the nature of the various elements that were being discovered. In 1869, however, the Russian chemist Dmitri Ivanovich Mendeléev listed the then-known elements according to the weights of their atoms. He showed that similar properties turned up at regular periods, so that the listing was called the periodic table. By use of it, Mendeléev predicted the properties of still-unknown elements, and lived to see those prophecies come true.

For thirty years the periodic table met every test. Then in 1896 uranium was found to give off strange forms of radiation. In doing so it broke down to another substance, which broke down to still another, and so on. The element thorium also behaved in this fashion.

Chemists located a series of breakdown products, over forty of them all told, each with its own properties, but there was simply no place in the periodic table to put these new elements. For instance, one substance was 'radium G'. Chemically it behaved like lead, but it gave off radiations, whereas ordinary lead did not. It fitted nowhere in the table. Then, too, three different gases were discovered as products of these breakdowns, but there was only one place in the periodic table to put them.

Several scientists, and notably the British chemist Frederick Soddy, solved this dilemma in 1913. They suggested that a particular element might have different types of atom. It has since been definitely proven that this is so. All atoms of a given element have the same number of protons, and are therefore alike in ordinary chemical ways. They may vary, however, in the number of neutrons, and will behave differently in other ways, in whether they break down or not, their particular manner of breakdown, and so on.

These different atomic varieties of a single element belong in the same place in the periodic table, and Soddy made that very point by calling such atomic varieties isotopes, from the Greek words *isos* (equal or same) and *topos* (place).

Jansky

KARL JANSKY was a radio engineer employed by Bell Telephone Laboratories. In 1931 they set him the task of studying the causes of static, which constantly interfered with radio reception and with radiotelephony from ship to shore. Static has a number of causes, from thunderstorms and aircraft to near-by electrical equipment.

Jansky, however, detected a new kind of weak static from a source which at first he could not identify. It came from overhead and moved steadily. At first, it seemed to be moving with the sun. It gained slightly on the sun, however, to the extent of four minutes a day. This is just the amount by which the vault of the stars gains on the sun. Consequently, the source seemed to lie beyond the solar system.

By the spring of 1932 Jansky had decided the source was in the constellation of Sagittarius, the direction in which the centre of the Galaxy was located.

He published his findings in December 1932, and this represented the birth of radio astronomy. It was the first indication that there were vast sources of radio-wave energy in the universe. Astronomers learned to receive and interpret microwaves (the shortest radio waves) in addition to light waves. Microwaves penetrate dust-clouds that light waves cannot, so that radio telescopes could receive information about objects totally invisible to ordinary optical telescopes. Furthermore, very distant or unusual objects—which could not be made out by ordinary telescopes—revealed themselves through radio waves.

Jansky himself did not continue to work with radio astronomy, for he was more interested in his engineering. He died in 1950 of a heart ailment while still only forty-five, so that he lived to see only the infancy of the science he had founded. He is, however, not forgotten. The unit of strength of radio-wave emission is now called the jansky in his honour.

Janus

ONCE THE TELESCOPE was invented it was soon found that planets had small companion bodies circling them. In 1610 the Italian scientist Galileo Galilei discovered four satellites circling Jupiter.

Other such discoveries were made, with the larger satellites of the closer planets detected first. In 1892 the American astronomer Edward E. Barnard (see BARNARD'S STAR) discovered Amalthea, a fifth satellite of Jupiter, far smaller and also closer to its planet than the other four. It was the last satellite to be discovered by eye. (Amalthea was the goat that suckled the infant Zeus, or Jupiter, or perhaps the nymph who fed him with goat's milk.)

By 1950 the resources of photography had raised the number of satellites to thirty-one. Earth had one, Mars two, Jupiter twelve, Saturn nine, Uranus five and Neptune two. Astronomers were sure that further satellites existed, but felt they were probably too small or too distant to detect except by space probes.

Saturn offered an interesting problem, though. In addition to its nine satellites, it had a brightly shining set of rings. The innermost known satellite was Mimas, but if there were any still closer it might well be lost in the glare of the rings. Every fourteen years, however, Saturn and Earth are so situated that the rings of the former are presented edge-on to the latter. The rings are so thin that they then cannot be seen.

The year 1966 was one of those in which the rings of Saturn were not seen from earth, and the French astronomer Audouin Dollfus then carefully investigated the close neighbourhood of Saturn. His search was rewarded. Only 6400 kilometres outside the outer edge of the ring system there was a satellite of respectable size (483 kilometres in diameter). It would certainly have been discovered long ago were it not for the rings. At the tenth and latest of the satellites of Saturn to be discovered, and the first in order counting out from the planet, he called it Janus, after the Roman god with two faces, one looking forward and one backward. What better name to call a satellite that was both last and first?

Jet Plane

ONE OF THE FUNDAMENTAL LAWS of the universe is the law of conservation of momentum. This says that if an object at rest is to begin to move in a particular direction, something else must move in the opposite direction. Thus, when we walk our feet push earth in the opposite direction, but only unimaginably slightly, because momentum is velocity times mass, and earth's mass is much greater than our own.

This holds true for all means of propulsion. Car tyres push earth in the opposite direction. Oars, paddle-wheels, screw propellers, all push water backward so that a ship can go forward. Aeroplane propellers push air backward so the plane can go forward.

There is a limit to how fast aeroplanes with propellers can go, because there is a limit to how fast propellers can turn without flying apart. In that case why burn fuel in order to turn a motor, in order to turn a propeller, in order to throw air backward? The burning fuel turns into hot gases that exert great pressure in expanding. If you give those gases an opening behind they will expand through that opening and be themselves thrown backward. The backward movement will force the aeroplane forward, and we can go straight from burning fuel to the desired motion without going through all the other steps.

The hot exhaust gases which so to speak throw themselves through the opening behind, are a jet (from the French word for to throw). A plane which pushes forward because of the jet pushing backward is a jet plane. In 1939 an Englishman, Frank Whittle, flew a reasonably practical jet plane, and the invention received impetus during the Second World War.

Since the War jet planes have come into commercial use and world travel has been revolutionized as large planes carry great numbers of passengers farther and faster than would have been possible in the days of propeller planes.

Jet Stream

MEN HAVE ALWAYS BEEN AWARE of moving currents of air, the winds. Naturally, they have known them best where they could directly experience them—on or near the surface of earth.

After the aeroplane was invented new kinds of winds were experienced. During the Second World War high-flying American planes, making their way westward toward the war zones near Asia, were astonished to find themselves making poor time and using up a surprising amount of fuel. They were fighting a head-wind, apparently, and a very powerful one.

When the phenomenon was studied it turned out that there was a steady, very strong, west-to-east wind blowing at a height of 13 kilometres or so above the surface of earth. Its speed sometimes reached up to 320 kilometres an hour. Because this stream of air moved with the force of a jet of fluid emerging from a narrow nozzle, it was called the jet stream. Actually, it turned out, there were two jet streams, located between the mainly cold arctic mass of air and the mainly warm tropic mass of air, north and south of the equator. The northern jet stream circles earth in the general latitude of the United States, the Mediterranean Sea and northern China. The southern jet stream moves around earth in the general latitude of Argentina, Chile and New Zealand.

The jet streams meander, often curving into eddies far north or south of their usual course. Aeroplanes now take advantage of the opportunity to ride on the swift winds. More important still is their influence on moving air masses nearer earth's surface. These control the surface weather so that studying the variations in the jet-stream course makes it possible to be a little surer about the long-range prospects in weather forecasting.

So far the southern jet stream has been far less studied than the northern one, but it is likely that these two streams are not very different.

Juvenile Hormone

INSECTS HAVE a tough cuticle made of chitin (from a Greek word for a coat of armour). While the insect is growing the cuticle presents a problem. The chitin of a growing insect larva must periodically split. The larva wriggles out of the split cuticle and quickly grows a new one. This is called moulting. After a certain number of moults, the larva undergoes more radical changes to become an adult insect.

The process of moulting seems to be an automatic one controlled by the production of an insect hormone called ecdysone, from a Greek word meaning to slip out from or moult. But then what stops the moulting and suddenly causes the larva to undergo the other changes on the way to adulthood? Apparently a second hormone must be involved.

In 1936 an English biologist, Vincent Wigglesworth, cut off the head of a certain species of insect to see what would happen in the absence of any hormones that might ordinarily be produced in the head. The insect lived (a head is not as important to an insect as to a vertebrate), but it at once began the changes leading to adulthood.

Apparently the head produced a hormone that kept the insect a larva, and moulting. When no head hormone existed the changes toward adulthood began. Since the head hormone acted to keep the insect in the immature state, it was called juvenile hormone.

The American biochemist Carroll Williams began to study the juvenile hormone in the 1950s. He found that juvenile hormone, placed on insects that had already begun the change toward adulthood, was absorbed through the skin. It then acted to stop the change and the insect died.

In short, the juvenile hormone, or any synthetic compound that worked similarly, could possibly be a very useful kind of insecticide. It works only on insects and on no other organism, and usually on only one group of insects, not others. It would thus lack the disadvantages of other insecticides.

K-Capture

ATOMIC NUCLEI are only stable when they possess protons (with a positive electric charge) and neutrons (with no electric charge) in a certain ratio. When there are too many neutrons one of those neutrons must change into a proton. Suppose a neutron has no charge because it possesses equal amounts of both positive and negative. It can become a proton by getting rid of the negative charge. For that reason an atomic nucleus with too many neutrons can become stable by emitting the negative charge in the form of an electron.

In similar fashion, an atomic nucleus which is unstable because it has too many protons can convert a proton to a neutron by getting rid of a positive charge. Such nuclei become stable by emitting positrons (electron-like particles that carry a positive electric charge).

In 1936 the Japanese physicist Hideki Yukawa showed that it was theoretically possible for a nucleus to capture one of the electrons in the outer reaches of the atom. Capturing a negative charge was equivalent to emitting a positive one; therefore nuclei that become stable by emitting a positron might capture an electron instead. The American physicist Luis W. Alvarez detected actual cases of this in 1938.

The electrons are grouped outside the nucleus in a series of shells, up to seven in number in very complicated atoms. This was found to be so originally because under certain circumstances atoms emit X-rays at various levels of frequency, one for each electron shell. The X-ray frequencies are labelled in alphabetical order, starting arbitrarily with K, so that there were K, L, M . . . frequencies in order of decreasing energy. The electron shells were correspondingly labelled K, L, M . . . in order of increasing distance from the nucleus.

The K-electrons (those in the K-shell) are nearest the nucleus and are most apt to be captured. The phenomenon of electron capture is frequently called K-capture for that reason.

Kaon

AT FIRST physicists were aware of only a few sub-atomic particles. Gradually they learned to work with larger energies and to make use of more delicate detection devices. As a result they learned to produce, identify and study many additional particles.

In 1944 French physicists noted a cloud-chamber track in their study of cosmic rays that indicated a particle about a thousand times as massive as an electron, and therefore about half as massive as a proton (see HYPERON). Such tracks were found again, usually in association with a particle called a pion. The unknown particle and the pion were apparently formed together by the impact of cosmic-ray particles on atoms. The two particles, after formation, would then move off in different directions, leaving a V-shaped track. This was therefore called a V-event, and the heavy particle was called a V-particle.

Eventually, though, it turned out that V-events were fairly common and that not all of them involved this particular particle. A new name, therefore, had to be found. Since the mass was intermediate between that of an electron and a proton, it belonged to the meson family. To distinguish it from other mesons, it was called the K-meson, and this name is frequently shortened to kaon.

The kaon is very unstable, and only endures for about a hundred-millionth of a second before breaking up in any of six different ways to form (usually) smaller mesons.

The different ways in which kaons break up proved important, for it turned out that a kaon could break down in such a way that a certain sub-atomic property called parity could be either odd or even. Until then it had been thought that parity had to stay odd, or had to stay even, in any particle change. The finding that it could be either odd or even in this case led to an interesting change in physical theory.

Karyotype

EVERY CELL NUCLEUS contains a group of chromosomes that can be arranged in pairs. The chromosomes resemble a mess of stubby strands of spaghetti all tangled up, and it is extremely difficult to count the number per cell. For instance, it was long thought that human cells each contained forty-eight chromosomes, in twenty-four pairs. It was not until 1956 that a very painstaking count showed the true number to be forty-six, in twenty-three pairs.

Fortunately, this problem no longer exists. A technique has been devised whereby treatment with a low-concentration salt solution in the proper manner swells the cells and disperses the chromosomes. They can then be photographed and that photograph can be cut into sections, each containing a separate chromosome. If these chromosomes are matched into pairs and then arranged in the order of decreasing length, the result is a karyotype (from Greek words meaning picture of the nucleus).

The karyotype offers a new tool in medical diagnosis, for it

212

shows clearly whether there are any missing chromosomes, extra chromosomes, or damaged chromosomes. Such imperfections, by adding or subtracting or distorting a whole series of genes (which control the chemistry of the cell), can produce serious birth disorders.

There is for instance a disease called Down's syndrome (because it was first described in 1866 by the English physician John Down). Babies born with it are mentally retarded, and have physical deficiencies as well. (The disease is sometimes called mongolism or mongolian idiocy because one of the symptoms is an eyelid shape that looks like those common in East Asia. This is a poor name, though, since the disease has nothing to do with East Asians.)

In 1959 three French geneticists, **Jérome** Lejeune, M. Gautier, and P. Turpin, found that cell nuclei in Down's-syndrome patients had forty-seven chromosomes, not forty-six. There was an extra chromosome at position 21.

Ketone

IN MANY organic compounds there occurs a carbon-oxygen combination which can in turn be attached to two other atoms. One of these other atoms is often hydrogen.

Compounds containing such a carbon-oxygen-hydrogen combination can be obtained by removing two hydrogen atoms from an alcohol molecule—i.e., by dehydrogenating it. In 1835 the German chemist Justus von Liebig recognized this fact and suggested the name aldehyde for such compounds as an abbreviation for *al*cohol *dehyd*rogenated. The name was accepted, and the official suffix for the names of such compounds is -al now.

The carbon-oxygen combination need not be attached to a hydrogen atom, however, but to two carbon atoms. Again, it was von Liebig (in 1831) who worked out the actual molecular structure of the simplest of the compounds containing such a combination. He suggested the name which in English is acetone. The acet- part comes from acetic acid (see ACID), from which acetone can be formed. The -one ending was originally used to indicate a chemical that was weaker than the one from which it was prepared, and certainly acetone is a milder substance than is acetic acid.

Now, in English c before e is pronounced s, so that acetone is pronounced asetone. But in Latin c is always pronounced k, and in German the letter c is never used except in foreign words anyway. Liebig's name for acetone was therefore *Aketon,* which is actually more correct.

In 1848 the German physiologist Leopoldus Gmelin wanted a word to express all compounds that contain the atom combination that was found in acetone. To obtain such a word, he merely dropped the a from aketon and used what was left. Consequently, our word for compounds of the acetone type is ketone to this day. Furthermore, the -one suffix in organic chemistry has lost its original meaning and is now used to indicate that the chemical in question is a ketone.

Krebs Cycle

THE CHIEF BODY FUEL is a six-carbon sugar called glucose, which is always present in the blood-stream and is picked up by every cell in the quantities needed. A little energy can be obtained by breaking glucose molecules into two-carbon fragments. Fat and protein molecules can also be broken down to two-carbon fragments.

The bulk of the energy available to living tissue then arises from the combination of the two-carbon fragments with oxygen to form carbon dioxide and water. Naturally, biochemists were very interested in discovering the details of this process.

One way of doing this was to chop up muscle tissue and measure the rate at which it would take up oxygen. When the rate of oxygen uptake falls off one or another organic molecule is added. Those which cause the oxygen uptake to jump again must play a part in the oxidation chain. In 1935 the Hungarian biochemist Albert Szent-Györgyi showed that four closely related four-carbon molecules were involved. The German biochemist Hans Krebs added additional compounds to the list, and by 1940 was able to complete the chain.

The two-carbon fragment was added to a four-carbon compound, with the formation of a six-carbon compound as a result. This six-carbon compound (citric acid, because it occurred in citrus fruits) was gradually broken down to the four-carbon compound by steps that formed carbon dioxide and water and liberated energy. The four-carbon compound was then ready to pick up another two-carbon fragment.

The reactions moved in a cycle from four-carbon to six-carbon and back to four-carbon, grinding up a two-carbon fragment at each turn. This is sometimes called the citric acid cycle because it begins with the formation of citric acid. It is more often called the Krebs cycle, after the man who worked it out.

Laser

IN 1953 the American physicist Charles H. Townes designed a device that would absorb a weak beam of microwaves and emit a strong beam of exactly the same sort (provided it had the proper energy supply). This process was microwave amplification by stimulated emission of radiation or, by taking the initials of the important words, a maser (see MASER).

Why should only microwaves be amplified in this way? The same system could be used for other wave-lengths, for those of visible light, for instance. In 1960 the American physicist Theodore H. Maiman used a bar of synthetic ruby for the purpose. Its molecules were energized to a high-energy level and then, when a feeble beam of red light of the proper wave-length was allowed to fall upon it, an intense beam of the same wave-length of red light emerged.

This was light amplification by stimulated emission of radiation, or a laser. (It might also be called an optical maser, but the one-word version has been universally adopted.)

The light emerging from a laser consists of waves that are all travelling in exactly the same direction and are, so to speak, all moving up and down in unison. In ordinary light of all other kinds waves are moving in all sorts of ways, possess all sorts of energies, and move up and down without relationship to each other. The laser light does not therefore spread out as ordinary light does, but remains a tight beam which spreads out only slightly even if it travels all the way to the moon (as has been tried). This light beam, sticking together as it seems to do, is called coherent light.

Because the light beam sticks so tightly together, a great deal of energy is concentrated in a small area, and at the point of impact a laser beam can produce temperatures far higher than that of the surface of the sun.

Lawrencium

BEGINNING IN 1940, physicists learned to manufacture elements with atoms more complicated than those of uranium, with its atomic number of 92. By 1960 ten of these elements, from 93 to 102, had been formed. One way of forming them was to bombard the atoms of elements already produced with small atomic nuclei. There might be coalescence.

In 1961, for instance, an American team under Albert Ghiorso bombarded californium (98) with nuclei of boron (4). When the two nuclei coalesced they occasionally underwent radioactive changes, and these produced a few atoms of element 103. These atoms were unstable, and half of them had already broken down within eight seconds, but while they existed they were a new element.

Previous elements that had been synthesized had been named after deceased scientists whose work had significance in nuclear science. Element 99 is einsteinium, after Albert Einstein, who first showed that mass could be converted to energy; 100 is fermium, after Enrico Fermi, who first bombarded uranium with neutrons; 101 is mendelevium, after Dmitri Mendeleev, who first developed the periodic table of elements. Following this precedent, 103 was named lawrencium, after Ernest O. Lawrence, who invented the cyclotron.

In 1965 a Soviet team under Georgy Flerov bombarded plutonium (94) with nuclei of neon (10) and obtained element 104, which they named kurchatovium after Igor Kurchatov, who led the group that developed the Soviet nuclear bomb. The Soviet method of forming 104 has not been confirmed elsewhere but the element has been formed in other ways. In the West it is called rutherfordium after Ernest Rutherford, who carried through the first man-made nuclear reactions. In 1970 Ghiorso and his team produced 105, which they named hahnium after Otto Hahn, who first discovered uranium fission. (See also TRANSURANIUM ELEMENTS.)

Lemur

PRIMITIVE PEOPLE, to whom many aspects of the physical world are strange and terrifying (because not understood) fill their environment with imaginary gods, demons, ghosts and monsters. The Greeks and Romans were no exception, and some of their words for ghosts are hidden in scientific terminology (see INSECT).

The ancient Romans, for instance, talked of horrible night-prowling spirits called *lemures*. The word may come from older words meaning an open, gaping mouth, so you can imagine the kind of spirit it was.

Early explorers of Madagascar came across small animals that prowled the night. What's more, they were so timid and elusive one caught only glimpses of them, as though they were ghosts. So, despite the fact there was nothing of the devouring horror about them, they were called lemurs.

The lemurs, which are primitive monkeys, exist mainly in Madagascar, but also in a few places in south-east Asia. Some romantic people imagine that there was once a continent in the

Indian Ocean, connecting Madagascar and Malaya, and that most of it sank, marooning its lemur inhabitants at the two ends which remained above water. This imaginary continent, which was named Lemuria, has entered fiction as a kind of second Atlantis.

One type of lemur has the bone (the tarsus) of its foot very elongated, so that it seems to sit on stilts. It is called the tarsier for that reason. It is a little night-creature with enormous eyes (for its size) in the front of its head. It is completely harmless, but seeing those large eyes suddenly in the dark, staring solemnly, must give one quite a shock. So it is called the spectral tarsier, or even the spectral lemur. Since a spectre (from the Latin *spectrum,* meaning vision) is again some kind of ghost, a spectral lemur is obviously a ghosty ghost indeed.

Lepidoptera

THE INSECTS are the one form of invertebrate life that have developed wings. Insect wings are membranous films that differ from the wings developed by vertebrate creatures in not being converted fore-limbs. It was the wings that apparently impressed those who classified the insects into separate groups, for almost all the main divisions have names that are derived in part from the Greek word *pteron* (feather or wing). Some primitive insects, of course, and some parasites, do not have wings, but they do remain a typical adult insect structure.

For instance, the most familiar insect, the housefly, differs from most insects in having two wings rather than four. It belongs to the order Diptera. The Greek prefix *di-* means two, so you see the order includes the two-winged insects.

The beetles don't look as though they have wings at first glance, but many do. The hind-wings are ordinary wings but are folded beneath the fore-wings which have evolved into horny opaque sheaths that fit closely over the body and hide and protect the gauzy hind-wings. Beetles belong to the order Coleoptera, from the Greek *koleon* (sheath), hence the sheath-winged insects.

The most spectacular wings of the insect world belong, of course, to the butterflies and moths. These are large wings covered with minute scales that come off as a dust when the wing is handled. The only possible name would seem to be Lepidoptera, from the Greek *lepis* (scale), or the scale-winged insects. This name was given them by Carl Linnaeus in 1735. He first classified the insects, and this was one of the names that was never changed.

The colours of the Lepidoptera wings are often startlingly beautiful, and in the early 1940's the chemical constitution of the compounds giving rise to these colours was worked out. The compounds were found to contain a double ring of atoms made up of six carbons and four nitrogens. Compounds containing this particular ring system are called the pteridines after their source. The most important pteridine is a rather complicated one called pteroylglutamic acid. This is one of the B vitamins essential to all life, and a Greek scholar might well wonder what wing was doing in the name.

Lepton

IN THE 1860s it was discovered that electricity could be forced through a vacuum, producing a new type of radiation. This was called cathode rays, because the radiation seemed to emerge from that part of the electric circuit called the cathode. For a generation these cathode rays were studied, and were eventually found to be made up of particles carrying a negative electric charge.

It turned out that atoms could be made to carry a negative electric charge, and that atoms of different mass could carry exactly the same charge. Indeed, the electric charge seemed to come in packets of fixed size, all of which were exact multiples of a certain minimum. If atoms of different size carried this minimum electric charge a given mass of small atoms would carry more than the same mass of large atoms.

In 1897 the English physicist Joseph Thomson found that a given mass of cathode rays carried many times as much electric charge as the same mass of even the smallest atoms (those of hydrogen). The conclusion was that the individual cathode ray particles were much less massive than even the smallest atom. These particles were named electrons and turned out to be 1/1836 as massive as a hydrogen atom.

The electrons are far smaller than the other particles making up atoms of matter. No one knows why they are so much smaller than the other two, the proton and the neutron. There are other particles, not commonly found in ordinary matter, that are as small or smaller. In fact, the neutrino seems to be a particle with no mass at all.

The electron and certain other particles are subject to weaker forces than those involved in the considerably more massive particles, and react more slowly. These particles, subject to weak forces, have in recent years come to be grouped together as leptons, from the Greek *leptos,* meaning weak.

218

Lever

THE LEVER is really a lopsided seesaw (see MOMENT). Imagine a seesaw four metres long with a fulcrum placed half a metre from one end. A force of one kilogramme at the long end has as much of a turning effect on the seesaw as a force of seven kilogrammes at the short end. The moments are equal, since 1 (kilogramme $\times$ 3·5 (metres from fulcrum) = 7 (kilos) $\times$ ½ (metre from fulcrum).

A 7-kilo weight at the short end of this seesaw can therefore be lifted by pushing downward at the long end with a force of one kilo. If the lever were long enough and lopsided enough, you could, with a one-kilo force, lift 100 kilos or 10000 kilos. In fact the Greek mathematician Archimedes, who discovered this principle of moments, said, 'Give me a place to stand on and I can move the earth'.

Where does the extra force come from? At the expense of distance. The long end of the lever in the case described here moves downward seven times the distance that the short end moves upward. Force times distance is the same at both ends.

When you're prying up a huge boulder, after all, it may be that you need only pry it up a few centimetres to send it hurtling down a hill at your enemy. So you sacrifice distance by forcing a long lever under one end of the boulder, propping the lever on a stone near the boulder as a fulcrum, then putting your weight on the long end, not minding if you have to push down a metre if only you can get that tonne of rock to move up the crucial few centimetres.

You might also sacrifice force for distance by using a lever wrong way round, as a catapult. Then, by dropping a heavy weight at the short end of a lever and moving it down a short distance, a lighter stone at the long end will sweep up through a long arc and go sailing off against a fortress. It is because levers are so often used to lift something against gravity that they are called by that name, since the Latin word for to lift is *levare*.

Liberal Studies

THE SUBJECTS taught in colleges are extremely varied. They include a large number of sciences, foreign languages, history, literature, economics and so on. All such subjects are spoken of as liberal studies, as opposed to the technical or professional subjects studied in trade schools, law schools, medical schools and others.

In ancient times the liberal studies were exactly seven in number, and were sometimes called for that reason the seven liberal arts. They were: arithmetic, geometry, astronomy, music, grammar, logic and rhetoric. These were the higher arts, and not everyone might have access to them. Slaves, for instance, might need a rudimentary education in order to be properly useful but the higher arts were for free men only. In Latin 'free men' is *liberi,* so these were the liberal arts.

The seven liberal arts were divided into two groups. The first four—arithmetic, geometry, astronomy and music—made up the quadrivium, which in Latin means a place where four roads meet, from *quattuor* (four) and *via* (road). In the mind of the student studying the quadrivium four roads to learning met, so to speak.

The remaining arts—grammar, logic and rhetoric—were, naturally, the trivium (a place where three roads meet, from *tres,* meaning three, and *via,* road). This idea of a common place (hence commonplace) is at the root of the word trivial. The trivium was considered less important, and the student in a hurry could omit it and study only the quadrivium. But it would be a mistake to consider the trivium trivial!

A student who mastered the seven liberal arts was, obviously, a Master of Arts, a degree still awarded by universities. Master comes from the Latin *magister,* which in turn comes from *magnus* (great). A less thorough mastery entitled one to be a Bachelor of Arts, bachelor being a word which originally referred to a knight's squire, a man still young enough to have to follow another, and who had not yet attained to being his own master.

Libration

THE MOON travels once about the earth in 27·321 66 days. It also rotates about its axis once every 27·321 66 days. (This is not a coincidence but is a situation that is forced upon the moon by the earth's tidal drag.)

Since the periods of rotation and of revolution are equal, the motion of the moon around earth and around its own axis is such that it always presents the same face to the earth. Or at least it would do so if both motions were exactly even.

The moon's rotation is quite steady and even to be sure, but its velocity of revolution about the earth varies with the distance between them. At one point in its orbit it is as close as 355 810 kilometres to the earth, and at another point as far as 405 720. Between those points it is at intermediate distances.

When the moon is relatively close to the earth it moves more quickly than usual in its orbit, and revolves a little too quickly for its period of rotation. The result is that the moon seems to turn a little from east to west so that we can see to some degree about its eastern edge. This is made up for in the other half of the moon's orbit, where it is relatively distant from earth and moves more slowly than usual, so that the moon turns a little from west to east (apparently) and we can see a small way about the western edge.

To a person watching the moon night after night, then, it seems to oscillate slowly about its axis, first to the east for two weeks, then to the west, repeating over and over again. It behaves something like the scales of a balance, slowly oscillating as they come to rest (except, of course, that the moon is not coming to rest but continues to oscillate indefinitely). The Latin word for a balance is *libra,* and this apparent oscillation of the moon about its axis is called libration (east-west).

The moon's axis is slightly tilted with respect to the earth, so that sometimes we can see over the northern edge and sometimes over the southern. This is libration, too, of the north-south variety. If we count the maximum librations of both types we can at one time or another see 59 per cent of the moon's surface altogether.

Lichen

LICHENS ARE simple plants, but in one respect are very unusual. They offer the best example in existence of two life-forms in equal partnership. A lichen consists of two parts: an alga and a fungus (see PLANKTON). The alga, possessing chlorophyll, is capable of manufacturing food and tissue components by use of the energy of sunlight. The fungus for its part provides water, salts and anchorage. Each is better off for the association, and such an association is called symbiosis, from the Greek *syn-* (together) and *bios* (life); a life together, in other words. It was the German botanist Heinrich Anton de Bary who in 1873 first suggested the word.

Lichens grow everywhere, spreading over soil, over the bark of trees, over leaves, even over naked rock. In fact, the observation of life in the form of lichen working its stubborn way up a rock, like a fire licking upward along a tree-trunk, may be what gave lichen its name, since the Greek *leichein* means to lick, and the word for moss is *leichen.*

Lichens may be divided into two types, depending on the nature of the algal component. The alga is sometimes of the blue-green type, and then the lichen is a phycolichen, from the Greek *phykos,* which means seaweed and is therefore equivalent to alga, which after all is just Latin for seaweed. More often the alga is of the bright green type, and then the lichen is an archilichen, from the

Greek *archaios,* meaning from the beginning, since these lichens are the more primitive.

There is one highly dramatic possibility in connection with lichens. There are on the surface of the planet Mars green patches which advance and retreat with seasons. In the northern summer, when the northern icecap melts, green patches advance in the northern hemisphere and retreat in the southern. Half a Martian year later the situation is reversed. The green colour resembles that of some lichens, and lichens, moreover, are the one form of life that may be sturdy enough to survive the severe Martian conditions. Perhaps when we land on Mars we'll find that lichens (or something like them) are licking along the Martian surface too.

Linac

FROM 1930 ONWARD physicists have constructed devices to accelerate sub-atomic particles, making them gain more and more energy. The more energetic they are the more effectively they will smash into atomic nuclei, and the more they will tell us about nuclear structure.

Some devices use a strong magnetic field to whirl the particles in a circle. Another method is to use an alternating electric field to push a particle in a straight line through a series of tubes in each of which the electric field changes its direction as the particle enters. As the particle speeds up, with each fresh push in each fresh section of tube, the section has to be made longer than the one before. The great length of such a device is a disadvantage when compared with the much more compact method of whirling the particles in a circle. Then, too, it proved difficult to keep the electric field alternating in exact rhythm with the particles passing from one tube to the next.

As a result the linear accelerator (or linac for short), first tried in 1931, passed into disuse. And yet it had its advantages, too. When particles move in a circle the necessity of constantly changing the direction in which they travel uses up energy which might otherwise go into speeding them up. In the case of massive particles like protons, this can be borne; but for light particles like electrons, this energy loss sets a sharp limit to the total energy they can receive.

In a linear accelerator, electrons are not forced to lose energy through direction change. Once techniques were worked out to keep the electric field alternating with greater precision, it became desirable to build a large linac for producing electrons with record energies. Stanford University has now constructed a linear accelerator for the purpose which is three kilometres long.

Liver

ALL GLANDS produce and liberate some chemical or fluid. Although the word gland (from the Latin *glans,* meaning acorn) was originally applied to small organs, some glands are quite large.

The largest is the liver, which weighs nearly two kilogrammes in man. It produces a juice that is delivered through a duct into the intestines. The word liver comes from the Anglo-Saxon, and may just possibly have some connection with the word life. (In German the word for liver is *Leber,* and for life is *Leben.*) Certainly the ancients, probably impressed by the sheer size of the organ, thought it the very centre of man's life and health.

We ourselves, for instance, usually talk as though we think the heart is the centre of life and the seat of emotions. We are broken-hearted or faint-hearted; we take heart or lose heart. Well, the ancients thought the liver was the seat of emotions, and used similar expressions with liver in them. Few of those old expressions survive now, but we still sometimes speak of cowards as being lily-livered. Their liver, in other words, is pale with fear, and so they themselves are afraid.

The second largest gland in the body is the pancreas, which is less than a tenth the size of the liver. Its name comes from the Greek *pan* (all) and *kreas* (flesh), because it is all flesh (that is, meat) without bone or fat. It is located just under the stomach, and when the pancreas of a calf or other animal is bought as an item of food at the butcher's it is called sweetbread. The word sweet is a corruption of an Anglo-Saxon word for glandular food in general, while bread is used here in its old general meaning of food as in 'Give us this day our daily bread'.

Another gland, the thymus, which occurs in the neck region of young animals (and which supposedly gets its name from a fancied resemblance to a bunch of thyme—a plant that the Greeks called *thymon*) has been called neck sweetbread by the butcher.

Lymph

THE VERY smallest blood-vessels have walls so thin that the watery portion of the blood can easily leak out. It does exactly this, bathing the cells and forming the interstitial fluid of the body. (It is called this because it fills the interstices between the cells. The word interstice comes from the Latin *inter* (between) and *sistere* (to stand); an interstice is something that stands between.)

The interstitial fluid has not left the blood-vessel permanently. It drains off into tiny vessels that join into larger ones and finally into two main ones that travel up the chest to the neck where they join veins, and it is there the interstitial fluid rejoins the bloodstream. The larger of the two main vessels is on the left, and is called the thoracic duct, from the Latin *thorax* (chest), which in turn comes from the Greek *thorax* (a corselet), and *ducere* (to lead).

The interstitial fluid is like the blood in most ways. The most conspicuous difference arises from the fact that the cells in blood which give it its colour cannot pass through the small vessels. The interstitial fluid is therefore colourless, and this is reflected in its name, for it is called lymph, from the Latin *lympha* (clear water)—whence we get the word limpid. The vessels which contain interstitial fluid are the lymph vessels.

Here and there along the course of the lymph vessels (particularly on the side of the neck, under the jaw, in the armpits, elbows and groin) are small swellings. The early anatomists thought the swellings resembled acorns in shape, and since the Latin word for acorn is *glans,* the swellings were called lymph glands.

The lymph glands form cells that are called lymphocytes. (The suffix -cyte or the prefix cyto-, when used in biological terms, mean cell. They come from the Greek *kytos,* meaning a hollow place, so we have here the same sort of poor derivation from which the word cell itself suffers. See PROTOPLASM.) Lymphocytes are primarily bacteria-fighters, and during an infection lymph glands become more active. They swell and become painful, and mothers these days keep a sharp eye out for swollen glands whenever children grow feverish.

Mach Number

ERNST MACH was an Austrian physicist whose most important work was in the philosophy of science. In the nineteenth century scientists were self-confident indeed, and charged forward in their search for greater knowledge without worrying too much over the fine points of the methods of reasoning which they were using. By and large, they blundered on correctly, but Mach began to insist they stop to analyse what they were doing. His views were unsettling and unpopular, and he did not have much influence in his own time. He did, however, impress Albert Einstein, who went on to revolutionize science with his theory of relativity (which Mach in his last years, ironically enough, refused to accept).

One line of practical investigation initiated by Mach consisted of experiments in airflow, which he published in 1887. He was the first to take note of the sudden change in the nature of the airflow over a moving object as it reaches the speed of sound.

Sound travels through any medium by means of the natural motion, back and forth, of the atoms or molecules in that medium. When a material object forces its way through the medium the molecules get out of the way by means of their natural swing. As the material object approaches the speed of sound, it is approaching the speed of the natural motion of the molecules of the medium. Those molecules can no longer get out of the way, and their behaviour changes.

After the Second World War aeroplanes began to move at speeds approaching that of sound, and the behaviour of the airflow became of great interest. The ratio of the velocity of the planet to the velocity of sound in air, at a given temperature and density, was called the Mach number. If the planet moved at exactly the speed of sound it was moving at Mach 1; if it was moving one and a half times the speed of sound, that was Mach 1·5, and so on. (The speed of light, the fastest speed possible, is about Mach 900000.)

Macromolecule

AFTER THE ENGLISH CHEMIST John Dalton had advanced the atomic theory in 1803 it was soon found that each substance was made up of a combination of atoms which was held together firmly to make up a molecule (from a Latin word meaning a small mass).

The first molecules studied consisted of only a few atoms each. Water molecules, for instance, consisted of three atoms each, two hydrogens and one oxygen (H_2O). Common salt could be considered as made up of combinations of one atom of sodium and one of chlorine ($NaCl$). Sand molecules are made up of three atoms (SiO_2), while other common substances of the earth's rocks, limestone ($CaCO_3$), alumina (Al_2O_3) and haematite (Fe_2O_3), are made up of five each.

These are all molecules characteristic of inanimate nature and are therefore inorganic (from Latin words meaning not life). The situation is different in the case of organic substances, however, those that are characteristic of matter that is living, or that was once living.

Organic molecules are combinations of greater numbers of atoms, but of a smaller variety of different atoms. Most organic molecules consist of carbon, hydrogen and oxygen (plus nitrogen or a scattering of others sometimes). Acetic acid molecules are made up of eight atoms ($C_2H_4O_2$); citric acid molecules of twenty-one atoms ($C_6H_8O_7$); table sugar molecules of forty-five atoms ($C_{12}H_{22}O_{11}$); and so on.

225

That was only the beginning. The substances most characteristic of living tissue, and most important, are giant molecules made up not merely of dozens, but of hundreds or thousands or even millions of atoms. These macromolecules (macro- is from a Greek word meaning large) include such substances as the proteins and nucleic acids which are the keys to life, so that the study of macromolecules is really the study of life.

Maffei 1

IN THE DISTANT REACHES of the universe there are galaxies by the millions. The individual galaxies do not usually exist singly, but tend to group into galactic clusters. As many as ten thousand galaxies may cling together under mutual gravitational influence.

Is our own Milky Way Galaxy also part of a cluster? Two cloudy patches visible in the sky of the Southern Hemisphere are called the Magellanic Clouds (because they were first described in 1521 by the chronicler accompanying Ferdinand Magellan on the expedition that first circumnavigated the globe). These are two galaxies, much smaller than our own and quite near by. They might almost be considered satellite galaxies of ours.

In addition, there is the Andromeda Galaxy (so named from the constellation in which it is to be found) which is probably even larger than our own. It is the closest of all the large galaxies, not much more than two million light-years away. At distances between those of Magellanic Clouds and the Andromeda there are about twenty other galaxies, all small ones. All these form the Local Group, of which we and Andromeda seemed to be the only two giant members.

But not all parts of the sky can be clearly seen. In the zone of the Milky Way itself, there is so much dust that the sky is obscured. In 1968 an Italian astronomer, Paolo Maffei, was searching the sky and found patches of infra-red in one of the dustiest regions of the Milky Way. (Infra-red will penetrate dust somewhat more efficiently than visible light will.) Studying the regions closely, he discovered two galaxies with only 1 per cent of their light filtering through the dust. They were close enough to be part of the Local Group and large enough to be giants. They have been named Maffei 1 and Maffei 2 in honour of the discoverer, and are our new (and unsuspected) neighbours.

Magic Number

IN 1916 it was found that the electrons in an atom were distributed in groups that enclosed the nucleus like spherical shells. As one progressed outward from the nucleus the shells were larger, and could hold more electrons. The innermost shell could hold only two electrons, the next eight, the next eighteen, and so on. Each shell was divided into sub-shells and it was found that the chemical properties of elements depended on the distribution of the electrons in their atoms among the shells and sub-shells. When certain sub-shells were filled, for instance, an atom was particularly inert, and did not engage in chemical reactions.

In the early 1930s, when it was found that the atomic nucleus was made up of protons and neutrons, physicists began to wonder at once if those particles were arranged in shells as electrons were. It took a great deal of energy to probe the nucleus, however, so the matter could not be worked out easily.

One way of considering the question was to take note whether any atomic nuclei were particularly stable. That might indicate their protons or neutrons had filled certain shells or sub-shells. It turned out that when a nucleus contained 2, 8, 20, 50, 82 or 126 of either protons or neutrons, it was particularly stable. In 1949 the German physicist Hans Jensen called these magic numbers. Later, feeling perhaps that this was an over-dramatic name ill-suited to science, he used the term shell numbers instead, but the earlier term remains more popular.

In 1949 Jensen and (independently) the German-American physicist Marie Goeppert-Mayer worked out a system for the arrangement of shells within the nucleus, based on the magic numbers. It is more complicated than the electron shells and not as well established, but it is a beginning.

Magnet

THOUSANDS OF years ago, men were amazed at finding pieces of black mineral that attracted iron. It is said that the early Greek thinker Thales of Miletus (a city in Asia Minor) first studied the phenomenon. He got samples of the material from Magnesia (another city in Asia Minor), so he called the mineral *magnes*. This has come down to us in the word magnet, while the mineral is today called magnetite.

The Roman naturalist Gaius Plinius Secundus (usually called Pliny the Elder) confused the *magnes* of Thales with another black mineral, which he also called *magnes*. In medieval times Pliny's books were copied by hand, sometimes by people who were not very careful, or perhaps not very learned. Distortions crept in. Pliny's mistaken *magnes* was mangled further, and misspelled as manganese.

In later times Pliny's *magnes* was used in glass-making to wash out the greenish colour that resulted from iron impurities in the raw materials. The mineral therefore was named pyrolusite, from the Greek words *pyr* (fire) and *louein* (to wash). (The fire referred to the strong heating required to make glass.)

In 1774 the Swedish mineralogist Johan Gottlieb Gahn isolated a new metal from pyrolusite and in naming it went back to the medieval, mistaken and misspelled name of manganese. That name has been accepted ever since.

In ancient times another mineral (a white one) had been discovered in the region of Magnesia. (This may not necessarily have been the same city as the first; there were three cities so named in ancient Greek territories.) The Romans called this mineral *magnesia alba* (*albus* means white) to distinguish it from *magnes*, which is black. In 1831 the French chemist Antoine Bussy isolated a metal from a chemical related to *magnesia alba* and named it magnesium. And so ancient Magnesia ended by having two metals and an important natural force named after it.

Magnetohydrodynamics

THE STUDY of the flow of liquids such as water under the impulse of various kinds of pressure is called hydrodynamics, from Greek words meaning motion of water. Gases can flow, too, but do so in a more complicated fashion because gases can easily compress and expand, whereas liquids cannot. Since both liquids and gases flow, they are grouped together as fluids, and the study of their flow is fluid dynamics. However, it has become customary to use the term hydrodynamics as a synonym for fluid dynamics, and to treat gas flow as well as liquid flow under that title.

If gases are heated to thousands of degrees the atoms that make them up are broken down into electrically charged particles, and these will interact with a magnetic field. Under such circumstances the magnetic field may exert pressure on them and cause them to flow. The study of such flow is magnetohydrodynamics.

The control of hot gas by a magnetic field is an essential part of attempts to achieve controlled fusions, so magnetohydrodynamics may be a key to the energy sources of the future. It may also be a key to the more immediate energy sources of today.

At present fuel must be burned; the heat must boil water to

steam; the steam must turn a wheel; the turning wheel must generate electricity. At each stage there is loss of energy, and in the end the electricity formed represents only 30 to 40 per cent of the energy present in the fuel to begin with.

It is possible to use the fuel to heat gases to high temperatures of 5000 °C and send them through a magnetic field to generate a current directly. This magnetohydrodynamic method could conceivably raise efficiency to some 60 per cent and effectively double the energy we can obtain from our ordinary fuels of today.

Magnetosphere

IN 1600 the English physician William Gilbert showed, from the behaviour of the magnetic compass, that the planet earth must be a huge spherical magnet. Like any other magnet, earth has magnetic poles, a North Magnetic Pole in northern Canada and a South Magnetic Pole on the rim of Antarctica. Between the two lie imaginary magnetic lines of force, curving outward from one pole to another and being at their highest point above earth's surface halfway between the poles. These lines represent the direction along which magnetic attractions and repulsions are directed.

In 1957 a Greek amateur scientist, Nicholas Christofilos, advanced a theory according to which charged particles in the neighbourhood of earth would be trapped by the planet's magnetic field and would spiral back and forth from one magnetic pole to the other along the lines of force (the Christofilos effect, as it is now known). Near the poles the magnetic lines of force approach earth's surface and the charged particles there interact with the molecules in the upper atmosphere, producing the aurora.

Christofilos' work was ignored at first, but in 1958 a region of charged particles enveloping earth were discovered by rocket investigations and named the Van Allen belts, after James Van Allen, the scientist in charge. The particles in the Van Allen belts turned out to behave as Christofilos had predicted.

The Van Allen belts were soon renamed. The various layers about earth have names ending in -sphere because of their spherical shape. Atmosphere (sphere of vapour) is itself an example. Since the belts are involved with earth's magnetic field, the Van Allen belts are now called the magnetosphere.

Mammal

ALL MULTICELLULAR organisms are divided into two 'kingdoms', the plant and animal. It was only in recent times that plants were recognized as being as fully alive as animals. Animals could breathe and move, which plants could not. In fact, the word animal comes from the Latin *anima* (breath).

In common speech the word animal is often restricted to those with four legs and hair, such as dogs, cats and cows, so that people will speak of animals and birds. Actually, of course, a bird is an animal, and so is an oyster, a butterfly or an earthworm. The four-legged animals with hair (unlike those without hair, such as birds, reptiles and insects) bear their young alive and feed them on milk formed in the mother's body. The milk-forming glands are called in Latin *mammae,* and hence the hairy animals are called mammals.

There is one small group of primitive mammals, restricted to Australia and New Guinea, that, unlike other mammals, actually lay eggs. The best known of these has broad webbed feet like those of a duck and a snout that looks uncommonly like a duck's bill. It is therefore called the platypus (from the Greek *platys,* meaning broad, and *pous,* meaning foot) or the duckbill (derivation obvious) or the duckbilled platypus. It is also called the Ornithorhynchus, from the Greek *ornis* (genitive, *ornithos*) (bird) and *rhynchos* (beak), or bird-beak, which is just another way of saying duckbill.

The other egg-laying mammal (which occurs in two different genera) is the spiny anteater. (It has spines and eats ants, so that is a good name.) It is also called the echidna, which makes less sense. In Greek mythology Echidna was a monster that was half woman, half serpent. Perhaps the spiny anteater struck its discoverers as another kind of monster, half mammal (hair) and half reptile (eggs).

Because these creatures have one common exit for wastes (and eggs) they are called monotremes, from the Greek *monos* (solitary) and *trema* (opening).

Mare Cognitum

IN 1609 the Italian scientist Galileo Galilei first looked at the moon with a telescope and saw mountains and craters. There were also extensive regions which seemed to lack mountains and craters.

As the seventeenth century progressed other astronomers began to draw maps of the moon and to give names to its chief features. The flat regions without mountains and craters they called maria (a Latin word for seas). Some may have thought they were really regions of water. Actually, it was soon recognized that the moon lacked an atmosphere, and on an airless world open stretches of water are not at all likely. The flat regions are seas only in the sense that they are fairly flat seas of compacted dust.

Nevertheless, they not only retained the name of maria, but individual stretches kept the romantic names originally given them, such as Mare Humorum (Sea of Fluid), Mare Imbrium (Sea of Rain), Mare Undarum (Sea of Waves), Mare Spumans (Sea of Foam) and so on. Some names, to be sure, are more apt. Mare Serenitatis and Mare Tranquillitatis are Sea of Serenity and Sea of Tranquillity, and the surfaces of those dead seas are indeed serene and tranquil.

In the 1960s men began to get a closer look at the moon's surface than even the best telescope could afford. In 1964 the United States sent lunar probes directly towards the moon. These headed for a crash landing, taking photographs and sending them back to earth as they approached. On July 28, 1964, *Ranger 7* approached the surface of the moon, sending back 4316 photographs, the last from a distance only metres away. It struck near the north-west edge of Mare Nubium (Sea of Nubia) in the midst of a small flat stretch which was named Mare Cognitum (Sea of the Known) because, thanks to *Ranger 7*, it was better known than any other part of the moon—at that moment. Since then, of course, *Ranger 8* and *9,* and still more the American Apollo and Soviet Luna programmes, have accumulated knowledge.

Mariner Programme

THE FIRST OBJECTS sent into space in 1957 and 1958 were artificial satellites that circled earth in as little a time as ninety minutes. Even satellites sent into orbits around earth that stretched hundreds of kilometres away completed those orbits in a matter of hours.

It is different for objects sent into space at speeds greater than earth's escape velocity. They do not circle earth, but recede indefinitely and go into an orbit about the sun. Such orbits are not completed for many months.

The first space probes (objects sent away from earth towards other worlds), launched in 1959 and 1960, were aimed at the moon. Some passed around the moon and back towards earth; some passed on and disappeared into the emptiness of space. In either case scientists were interested only in the reports of conditions in the neighbourhood of earth and the moon.

It takes only three days for a probe to travel from earth to the

231

moon, and this is not a long journey. Men have made the round trip safely a number of times since then.

In 1962, however, space probes were being sent towards the planet Venus. Venus is, at its closest, forty million kilometres from earth, a hundred times as far away as the moon is. A Venus probe takes about four months to reach Venus. The thought of that long, lonely voyage through space was very impressive.

The planetary probes were therefore part of the Mariner programme. *Mariner 2* was launched in August 1962 and successfully passed Venus in December. In November 1964 *Mariner 4* made the even longer voyage to Mars, which it successfully passed in July 1965. *Mariner 5* in June 1967 passed some ten times nearer to Venus than had *Mariner 2*, and then in February and March of 1969 Mariners 6 and 7 sent back a marvellous series of photographs of Mars.

Marsupial

AUSTRALIA WAS cut off from the remaining land areas of the world at a time when the only living mammals were extremely primitive and were either egg-laying (see MAMMAL) or if bringing forth young alive were doing so while these were at a very early stage of development. What's more, Australia remained cut off, so that these primitive mammals could develop and multiply, while in the rest of the world more efficient mammals evolved and took over.

The best-known native mammal of Australia is the kangaroo. The story is that when the ship of the English explorer James Cook first landed in Australia in 1770 and the men saw a strange leaping animal they asked the inhabitants, "What's that?" The latter answered, 'Kangaroo', meaning 'What are you saying?'—and that was it.

In the kangaroos and related animals the young are born after only a very short development. They emerge into the world minute, with the bare ability to crawl through the mother's fur and into a special pouch where they feed on milk and remain until grown large enough to be able to assume independent life.

This pouch within which the tiny newborn animals develop is the most typical feature of the kangaroo and its relatives. The Latin word for pouch is *marsupium,* so these animals are called marsupials.

The higher mammals allow their young to develop a longer time within the body before birth. In the case of man the period of gestation (that is, the time during which the young are carried within the body before birth, from the Latin *gestare* meaning to carry) is nine months; in elephants and whales, two years.

This is made possible by the fact that the mother bearing the child develops a special organ called the placenta, which is penetrated by blood-vessels of both the mother and the developing baby, so that food and air can cross in one direction and wastes in another. (The blood-vessels approach one another but do not actually join.) The word placenta is a Latin one, meaning a kind of flat cake, which is what the organ looks like. The higher mammals are called placental mammals for this reason.

Mascon

THE INTENSITY of the gravitational field increases with mass, so that gravitation shapes the form of large bodies. These become spheres under the pull of their own gravity. They can have unevennesses on the surface, so that mountains and valleys are present, but these are very small compared to the whole body. A quickly rotating body will bulge at the equator, but a slowly rotating one like the moon is an almost perfect sphere. Gravity also makes the density of a body vary from centre to surface in the same way in all directions. This means the gravitational field should be the same at a given distance from the body's centre in all directions. It was assumed this was true of the moon.

Between 1966 and 1968 a number of satellites were placed into orbit about the moon (and hence were called lunar orbiters). Assuming an even distribution of gravity about the moon, those vessels were expected to follow certain paths. They deviated slightly from those paths, however, and it appeared that over certain sections of the moon's surface the gravitational field was a trifle stronger than over others. The field was stronger over the flat 'seas' than over the mountains and craters, and it had to be assumed then that the density of the moon's crust was higher in the sea regions than in the mountainous areas.

There had to be concentrations of mass under those seas (the only reasonable explanation) and the term mass concentrations was quickly abbreviated to mascons. One possible reason for the mascons is that the seas were formed by the collision of large, dense meteors with the moon, and that those meteors (iron, perhaps) still lie buried under the seas. Or else the seas contained water once, and the dry surface is now covered by thick and dense layers of sedimentary rock.

Perhaps we may one day find out which, if either, of these explanations is correct.

Maser

THE NEW KNOWLEDGE concerning atomic structure which came with the opening of the twentieth century made it clear that a particular atom or molecule could absorb a photon containing a fixed quantity of energy and enter a higher-energy state. In 1917 Albert Einstein showed that if a photon of just the right size struck an atom or molecule that was already in the higher-energy state, it would drop to the lower, emitting a second photon just the size of the first and travelling in the same direction.

If all the atoms or molecules in a particular volume of matter were in the higher-energy state a photon of the right size would bring down one molecule. Then there would be two photons, each of which would bring down one molecule, producing four photons, and so on, until in a very short time a gigantic flood of photons would be emerging, all moving in the same direction.

In 1953 the American physicist Charles H. Townes devised a method for pumping a quantity of ammonia molecules into a higher-level state. When he then allowed a very weak beam of microwave photons, representing the proper energy difference between the high-energy and low-energy levels of the molecule, all the molecules were dropped to the low-energy level, and a large flood of microwaves resulted.

By stimulating the emission of radiation through the original weak beam, that beam was amplified into a large one. This was microwave amplification by stimulated emission of radiation. Taking the initial letter of the key words of this phrase, we have maser.

Since Townes's first maser, the Dutch-American physicist Nicolaas Bloembergen has devised a system with three levels so that pumping and emission can go on at different photon sizes. Emission is then steady, and the first continuous maser invented.

Mathematics

MATHEMATICS IS that branch of knowledge that deals with quantities, their measurement and their interrelationship. It is derived from the Greek *mathein* (to learn), which implies that it is the subject which, above all others, must be learned. We admit that partially ourselves, when we speak of the three R's as the absolute fundamentals of learning.

And arithmetic (from the Greek *arithmos,* meaning number) may well have come first. People even in advanced cultures can get along (though rather poorly) if they cannot read or write, but even illiterates find it useful to add. Some primitive tribes are supposed to have no names for numbers over two, but any tribe that gets to the stage of making stone axes must want to count higher than that, if only to make sure no one has been appropriating an axe that wasn't his.

Mathematics is generally not counted among the natural sciences (see SCIENCE). It is admitted to be the most important tool of many of the sciences, but the point is made that a natural science begins with the observation of nature and the collection of facts. These are then brought into order by the use of mathematics. Mathematics, itself, however, is apart from nature and is a creation of pure mentality, being built up step by step by the mathematician with closed eye and thoughtful mind from a minimum number of fundamental and obvious principles called axioms. This comes from the Greek *axios,* meaning worthy, perhaps because axioms are principles which all people will agree are worthy to serve as foundations.

An example of an axiom is: A straight line is the shortest distance between two points.

One can't help but wonder how axioms arose in the first place. Did they spring up, pure and unsullied, in the mind? Or did some early thinker observe two points and draw various connecting routes before deciding by observation and experiment that it looked as though a straight line were probably the shortest connecting route between them?

Melanin

IN MAN the most noticeable natural colours are those of the blood and those of the hair, skin and eyes. The coloured substance of blood, as it happens, is not named after its colour (see PORPHYRIN; HAEMOGLOBIN), but that of the hair, skin and eyes is another story. (Those three are lumped together, because the same substance is involved in each.)

The colour of hair, skin and eyes is due mostly to a blackish pigment, produced in varying amounts by practically all human beings. This pigment is called melanin, from the Greek *melas* (black). The few people who cannot form this substance have very fair skins, white hair and eyes that would be colourless except that the blood vessels show through, giving them a pink appearance. Such people are albinos, from the Latin *albus* (white).

People who can form a little melanin are yellow-haired and blue-eyed. (The blue of blue eyes is not the result of a blue pigment but the effect of tiny particles of melanin which scatter blue light the way dust particles in the air scatter it to form the blue sky.)

Such people are blond, a word which is of ancient German origin meaning light or fair.

People who can form a fair amount of melanin have brown hair (or even black hair) and brown eyes. There is enough melanin in the skin to give them a swarthy complexion. (Swarthy comes from the Anglo-Saxon *sweart,* which is analogous to the German *schwarz,* meaning black.) Such people are brunette, a French word originating from *brun* (brown).

If the amount of melanin formed is high enough the skin itself is distinctly brown, sometimes quite dark brown. The African Negro and his descendants are an example of this. The word Negro is a Spanish word meaning black, derived from the Latin *niger* (black). (The Spaniards and Portuguese were the first of the modern Europeans to make contact with the dark-skinned inhabitants of West Africa, which is why a Spanish word has come into use in this connection.)

Mercury

THE ANCIENTS knew seven metals: gold, silver, copper, iron, tin, lead and mercury, and of these mercury was by all odds the most remarkable. It was the only liquid among them, and it was such a heavy liquid, too; only gold was heavier. The Greeks called the metal *hydrargyros,* from *hydor* (water) and *argyros* (silver); i.e., liquid silver. That name still lingers today in the chemical symbol for mercury, which is Hg.

The notion of mercury's being a special kind of silver lingers also. Unlike ordinary silver, which just lies there in a lump, mercury, being liquid, moves and vibrates and if spilled breaks into thousands of silvery droplets that race about. There is something seemingly alive about it, and one of its names in English is quicksilver, where quick has its older meaning of alive, as in the phrase the quick and the dead.

The medieval alchemists, who were great for masking what knowledge they had in mystical gobbledygook, were struck by the fact that there were seven known metals and seven known planets. They thought there might be some mysterious significance to this, and matched up the two sets.

Gold they referred to as Sol (Latin for sun) and silver as Luna, (Latin for moon) because they were the most precious, and should be named after the most prominent 'planets'. Copper, as next most precious, was named Venus, which was the next most prominent. Iron was a natural for Mars because of the use of iron weapons in war, while the heavy and dull lead seemed appropriate for the slow-moving Saturn. Tin was Jupiter, by elimination, probably, since hydrargyros, the one remaining metal, could only be tied up with one planet, Mercury.

Just as hydrargyros was the most alive of the metals, being liquid and mobile, so Mercury was the most alive of the planets, since its apparent motion across the sky was most rapid. So the liquid metal received the name mercury, for, after all, Mercury was the wing-shoed messenger of the gods, fleet as thought, and of all the planet-named metals, only that one kept its planet name—until the fashion was begun again in modern times (see URANIUM).

Mercury, Project

AFTER THE SOVIET UNION launched its first artificial satellite, *Sputnik 1,* the United States began to make plans at top speed for the exploration of space. While unmanned satellites are important, the American Government was sure that it would be necessary to send a man into space eventually. The National Aeronautics and Space Administration (NASA) was established in 1958 to oversee this.

The first men in space would merely orbit earth. To do this, however, the rocket carrying them would have to travel at a speed of eight kilometres per second, and at that rate earth would be circumnavigated in ninety minutes. (When Magellan's men made the first circumnavigation of the earth in 1522 it took them three years.)

The speed at which men in orbit could go round the planet was reminiscent of the winged messenger of the gods in the Greek myths, who moved so quickly in his journeys that he was pictured with wings on his sandals and cap. He was Hermes, and the Romans called him Mercury. The plan to put men in orbit was therefore called Project Mercury.

In 1961 two Americans were sent up in rockets and returned safely, but they did not go into orbit. On February 20, 1962, however, John H. Glenn was sent into orbit and circled earth three times before landing safely. (Two Soviet cosmonauts had been placed into orbit before that.)

Others followed Glenn, until on May 15-16 L. Gordon Cooper was placed in orbit and circled earth twenty-two times, remaining in space for thirty-four hours before coming back safely. Project Mercury had proved a resounding success, and the United States was then ready to go on to the more ambitious programme of Project Gemini.

Meridian

ANY LINE drawn due north and south on the surface of the earth (or against the vault of the sky), passing from pole to pole, is called a meridian. Wherever you are you are standing on (and under) one such line. About noon (time by clock can vary several minutes from the time marked by the sun) the sun passes across the meridian on which or under which you are standing, and that is the middle of the day for you, exactly halfway between sunrise and sunset. The Latin word *medius* is middle and *dies* is day; consequently *medidies* means midday. *Medidies* was corrupted into *meridies,* and from that comes the word meridian.

(Strangely enough, the word noon comes from the Latin *nonus* (nine). Originally it referred to the ninth hour of the day counting from sunrise, which, on the year-average, is at 6 A.M. This put *nonus* at 3 P.M., or halfway between midday and sunset. Thus what was originally the middle of the afternoon has come to mean the middle of the day itself.)

The position of a particular point on the earth's surface can be given, at least in an east-west sense, by choosing some particular meridian as a zero point, then marking off the rest of the earth in equally spaced meridians.

At first most nations chose their own capital (or some other point particularly convenient to themselves) as starting-point and figured east and west from that. This aroused confusion in sea travel, where a knowledge of accurate position was more important than on land, and where ships of all nations, each with its own system, found things a mess.

In the middle of the nineteenth century Great Britain had more ships and a larger seagoing trade than any other. It was natural, then, for the Washington Meridian Conference in 1884 to adopt Great Britain's standard internationally. The meridian passing through the Greenwich Astronomical Observatory in London is accepted as the zero point the world over now, and that is the Prime Meridian (from the Latin *primus*, meaning first). Greenwich is thus about 51½°N 0°E (or W).

Meson

IN THE 1920s the only sub-atomic particles known were protons and electrons, and it was supposed that the atomic nucleus had some of each. Protons repel each other, but protons attract electrons, so the electrons were thought to serve in the nucleus as a kind of cement, holding it together.

In 1930 the neutron was discovered, and it was soon realized that the nucleus had to be made up of protons and neutrons only. But what kept the protons from repelling each other, then, and breaking up the nucleus?

A Japanese physicist, Hideki Yukawa, suspected a hitherto unknown nuclear force might exist which held the protons and neutrons together. In 1935 he worked out the requirement to make an attractive force strong enough to hold the nucleus together and yet be felt only over the width of the nucleus.

It turned out that certain particles had to be endlessly and rapidly exchanged between the protons and neutrons of the nucleus to make such a force work. Yukawa calculated they would be about 270 times as massive as electrons, and therefore about $1/7$ as massive as a proton or neutron. No such particles, intermediate in size between electrons and protons, were known.

In 1936, however, an American physicist, Carl D. Anderson, who was studying the tracks of cosmic rays in cloud chambers, came across tracks that had to be formed by particles of such intermediate mass. He labelled the particle mesotron (meso- coming from a Greek word meaning intermediate). This was eventually shortened to meson.

A number of different kinds of particles in this mass range were eventually discovered, and the term meson has been used for the entire group of particles that are more massive than electrons but less massive than protons and neutrons.

Messenger-RNA

BEGINNING IN THE 1940s, biochemists could demonstrate that the nucleic acid molecules of the chromosomes guided the synthesis of specific enzyme molecules. Yet the nucleic acids of the chromosomes were confined to the nucleus, deep within the cell, while enzyme molecules were manufactured in the cytoplasm, outside the nucleus. Somehow the information concerning the structure of the nucleic acid molecule had to be carried outward to the cytoplasm. There had to be a messenger.

The nucleic acid of chromosomes is a particular variety called deoxyribonucleic acid, which is frequently abbreviated DNA. This is not, however, the only kind of nucleic acid in the cell. Another kind, just as complicated as DNA and made up of very similar units, is ribonucleic acid, or RNA. Both have as a portion of their structure a sugar called ribose, but in the case of DNA there is an oxygen atom missing in the ribose portion, hence the deoxy- prefix.

RNA differs from DNA in that the former is located both in the nucleus and in the cytoplasm. Was it possible, then, that there might be a particular RNA molecule which could be synthesized in the nucleus, using one of the DNA molecules in the chromosomes as its model? This RNA molecule, carrying a copy of the structure of a particular DNA, could then move out into the cytoplasm where it could serve as a guide for the synthesis of an enzyme molecule. Such a RNA molecule would carry information from nucleus to cytoplasm. It would be the sought-for messenger, and was almost inevitably called Messenger-RNA.

Messenger-RNA was first identified in bacteria, but in 1962 two American biochemists, Alfred E. Mirsky and Vincent G. Allfrey, demonstrated its presence in mammalian cells. It is now recognized as a universal part of the mechanism of the inheritance of physical characteristics from cell to cell and from generation to generation.

Metal

FOR HUNDREDS of thousands of years mankind had been using stone and wood for various tools. Wood was easier to handle but stone could take more punishment. Both were useful.

And then, not more than six thousand years ago, men discovered a new kind of material altogether. Perhaps they first found small nuggets of yellow gold. Or perhaps somebody built a charcoal fire on some green rock in the Sinai peninsula and found ruddy drops of copper under the ashes afterwards. In any case, copper, gold and silver date back to nearly 4000 B.C.

The qualities that struck men first about these substances must have been those in which they differed from stone and wood. Take gold as an example. It reflects light beautifully and shines, instead of having the dull appearance of stone. It shows lustre, from the Latin *lucere* (to shine).

Gold can also be hammered into various shapes or into very thin sheets, whereas stone so treated will simply powder. Gold is therefore malleable, from the Latin *malleus* (hammer).

Gold can be drawn out into a thin wire, whereas again stone so treated would break and powder. Since one end of a heated strip of gold will seem to follow the pincers that are drawing it out and leading it, so to speak, away from the other end, gold is said to be ductile, from the Latin *ducere* (to lead).

Then too gold sheets or gold wire can be bent to any shape without breaking. (Can you imagine stone bending?) Gold is therefore said to be flexible, from the Latin *flectere* (to bend).

For all these reasons, appearance plus the way these substances could be bent and drawn and hammered, they were ideal for the fashioning of jewellery and ornamentation, so they were eagerly desired even before more practical uses turned up. Since they were much rarer than wood or stone, they had to be carefully searched for. The Greek word *metallon* means mine, and also metal. But this probably comes from the Greek word *metallan,* meaning to search for. Anyway, the new class of materials received the name of metal.

Metalliding

THERE ARE DOZENS of different metals that exist, but most of them in pure form are of only limited use. Gold, for instance, would seem perfect for jewellery, but pure gold is so soft it would lose its shape and wear away too easily. Add a little copper, though, and it becomes hard enough. Again, pure copper is too soft to use in tools and weapons or for knives and spears. Add tin and you get bronze, which is hard enough. Iron will rust, but the addition of some other metals will make stainless steel, a non-rusting metal that is mostly iron.

Mixing metals to form alloys will produce a myriad useful substances. The simplest method is to melt the two metals and stir them together. The alloy that results is a mixture all the way through. Another way is to plate a thin layer of one metal on to the surface of a bulky piece of another. This can be done mechanically or by means of an electric current which will layer a metal out of solution (electroplating). In that way a small quantity of gold layered over another, much cheaper metal will make the whole object as beautiful and as resistant to rust as though it were solid gold.

In 1968 the American metallurgist Newell C. Cook tried to plate a silicon layer on a platinum surface, but found that under the conditions he used, silicon atoms worked their way below the surface. Instead of plating one substance on to another, he had prepared a metal with an outer skin that was an alloy, a situation more stable and permanent than plating. Cook called the process metalliding.

This process shows promise of economy. For instance, copper alloyed with 2 to 4 per cent of beryllium becomes extraordinarily tough. This can be achieved if copper is beryllided so that only the outer skin is alloyed. The entire piece of copper is just about as tough as though it were all alloyed, but much less beryllium need be used.

Metre

IN 1791 the French nation, in the midst of a revolution, wished to break with the past, especially with those aspects of the past which they considered illogical and useless. For instance, the system of weights and measures then in traditional use was far too complicated and, moreover, varied from place to place.

The French began by trying to set up a unit of distance equal to one forty-millionth part of the earth's circumference. Unfortunately, later measurements showed that the unit they had chosen was not exactly that fraction, and today the unit is defined simply as the distance between two marks on a platinum-iridium bar kept in Sèvres, a suburb of Paris. This unit they called the *mètre* (metre in English), from the Latin word *metrum,* meaning measure. The whole system of measurements which began with the metre is called the metric system, and is today used the world over.

The metric system is built up in units of ten, using Greek prefixes for multiples and Latin prefixes for fractions. Thus ten metres is a dekametre, a hundred metres is a hectometre, a thousand metres is a kilometre, and ten thousand metres is a myriametre—from the Greek words *deka* (ten); *hekaton* (hundred); *chilioi* (thousand) and *myrioi* (ten thousand).

On the fraction side, a tenth of a metre is a decimetre; a hundredth of a metre is a centimetre; and a thousandth of a metre is a millimetre—from the Latin words *decem* (ten); *centum* (hundred); and *mille* (thousand).

The Greeks had no word for a number larger than ten thousand, and the Romans no word for a number larger than one thousand. However, the system was extended by using less specific words. For instance, a megametre is one million metres, from the Greek word *megas,* meaning simply large; while a micrometre is a millionth of a metre, from the Greek word *mikros,* meaning simply small.

In 1958 new prefixes were adopted internationally. The prefix giga- from the Greek *gigas* (giant) stands for a thousand million and tera from the Greek *teras* (monster) stands for a billion, so that a gigametre is a thousand million metres and a terametre is a billion metres. On the other hand, nano- from the Greek *nanos* (dwarf) stands for one thousand millionth and pico for one billionth, so that a nanometre is a thousand millionth of a metre and a picometre is a billionth of a metre.

Microbe

ONE DUTCHMAN first used lenses to penetrate the incredibly distant (see TELESCOPE); another Dutchman, Anton van Leeuwenhoek, used lenses to penetrate the incredibly small. Leeuwenhoek produced lenses so powerful in magnification that he could see single cells. These were the first microscopes, from the Greek *mikros* (small) and *skopein* (watch). With them Leeuwenhoek indeed watched the small.

In 1675 he described Protozoa, from the Greek *protos* (first) and *zoon* (animal), which are single-celled animals and which in some form were indeed the first animals to exist on earth. In 1683 he discovered single-celled organisms smaller still, which were not animals nor yet quite plants.

These are today known by a variety of names. The most general name is germ (from the Latin *germen,* meaning bud—i.e., something small that yet contains the beginning of life). Germ is also the name of cells from which complex organisms develop. The human ovum and sperm are, for instance, referred to as germ cells. The part of a plant seed that develops into the plant (i.e., that germinates) is also a germ, so that we may speak of wheat germ.

Another name for the small creatures is microbe. The -be ending is what is left of the Greek *bios* (life). But small life really includes the Protozoa, too, and still other forms. Because microbe has been applied so often to just a particular type of cell, the modern term for small life in general is micro-organism.

The name most used today for Leeuwenhoek's creatures is bacteria, from the Greek *bakterion* (a little rod), since a number of them do indeed have the appearance of tiny rods. The study of such creatures is now known as bacteriology. (The suffix -logy comes from the Greek *logos,* meaning word; a bacteriologist is someone who speaks words about bacteria.) The study of micro-organisms in general is microbiology (words about small life).

Microwaves

VISIBLE LIGHT with the longest waves is seen by us as red. Light with still longer waves is invisible, and is called infra-red (infra- coming from a Latin word meaning below). The wavelength of infra-red light is anything up to a millimetre.

In 1887 the German physicist Heinrich Rudolf Hertz first detected radiation of extremely long wave-length, far longer than infra-red, generated from the spark of an induction coil. They were called Hertzian waves by some and radio-waves by others. The latter name, which merely means waves that radiate (which all waves do) is quite meaningless, but it won out.

The term radio-waves is sometimes applied to the entire range of waves, from a millimetre to many kilometres. At first it was the longer radio-waves that proved most useful. They were reflected by layers of charged particles in the upper atmosphere, so that they would bounce between earth and sky, and therefore follow the curvature of earth for long distances. It was this which made long-distance radio communication possible.

On the other hand, the very short radio-waves went right through the charged layers, and so not much attention was paid them. From the 1930s onward, however, they became important. They could penetrate clouds and fog and bounce off obstacles much better than longer radio waves could. The short radio-waves were therefore used to detect incoming aeroplanes during the Second World War (radar). Then it was discovered that short radio-waves were emitted by many astronomical bodies, and radio-astronomy was developed.

The short radio-waves soon received a name of their own, microwaves (micro- coming from a Greek word meaning small). Microwave wave-lengths are from 1 to 16 millimetres.

Mid-Oceanic Ridge

IN 1853, when the Atlantic Cable was being laid, it seemed to those who were making soundings that there was a plateau in mid-ocean. The Atlantic Ocean was shallower in the middle than at either end.

After the First World War, soundings were made by listening to ultra-sonic echoes. This gave much more detail of the rise and fall of the bottom than one could get by dropping a line. By 1925 it seemed there was a vast under-sea mountain range winding down the Atlantic's centre.

Later soundings elsewhere showed that the mountain range was not confined to the Atlantic. At its southern end it curves around Africa and moves up the western Indian Ocean to Arabia. In mid-Indian Ocean it branches, so that the range continues south of Australia and New Zealand and then works northward in a vast circle all around the Pacific Ocean.

At first the mountain range had been called the Mid-Atlantic Ridge, but when its world character was understood it came to be called the Mid-Oceanic Ridge.

After the Second World War, the details of the ocean floor were probed with renewed energy by the American geologists Maurice

Ewing and Bruce C. Heezen. In 1953 they discovered that a deep canyon ran the length of the ridge and right along its centre. This was eventually found to exist in all portions of the Mid-Oceanic Ridge, so that sometimes it is called the Great Global Rift. There are places where the rift comes quite close to land. It runs up the Red Sea between Africa and Arabia and it skims the borders of the Pacific through the Gulf of California and up the coast of the state of California.

Through the rift hot molten rock from below pushes upward and spreads the sea floor apart. This moves great plates of earth's crust this way and that, and is the cause of continental drift.

Million

THE LARGEST number-word among the Romans was *mille*, meaning thousand. It has come down to us in a number of ways. A millipede is an arthropod with perhaps a thousand feet (see CEN-TIPEDE) and a millimetre is a thousandth of a metre (see METRE). The Roman used as a measure of length the distance covered by one thousand paces of their marching legions. This they called a *milia*, and it has come down to us as mile.

In medieval Italy, with the growth of finance and the increasing prosperity of Europe, it became handy to have words for numbers larger than a thousand. So for the number 1 000 000 (a thousand thousand) someone tacked the ending ione on the *mille*. This suffix implied something of unusual size, so that *millione*, or million in English, is a king-size thousand, so to speak.

In the fifteenth century still larger words were needed and the French invented billion for a thousand million (1 000 000 000). The ending -llion had already become established as implying a large number, and the bi- prefix came from the Latin *bis* (twice), perhaps because billion was the second made-up word of the sort.

By using additional Latin number prefixes still larger numbers could be invented. Trillion is a million billion (*tres* is Latin for three) and quadrillion is a million trillion (*quattuor* is Latin for four). This can continue indefinitely.

The French use these names for numbers at multiples of a thousand, England and Germany at multiples of a million. Thus a quadrillion is a thousand trillion in France, but a million trillion in England or Germany. America went along with the French system after the American Revolution when France was popular and England was not. Oddly enough, the French dropped the word billion. They now call 1 000 000 000 a milliard. The suffix -ard again implies something has gone all the way (a drunkard is one who drinks all the way) so that 1 000 000 000 is a veritable king-size thousand, so to speak.

Mitochondria

IN THE 1830s it was found that all living tissue was composed of tiny cells too small to be seen by the unaided eye, and each walled off from the others by a membrane.

It was supposed at first that each cell was a microscopic drop of homogeneous living fluid which, beginning in 1846, was called protoplasm. As microscopes improved, however, it became clear that the contents of the cells included dimly seen granules.

It was hard to see the details within the cell, because everything in it was more or less transparent. In the 1850s, however, chemists began to synthesize organic molecules, some of which were coloured. Using these synthetic dyes, biologists found that some would be absorbed by some parts of the cell contents only. This meant that these portions of the cell contents would stand out against the rest in bright colour.

Unfortunately, the dyes killed the cells, and sometimes coagulated the clear substances within and created apparent granules that would not have existed in the dye's absence. Thus when a German biologist, Richard Altmann, reported certain granules in the outer regions of the cells the observations were met with considerable scepticism. It was not till another German biologist, Carl Benda, repeated the observation in 1897 that scientists were convinced. The objects were called different names by different observers, but Benda had proposed the name mitochondrion (mitochondria is the plural form), and that was finally accepted.

The name isn't very descriptive. The granules as observed are like tiny threads or filaments, and mitochondria is from Greek words meaning filament granules. In the 1950s and 1960s electron microscope studies showed the mitochondria to have a complex structure. Biochemists found them to be the powerhouse of the cell, containing enzyme systems that supervised reactions liberating energy for the use of the body.

Mohole

THE STUDY OF EARTHQUAKE WAVES has shown that at the centre of earth there is a sphere of molten nickel-iron. Around this nickel-iron core is a layer of rocky substances called the mantle. The core and the mantle bear about the same position and proportions as the yolk and white of an egg.

In 1909, a Croatian geologist, Andrija Mohorovicic, was studying the manner in which earthquake waves travelled through the upper layers of earth. To account for the time it took them to reach various points on the surface he had to suppose that they bent sharply about thirty kilometres beneath the surface. This meant the mineral structure above that level was distinctly different from that below. It was a discontinuity and came to be known, in his honour, as the Mohorovicic discontinuity, or the Moho discontinuity for short.

Above the discontinuity is earth's crust (like the eggshell lying outside the egg white). It is of a different kind of rock than the mantle. It is possible that the crust was built up in the course of earth's evolution and that the top of the mantle represents the planet's original surface.

It would be useful to penetrate through the crust, therefore, to the top of the mantle. The Moho discontinuity lies farther below the surface on dry land than at sea. At sea the Moho discontinuity may be only thirteen kilometres below the surface, and most of that is just water. In places only five kilometres of solid crust would have to be penetrated.

Plans were made in the 1960s to drill down to the Moho discontinuity, and the drilling was slangily referred to as a Mohole. The prospect aroused considerable excitement among geologists, but it meant the expenditure of a lot of money which the Government decided to spend in other ways. For the time, therefore, the Mohole has been shelved.

Molecular Biology

BIOLOGY (from Greek words meaning study of life) is the science of living things. The word was first used in 1802 by the German naturalist Ludolf Treviranus. At first biology concerned itself only with organisms and parts of organisms large enough to be seen with the unaided eye. In the seventeenth century (long before the word biology, but not the study, was invented) microscopes came into use, and biologists could see what had hitherto been invisible.

It was discovered that all life-forms were made up of tiny units, invisible to the unaided eye, which were called cells. In the nineteenth century the inner structure of the cell was slowly worked out. Coloured chemicals were used, for instance, because some of these adhered to some structures within the cell but not to

247

others. The coloured structures could then be studied in some detail.

Chemists went further and began to study the molecules of those substances that characteristically made up cells. By the mid-nineteenth century, this science of biochemistry was flourishing.

As scientific techniques grew subtler chemists were able to study the complex molecules in greater detail. By the 1940s, for instance, methods were developed to break down large molecules into fragments that could be analysed and then put back together again. In this way the fine structure of large molecules could be worked out. By this method, and by studies of the behaviour of large molecules in beams of X-rays, in electric fields, and under strong centrifugal effects, there came to be a new understanding of life, with still more promised.

Biology came to deal more and more with the properties of the very large molecules of protein and nucleic acid. Since the Second World War, the term molecular biology (first used by the English biochemist W. T. Astbury) has come into fashion to describe this new study.

Molecule

THE LATIN word *moles* means a mass, and the diminutive form *moleculus* naturally means a small mass. The term molecule was therefore originally used to mean any tiny particle of matter.

In the study of gases, however, it early became obvious that they must be composed of tiny flying particles of matter—molecules—separated by gaps. In 1811, moreover, the Italian physicist Amadeo Avogadro suggested that a given volume of any gas at the same temperature and pressure contained the same number of molecules. This was Avogadro's Law, which showed that one could determine the comparative weights of different molecules by measuring the densities of gases.

It turns out that the molecules of some gases (e.g., helium and argon) consist of single atoms, but that in most cases gas molecules consist of two or more atoms. The molecule concept has been extended to liquids and solids also, and a molecule is now defined as the smallest particle into which a compound can be subdivided without change in chemical properties. Molecules are now known that are made up of millions of atoms.

The molecular weight of a substance is the sum of the weight of the individual atoms composing its molecules. Each atom has an atomic weight relative to that of the oxygen atom, which was set, arbitrarily, at exactly 16. According to this scale, carbon weighed just a little over 12 and hydrogen a little over 1. (See PROTON.) A molecule of ethyl alcohol (with two carbon atoms, six hydrogen atoms and an oxygen atom) has therefore a molecular weight of 24 plus 6 plus 16, or 46.

If the molecular weight is expressed in grammes, then the result is a gramme-molecular weight, which is usually abbreviated as mole (or sometimes mol). Thus, 46 grammes of ethyl alcohol (46 being its molecular weight) is the equivalent of one mole of ethyl alcohol. It is convenient for chemists to talk about moles rather than grammes sometimes, because it turns out that one mole of any substance contains as many molecules as one mole of any other. The total constituent number of molecules in a mole is 602 000 000 000 000 000 000 000 (or Avogadro's number, since the existence of this number was deduced from Avogadro's Law).

Moment

A SEESAW is a flat board supported in the centre. The support is the fulcrum of the seesaw, from the Latin *fulcrum* (a bedpost), which in turn comes from *fulcere* (to prop up).

If two children of equal weight sit at opposite ends of the board and one pushes himself up the other will come down; the other pushes and the situation reverses itself. The downward pull of gravity on one child is transmitted as an upward lifting force on the other. The seesaw is thus an example of a simple machine, which may be defined as any device that will allow force applied at one point to have its effect at another point or in another direction. Machines are generally used as a means of doing work that would be more difficult without such transmission or direction-change of force; the word is derived from the Greek *mechos*, which signifies means or expedient.

But suppose one child on the seesaw is twice the weight of the other. The lighter child, once in the air, will not be heavy enough to lift the other. The seesaw will come to a halt. In order to rebalance the seesaw, it is necessary for the heavier child to sit closer to the fulcrum. The ability of a force to turn a board about a fulcrum depends not only upon the size of the force but also upon the distance from the fulcrum of the point at which it is applied. For a well-balanced seesaw the weight of one child times his distance from the fulcrum must equal the weight of the other times his distance.

When force times distance on one side of the fulcrum is equal to force times distance on the other we say the torques are the same.

249

(The word comes from the Latin *torquere,* meaning to twist.) Or we can say the moments are the same. This comes from the Latin word *momentum,* which in turn is just a shortened version of *movimentum,* meaning movement. The two words refer to the twisting or moving effect of a force.

The word moment has also come to mean an instant of time. Since time is always measured by movement (of heavenly bodies, of pendulums, etc.), a small movement in space is a small moment of time.

Monopole, Magnetic

ABOUT 1870 the Scottish mathematician James Clerk Maxwell worked out a detailed theory which related electricity and magnetism in a manner so intimate that you could not have one without the other. As a result, we speak of an electromagnetic field. When such a field fluctuates, radiation is given off. Light is an example of such electromagnetic radiation.

Maxwell's theory could be improved if electricity and magnetism were analogous in every possible way. There remains one great difference, however. An object may have one of two kinds of electric charge, positive or negative, but both need not be present on a single particle. An electron carries *only* a negative charge; a proton, *only* a positive charge.

There are also two kinds of magnetic charge, concentrated at the north pole and the south pole. (These are so called because the most impressive magnet known in early modern times was earth itself, and its magnetic charge was concentrated in the polar regions.) Every body which possesses a magnetic field, however, always has a concentration of each kind—*both* a north pole and a south pole.

In 1931, the English physicist Paul Dirac suggested that to make electricity and magnetism completely analogous there ought to be particles containing only a north pole or only a south pole. These particles would be magnetic monopoles.

Physicists have been trying to detect monopoles ever since, but without success. If they exist they would have to possess huge energies, and could only be formed by use of such energies.

In 1969 the American physicist Julian Schwinger suggested that certain hypothetical electrically charged particles called quarks are also magnetic monopoles. He has suggested dyon as a name for this combination (with the prefix dy- from a Greek word meaning two).

Monosaccharide

PROTEINS, THE IMPORTANT giant molecules of living tissue, are built up from long strings of relatively small molecules called amino acids. In the process of digestion these long strings are broken up into smaller strings. Since the Greek word for cooking or digestion is *pepsis,* the short strings of amino acids resulting from digestion were named peptides.

Different peptides may be distinguished according to the number of amino acids in the chain. A peptide made up of two (or three or four) amino acids is called dipeptide (or tripeptide or tetrapeptide) using Greek number prefixes (see NEOPRENE). Proteins themselves, since they contain a large and indefinite number of amino acids (sometimes thousands), are polypeptides, from the Greek *polys* (many).

There are a number of kinds of giant molecules in living tissue, but in every case they are built up of smaller units joined together. For instance, starch is made up of a number of glucose units (see GLUCOSE) joined in a string. Starch is therefore an example of a polysaccharide (many-sugar), since glucose is a kind of sugar and the Latin *saccharon* (from the Greek *sakchar*) means sugar.

Glucose itself is a monosaccharide (one-sugar) the Greek *monos* meaning one. (However, there is no analogous monopeptide since a single amino acid is just called an amino acid. Even scientists are not always logical.) Two other important monosaccharides are fructose and galactose; each has the same number of the same kind of atoms as glucose, but the atoms are arranged differently. Fructose is found in a variety of sugar contained in fruits (the Latin *frux* means fruit), while galactose is found in the variety of sugar contained in milk (the Greek *gala* means milk).

Glucose and fructose will combine to form a disaccharide (two-sugar) called sucrose. This is directly from *saccharon,* since it is common everyday sugar, the kind we buy in the store. Glucose and galactose form a disaccharide called lactose, the sugar in milk (the Latin *lac* means milk).

Mössbauer Effect

CERTAIN ATOMS emit gamma rays. In theory these should be of a certain energy, but as the gamma ray emerges the atom emitting it recoils. The recoil takes up a certain amount of energy. When a number of atoms emit the gamma ray each may recoil by a slightly different amount, so that gamma rays of a fairly broad range of energies will be emitted.

In 1958 a German physicist, Rudolf Mössbauer, showed that when atoms are part of a crystal the recoil is sometimes spread out over the entire crystal. There is then hardly any recoil at all, since the crystal is so much more massive than a single atom. The gamma ray that issues is of sharply defined energy. This is the Mössbauer effect.

Gamma rays of the energy emitted by a particular crystal will be absorbed by another crystal of the same kind. (The gamma ray emitted by a particular crystal exactly fits its own nuclear structure, so to speak. If you make a key that fits a particular lock you can then use it to open another lock identical to the first.) Gamma rays of a different energy will not absorbed by the crystal.

An ordinary gamma ray beam, with a range of energies, cannot be dealt with sharply. Another crystal will absorb some of it and not the rest. A sharp Mössbauer-effect beam will, however, be absorbed entirely or not at all, a much more noticeable thing.

This has been used to check Einstein's general theory of relativity. According to Einstein's theory, a beam of gamma rays should lose energy very slightly if it must climb against the pull of gravity—or increase very slightly if it drops with gravity. If a beam of gamma rays, which a crystal can absorb, is allowed to drop from top floor to basement the crystal will no longer absorb it at the basement. The energy has increased—very slightly, to be sure, but enough to be detected by the crystal—and Einstein's theory is supported.

Muon

THROUGHOUT the early twentieth century sub-atomic particles were best studied by the tracks they left passing through cloud chambers. Particles with electric charge knocked electrons out of the atoms they struck and left a trail of charged atoms (ions) behind. The chamber was saturated with water vapour and tiny water droplets formed around each ion. In this way the track of the sub-atomic particle was made visible.

The nature of the particle could be determined from the manner in which the track curved in the presence of a magnetic field. Particles with a positive charge curved in one direction; those with a negative charge, in the other. Massive particles carrying a unit electric charge curved gently; light particles carrying the same charge curved sharply. An experienced physicist could easily read off the tracks at sight.

In 1935 the American physicist Carl D. Anderson was studying cosmic rays. These were composed of energetic particles that

252

struck atoms in the atmosphere and produced showers of sub-atomic particles of all kinds. Anderson was astonished to find that some tracks curved more sharply than those of protons but less sharply than those of electrons. They were particles of inter-mediate mass, the existence of which had been predicted a year earlier by the Japanese physicist Hideki Yukawa.

Anderson named the new particles mesotrons (from the Greek *mesos,* meaning intermediate), and this was shortened to meson. As time went on, though, it turned out that Anderson's particle was by no means the only kind of object of intermediate mass. Other kinds of mesons were discovered, and all had to be disting-uished from one another.

Since Anderson's meson was the first to be found, it was given the honour of the initial letter of the word (m) as its special symbol, but the letter was used in its Greek form, mu. The particle was therefore called the mu-meson, and this has now been shortened to muon.

Mutagen

ABOUT THE TURN of the century biologists noted that young were sometimes born which did not resemble either of their par-ents in a particular characteristic. This was termed a mutation by the Dutch botanist Hugo de Vries, from a Latin word for change.

In order to learn about the mechanism of heredity biologists studied these mutations and the way they were inherited. Notice-able mutations are usually quite rare, however, so methods were sought for increasing their frequency.

The American biologist Hermann J. Muller found that raising fruit flies at higher temperatures increased the incidence of muta-tions, but only slightly. It occurred to him to try X-rays. An energetic X-ray might hit a chromosome in some key spot, intro-ducing a change in the hereditary mechanism. By 1926 Muller had succeeded. X-rays produced many mutations, and they were now easy to study.

In 1937 the American botanist Albert F. Blakeslee discovered that the alkaloid colchicine interfered with the process of cell-division in plants, so that cells with abnormal numbers of chromo-somes were produced. These were the first mutations to be pro-duced by chemicals.

During the Second World War it was found that mustard gas somehow reacted chemically with the substances in chromo-somes, altered them, and produced mutations. The search was on, and a rather long list of other chemicals capable of doing the same was discovered. These chemicals came to be called mutagens, the Greek ending making the word mean capable of producing mutations.

It would appear that cancer results from some particular mutations (that in some cases may be virus-induced). A mutagen may therefore bring about cancer. In that case it is also a carcinogen, from Greek words meaning to produce cancer.

Mutation

DURING THE process of cell division each chromosome (see GENE) forms a duplicate of itself. When cell division is complete, and there are two cells where only one existed before, it is found that the original of each chromosome is in one cell and the duplicate in the other. Since the duplicate is usually identical with the original, and since the chromosomes govern the chemistry of the cell, the new cell is just like the old one.

The same principle governs the formation of the sex cells and the growth of the new individuals arising from them, so that giraffes have baby giraffes and elephants have baby elephants rather than vice versa.

However, the duplication is not always perfect. Whenever through some chance a new chromosome is formed that is not exactly like the old the new cell is not exactly like the old, and the new young creature is perhaps not exactly like its parents. In 1886 the Dutch botanist Hugo de Vries noted a group of plants in which some were radically different from the rest, though all seemed to have originated from the same stock. He applied the word mutation to such a sudden change in going from parent to offspring, from the Latin *mutare* (to change).

Actually mutations are an old story to herdsmen. Domestic antimals (and human beings, too) sometimes give birth to young that are so far out of the ordinary as to seem freaks of nature. The word comes from the Anglo-Saxon *frician* (to dance). A wild dance is characterized by startling changes in step and movement, so the word comes to mean any unexpected chance happening (as a freak wind or a freak bounce of a ball).

Freaks appear to be a kind of practical joke perpetrated by nature. It is almost as though nature is making grim sport of the herdsman expecting a normal calf or lamb, or even grimmer sport of a human mother, and freaks are also, in point of fact, called sports. On the other hand, a freak birth was considered an ominous warning by the diviners of ancient Rome, and from the Latin *monere* (to warn) they got *monstrum* (a divine warning of ill omen), and we get our word monster.

Nanometre

IN 1795 the French revolutionaries established a new system of measurement, so convenient and so logical that it has now been accepted (or is being accepted) by almost the whole world.

Essentially the system establishes a particular unit of measurement, the metre, which is a unit of length equal to about 39·4 inches. Other units of the same sort are then indicated by different prefixes. A kilometre is a thousand metres, the prefix kilo- coming from the Greek word for thousand. A millimetre is one-thousandth of a metre, the prefix milli- coming from the Latin word for thousand.

The original designers of this metric system did not make up prefixes for less than a thousandth, because at that time they scarcely seemed to be needed. Since then, however, scientists have been probing smaller and smaller measurements. They began to speak of microns, or micrometres, for instance, where a micrometre is a thousandth of a millimetre, or a millionth of a metre. The prefix micro- is from the Greek word *mikros,* meaning small.

That was not enough either. In 1958 the prefix micro- was made official and two prefixes, representing even smaller quantities, were established. A thousandth of a micrometre is a nanometre (from the Greek *nanos,* meaning dwarf) and a thousandth of a nanometre is a picometre. There is no Greek word related to pico-, but it may have originated by association with *picayune,* meaning very small or unimportant.

In 1962 it was established that a thousandth of a picometre is a femtometre and a thousandth of a femtometre is an attometre. Neither femto- nor atto- is related to any Greek word. All these new prefixes can also be used for units for measurements other than length.

Neoprene

THE FIVE-CARBON compound isoprene (see TERPENE) is used by living tissue (particularly plant tissue) as a unit out of which to build up a number of more complicated compounds. A molecule made up of two isoprene units is a terpene. Similarly, one made up of four isoprene units is a diterpene (two-terpene, from the Greek *dyo,* (meaning two); one of six isoprene units is a triterpene

255

(three-terpene, from the Greek *treis,* meaning three) and one of eight isoprene units is a tetraterpene (four-terpene, from the Greek *tettares,* meaning four). A molecule made up of three isoprene units is a sesquiterpene. The Latin prefix *sesqui-* is a running together of *semis* (half) and *que* (and) and means one and a half.

Certain important compounds are built up out of isoprene. Vitamin A, for instance, is a diterpene and carotene (see CHLOROPHYLL) is a tetraterpene. However, the most familiar substance built up out of isoprene units is rubber. The rubber molecule is a large one, made up of an indefinite, but high, number of such units. It is a polyterpene (or many terpenes, from the Greek *polys* meaning many).

As rubber grew more valuable, chemists tried to prepare artificial rubber in the laboratory. They began by working with isoprene, trying to force it to build up large molecules like those in rubber. Until recently they failed. The isoprene molecules joined all right, but not in the correct fashion.

They had better luck with compounds similar to isoprene. One was a compound with a molecule in which a chlorine atom replaced the one carbon atom branching off the main carbon chain in isoprene. This compound was named chloroprene (a combination of *chlor*ine and is*oprene*). In 1931 the American Du Pont Corporation successfully manufactured an artificial rubber built up out of chloroprene units. At first they called it Duprene, in honour of the company, but then they changed the name to Neoprene, the Greek prefix *neo-* coming from *neos* (new), and so the nonsense syllable prene (see TERPENE) has really branched out.

Nerve Gas

POISON GAS was first used in warfare on a large scale during the First World War. In 1915 chlorine gas was used for the purpose, but much more dangerous gases were quickly developed. The most effective gas used during the First World War was mustard gas (actually a readily boiling liquid, which was given its name by British soldiers from its smell). Mustard gas is a vesicant (from a Latin word meaning blister); that is, it irritates the skin, reddening it and eventually producing blisters and considerable pain. It is poisonous to breathe, and damages the eyes.

An even more damaging gas was synthesized during the War by an American chemist, W. Lee Lewis, and was named Lewisite. It wasn't produced in quantity before the end of the War, and was never used in combat.

In the Second World War poison gas was not used. During that war, however, German chemists, searching for compounds to serve as insecticides, came across certain organic phosphates that would kill insects.

Indeed, they were highly effective against any form of life, because they interfered with an enzyme called cholinesterase. They prevented cholinesterase from breaking down the compound acetylcholine. Nerve impulses passed from one nerve cell to another by way of acetylcholine, which had to be formed for one impulse and then broken down ready to be formed for the next. If cholinesterase stopped working, and acetylcholine didn't break down, various nerve impulses stopped, including those to the heart and lungs. Inhaling small quantities of these odourless organic phosphates meant death within five minutes. Because they interfere with nerve action, they are called nerve gases. They are by far the most terrifying poison gases yet developed, but they have never yet been used in actual warfare.

Neuron

THE BRAIN and spinal cord are composed of irregular cells that usually have a number of short branching fibrous extensions at one end and a single long fibrous extension at the other. The long fibrous extension is encased in a fatty sheath, and this is by far the most noticeable part of the cell. Without a microscope, in fact, it is the only part that can be noticed.

The Romans used the word *nervus* to apply to any fibre occurring in the animal body—a tendon, sinew or this extension of a brain cell. In time the word was used only for the last, which is now called a nerve.

By 1891 the German anatomist Heinrich Waldeyer, who first studied the nerve cells thoroughly, wanted to get away from nerve and its implication of the long fibre only, so he turned to the Greek equivalent, *neuron*, for the entire cell. (Of course, the Greek word also meant tendons and sinews originally, but in modern anatomy Waldeyer's meaning stuck.) It is from neuron that we get words like neuralgia for pain along the course of a nerve (the Greek *algos* means pain); neuritis for inflammation of a nerve (the Greek suffix *-itis* means inflammation); neurology for the study of the nervous system and its diseases (the Greek *logos* means word and studies always result in many words in lectures and books); and neurosis for a mental ailment or nervous condition (the Greek suffix *-osis* means a condition of).

The small branched extensions at one end of the neuron are the dendrites, from the Greek *dendron* (tree), which is what the branches resemble. The long extension, running as it does down the centre of the fatty sheath, was called the axis at first (see VERTEBRA), but this was later modified to axon in order to match the -on suffix of neuron.

The axon of one nerve cell branches at the end, and these branches usually mingle with the dendrites of another, though they do not join. Nerve impulses can cross this microscopic gap, which is called a synapse, from the Greek *syn-* (together) and *haptein* (to fasten); it is at this point that nerves are fastened together.

Neutrino

THE FIRST TWO SUB-ATOMIC PARTICLES discovered, the electron and the proton, carried electric charges. In 1932 a sub-atomic particle was discovered that carried no charge. It was electrically neutral (from Latin words meaning neither one nor the other). Physicists made use of the -on suffix of the electron and proton and named the new particle neutron.

Meanwhile in 1931 the Austrian-born physicist Wolfgang Pauli had suggested the existence of a new particle to account for the fact that there was some energy missing when a radioactive atomic nucleus broke down and emitted an electron. Pauli showed that such a particle would have to have no charge and very little mass.

What could this particle be called? The term neutron was already used by the time other physicists began to take Pauli's particle seriously. Pauli's particle had far less mass than the neutron, so the Italian physicist Enrico Fermi suggested it be called a neutrino, which in Italian means little neutron. This name was accepted.

For many years the neutrino was thought to be a very doubtful particle indeed—very useful in physical theory, but able to slip through matter so speedily and subtly that it could not be detected. If it could not be detected how could anyone ever tell if it existed?

In the 1950s, however, American physicists Clyde L. Cowan and Frederick Reines set up an experiment in which hordes of neutrinos (assuming they existed) would bombard large tanks of water. It was calculated that a very few neutrinos would be absorbed and produce effects that could be observed. The predicted effects were indeed produced, and, in 1956 Cowan and Reines were able to announce that the neutrino had been detected, and that it really existed. Since the 1960s physicists have been trying to detect neutrinos produced by the sun.

Neutron Activation Analysis

AS EARLY AS 1906 the New Zealand-born physicist Ernest Rutherford was bombarding matter with sub-atomic particles. If such sub-atomic particles struck the tiny nucleus of an atom they could bring about changes in the nuclear structure and change one type of atom into another.

The first sub-atomic particles that were available for bombardment carried a positive electric charge and were repelled by atomic nuclei which were also positively charged. As a result, fewer sub-atomic particles managed to strike a nucleus than one might hope.

The situation was changed when the neutron was discovered by the English physicist Sir James Chadwick in 1930. The neutron carried no electric charge, and it was not repelled by any part of the atom, either by the positively charged nucleus or by the negatively charged electrons that surrounded it. Quickly the neutron became a favourite bombarding particle, and almost every type of atom was exposed to its action.

Sometimes a particular stable nucleus would absorb a neutron and become a more massive but still stable nucleus, as when carbon 12 became carbon 13. Much more often, a stable nucleus would absorb a neutron and become radioactive. The new radioactive nucleus would break down to emit certain particles at certain energies. Every different radioactive nucleus would break down in its own special way, different from that of all others. The speeding particles it gives off can be detected with great accuracy. In this way one can detect what nuclei were present to begin with. This method of analysing is called neutron activation analysis.

Neutron activation analysis is very delicate; amounts as small as a billionth of a gramme of a particular type of nucleus can be detected. Tiny scraps of paint from a work of art can be studied to see if it is a fake. Even hair from Napoleon's century-and-a-half old corpse was studied, and found to contain suspicious quantities of arsenic.

Neutron Star

IN 1968 pulsars were discovered, objects that gave out very short, regular bursts of radio waves at intervals of a second or less. Naturally, the question was: what could produce such bursts? Some astronomical body must be undergoing some change at intervals rapid enough to produce the pulses. Something must be pulsating, rotating, or revolving about another object, and at every pulsation, rotation or revolution send out a beam of radio waves.

But the regular change must take place so quickly that the only way to account for it was to suppose some very small body was involved and that it was affected by a very intense gravitational field. The smallest, most intensely gravitational bodies known were white dwarfs, stars in which ordinary atoms had broken down into a mixture of electrons and nuclei. Without their electron shields the stars collapsed, the nuclei approaching each other far more closely than they would in ordinary matter.

A white dwarf would have the mass of the sun compressed into a ball not much larger across than earth. Even so, astronomers couldn't see any way of having a white dwarf pulsate, rotate or revolve quickly enough to account for the pulsars.

Could there be still smaller objects? Suppose the shattered atoms of a white dwarf collapsed further. The negatively charged electrons present would be forced by enormous pressures to combine with the positively charged protons to form uncharged neutrons. The neutrons would smash together to form solid neutronium. A star like our sun could condense into a small sphere, sixteen kilometres across, retaining all its mass. Such a tiny, but extremely dense neutron star could rotate about its axis rapidly enough to account for the pulsars. The theory that pulsars are rotating neutron stars is in fact generally accepted now.

Niacin

THE MOLECULE of nicotine (see NICOTINE) consists of two rings of atoms. It was found in 1867 that one of the rings could be destroyed by treatment with strong acid. What was left showed acid properties, and was therefore named nicotinic acid. The resemblance in names does not extend to properties. Nicotine is a violent poison, but nicotinic acid is not.

Now to change the subject. A certain disease marked by mental difficulties, a sore mouth, and an inflamed roughened skin has long been prevalent in Spain, Italy, and the southern parts of the United States. The disease is named pellagra, from the Italian *pelle agra* (rough skin). In 1915 J. Goldberger, a physician working for the United States Public Health department, showed by experiments on prisoners that he could cause the disease in healthy people by limiting the diet, and then cure it by adding such foods as milk to the diet. Pellagra, like scurvy (see ASCORBIC ACID), was a vitamin deficiency disease and not an infectious disease. Temporarily the vitamin that prevented pellagra was called the P-P factor (the P-P standing for pellagra-preventive).

Then in 1937 the American biochemist C. A. Elvehjem showed that nicotinamide (with a molecule consisting of nicotinic acid to which an amine group was attached) was the compound that would prevent pellagra. A few months later he showed that nicotinic acid itself would do the job too, since the body could convert it to nicotinamide easily.

This presented doctors with a problem. The name nicotinic acid could be easily confused by the general public with the name nicotine. People might get the idea they could get vitamins by smoking. Or a person seeing a label on a food product stating that it contained nicotinic acid might think it contained poison.

It was therefore decided to take the first two letters of nicotinic and the first two of acid, add suffix -in by analogy to vitamin, and end up with the made-up word niacin to prevent such confusion. And nicotinamide became in the same fashion niacinamide.

Nicotine

THE TOBACCO PLANT is a native of the western hemisphere, and Europe did not see it until samples were brought into Spain from America in 1558. Two years later, the French ambassador to Portugal, Jean Nicot, sent tobacco seeds to Catherine de Medici, mother of the French king. In Nicot's honour the group of plants to which tobacco belonged was eventually given the latinized name of Nicotiana.

Tobacco is one of a number of plants that get rid of some excess nitrogen by forming complicated nitrogen-containing organic compounds and storing them. These compounds possess weakly alkaline properties (see POTASSIUM), and are therefore called alkaloids. The suffix -oid means having the form of, from the Greek *oeides* (form).

Often the alkaloid is named after the Latin name of the genus (a Latin word meaning group or kind) to which the plant belongs. Invariably their names are given the -ine suffix which chemists reserve for nitrogen-containing organic compounds. The chief alkaloid of tobacco received, therefore, the name nicotine.

Other familiar alkaloids named in this fashion include strychnine, which is found in nux vomica, the seed of an Indian tree which belongs to the genus Strychnos. Coniine is found in poison hemlock, and is therefore the poison that killed Socrates. Hemlock belongs to the genus Conium.

Sometimes alkaloids are named for other reasons. Quinine, for instance, the world's first effective medicine for malaria, is found in the bark of a tree that originally grew in South America. The South American Indian word for bark was *quina,* hence the name. Morphine, which is obtained from the opium poppy, can be useful, under a physician's guidance, in relieving pain and lulling a suffering patient to sleep. Its name comes from Morpheus, the Roman god of sleep. (However, the opium poppy belongs to the genus Papaver—the Latin word for poppy—and another alkaloid from that plant, which contains a number of them, is named papaverine in consequence.)

Nimbus

THE FIRST weather satellites belonged to the TIROS series. They worked most successfully, but they did have certain limitations. Their cameras did not point to earth all the time, so that pictures of earth were taken only at intervals. Futhermore, the orientation of the TIROS satellites wobbled somewhat, so that it shifted from Northern Hemisphere to Southern and back again. Then, too, their orbits were such that they were always to be found over the tropical or temperate zones, and the polar areas were never properly viewed.

On August 28, 1964, the first of a new type of weather satellite was launched. It was sent into an orbit with a high inclination to the equator, so that it could take photographs of the polar regions too (these were very important in any consideration of global weather patterns). Furthermore, it was oriented in such a way that its television camera always pointed to earth as it moved, so that pictures could be taken continually.

The new satellite could also sense infra-red radiation given off by earth. This radiation is given off by the night side, and its amount varies according to whether a region of the surface is covered by snow, by ice, by clouds or by none of these. The infra-red data can therefore be used to study the night side of the planet as well as the day side.

Because of the efficiency with which the new satellite studied the cloud cover it was called NIMBUS 1, from the Latin word for cloud.

By 1966 there were satellites so designed that a picture of the entire planet earth and its cloud cover could be taken bit by bit, every twenty-four hours. The pictures can be put together to form a large mosaic, and weather stations now use satellite data routinely, as a further refinement in weather prediction. The later NIMBUS satellites, and the new TIROS series (called ITOS) have been even more sophisticated. The Soviet Union also has launched many weather satellites since the 1960s, mainly in the COSMOS range. (See also TIROS.)

Niobium

IN 1944 the United States received the honour of having an element named after it (see GALLIUM). It had, however, received the honour more than a century ago and lost it.

It came about in this way. John Winthrop the Younger, Governor of Connecticut in 1635, was an amateur mineralogist and came across a fragment of strange rock near his home in New London. Eventually this rock was sent by his grandson to London, where it is still preserved in the British Museum (Natural History). In 1801 the English chemist Charles Hatchett detected in the rock element number 41, which he named columbium, in honour of the country in whose territory the mineral was first discovered. (The United States is sometimes called Columbia.)

However, this did not end the story. In 1802, the very next year, the Swedish chemist Ekeberg (see TANTALUM) discovered the element tantalum. Columbium and tantalum are very similar chemically, and in 1809 the English chemist William Hyde Wollaston decided that the two were identical, and the chemical profession in general went along with the notion.

Of course, if they were the same element Hatchett was still first by a year and his name should have stuck, but Berzelius, Europe's leading chemist, thought Ekeberg's work more thorough and convincing, and in 1814 voted against columbium and for tantalum, and again the rest of the profession followed.

Finally, in 1845, the German chemist Heinrich Rose showed columbium and tantalum to be two different elements. However, because of their similarity, Rose called columbum niobium, after Niobe, the daughter of Tantalus (see TANTALUM).

For many years the element kept its two names, columbium in America and niobium in Europe. In 1949, however, an international conference of chemists decided to make niobium the official name of the element everywhere, and America lost the honour of the name.

Nitrogen

IT WAS in the 1770s that chemists first realized there were two substances in air, one of which was necessary to life and one of which could not support it. If animals were forced to remain in a closed container of air, or if wood were burned in one, the part of the air that was essential to life was used up. Substances would not burn in what was left, and animals would not live in it.

As a result this gas received a number of unpleasant names. The Swedish chemist Karl Wilhelm Scheele called the life-supporting fraction fire air, and the remainder foul air. (Scheele's fire air was oxygen, which he discovered two years before Priestley (see OXYGEN). Priestley gets the credit, though, because his results were published at once, while Scheele's were delayed.)

The British chemist Daniel Rutherford in the same year, 1772, named the foul air fraction mephitic air, from the Latin word *mephitis*, meaning poison gas. The French chemist Antoine-Laurent Lavoisier called it azote, from the Greek prefix *a-*, meaning not, and *zoe*, meaning life. Azote was therefore the gas that was without life. The Germans call it on the same principle *Stickstoff*, which means in German suffocating substance.

None of these names stuck in English, although azote gave rise to a number of names in organic chemistry for various substances that contain nitrogen atoms. For instance, there are azo compounds, diazo compounds, hydrazo compounds, azoxy compounds and so on.

In 1790 the French chemist Jean-Antoine Chaptal did the trick. The new substance was found to form part of the molecule of the common chemical known as nitre (one of the components of gunpowder). Since there was a fashion in those years to name new gases with the ending -gen, from the Greek suffix *-genes*, meaning born or produced, Chaptal called the gas nitrogen. It was something, in other words, from which nitre was produced.

Noble Gas Compounds

IN 1894 a Scottish chemist, Sir William Ramsay, and an English physicist, Lord Rayleigh, discovered a gaseous element which did not react with any other substance to form compounds. Its atoms would not join those of any other element to form molecules, and it was therefore inert. It was named argon from a Greek word meaning inert.

In the course of the next four years Ramsay, working with liquid air, located four more gases very like argon in properties: helium, neon, krypton and xenon. In 1900 still another gas, a radioactive one, was discovered as a breakdown product of radium by a German physicist, Friedrich Dorn. He called it radon.

The six new elements were lumped together as the inert gases.

They were also sometimes called the noble gases, the point of the adjective being that these gases were too haughty and aristocratic to mingle with the common herd of elements and form compounds with them.

In 1933, however, the American chemist Linus Pauling worked out detailed theories of the manner in which atoms behaved. The inert gases with the larger atoms, he maintained, were not entirely inert and the large atom of xenon ought to combine with the most active of all substances, fluorine.

Xenon was quite rare, and fluorine was dangerous to work with, so the suggestion did not come to anything for a while. During the Second World War, however, techniques for handling fluorine were improved, and in 1962 the Canadian chemist Neil Bartlett was able to form a compound with a molecule that included both xenon and fluorine atoms. Other such compounds followed with radon and krypton as well, and the term inert gases was dropped. Noble gases was used instead (although that was just as much a misnomer). Beginning in 1962, a new and rather unexpected branch of chemistry was founded, that of the noble gas compounds.

Notochord

IN THE HUMAN the skeleton develops in stages. In the early embryo it begins as a straight rod of cartilage down the middle of the back. In 1848 the English physician Richard Owen named this the notochord, from the Greek *notos* (back) and *chorde* (string). The remainder of the skeleton—first of cartilage, then of bone—develops in stages, the process of bone-formation not being complete until some years after birth. This happens not only in man, but in all mammals, birds, reptiles, amphibians and bony fishes. In sharks and their relatives the skeleton remains cartilage throughout life (see TELEOSTEI).

In some very primitive animals the skeleton never passes the notochord stage. There is, for instance, a little animal about five centimetres long, vaguely fish-like, whose only skeleton is a notochord running the length of its back. Its front and rear ends both end in similar pointed fashion, so that at a quick glance it is hard to tell which is front, which rear. It is therefore named Amphioxus, from the Greek *amphi* (both) and *oxys* (sharp); it is sharp at both ends.

There are animals still more primitive which never even develop a full notochord, only the beginnings of one; or develop one in early life and lose it in adulthood. There are the tunicates, for instance, so called because in adult form they are covered with a hard layer or tunic. They live as motionlessly as oysters, and have no trace of any internal skeleton. In fact, for many years they were classified as molluscs, the group to which clams and oysters belong. The young tunicate, however, fresh from the egg, swims about freely, and has a notochord.

To a zoologist the development of a notochord, however imperfect or temporary, shows a relationship to other animals with internal skeletons—a relationship that animals such as lobsters and beetles, which never form the slightest trace of internal skeleton at any time, do not have. Consequently, all animals that possess even a notochord, including the tunicates, are put in the phylum Chordata. Man, of course, by virtue of his adult skeleton, is assigned to the immense sub-phylum Vertebrata.

Nova

IN 1572 a very unusual thing happened in the heavens. Among the stars there appeared a new one, in the constellation Cassiopeia. It grew so bright it could be seen in daylight, and then it faded away. To a world which was used to considering the heavens perfect and unchanging (see ALGOL and COMET) this was a noteworthy thing. One observer was a young Danish scientist named Tycho Brahe who the next year published his observations under the title *De Nova Stella* (Concerning the New Star). And from that day a new star became known as a nova (the feminine form of the Latin *novus,* meaning new).

In 1604 another nova appeared, and was this time observed by Brahe's pupil Johann Kepler, who also wrote a book with a title *De stella nova in pede Serpentarii.*

Of course, novae, as we now know, are not new stars but old ones that have exploded. There are about 25 novae occurring every year in our own galaxy. Most of them are unspectacular, stars that increase a few hundredfold in brightness, then fade. (Of course, if the sun were to do this life on earth would be destroyed, but astronomically such a phenomenon is still rather mild.)

Occasionally, though, a star tears itself apart completely and increases in brightness a few thousands of millions of times and outshines an entire galaxy of stars. Such a star is a supernova (the Latin *super* meaning above or beyond).

Only three supernovae are known to have occurred in our own galaxy. One was Brahe's nova and one Kepler's. Two supernovae, in other words, occurred in a single generation, and none since, to the great frustration of astronomers who now have telescopes and cameras.

The third supernova occurred in 1054, but it was observed by Chinese and Japanese astronomers only. At the point indicated by them there now exists a small mist of light that is obviously the cloudy remnant of a super-explosion which because of its shape is called the Crab Nebula (see GALAXY for meaning of nebula).

Nuclear Reactor

THE FIRST self-sustaining nuclear reaction (see NUCLEUS) was set up at 3.45 P.M. on December 2, 1942, under the stands of a football stadium at the University of Chicago. The reaction took place within a huge cube of uranium and carbon. The uranium atoms were split by slow-moving neutrons, and that (see FISSION) liberated the energy. The carbon was needed to slow the neutrons to the proper speed, and was therefore the moderator, from the Latin *moderare* (to regulate), since it regulated the neutron speeds.

Because the cube was made by first forming a layer of uranium, then piling a layer of carbon on it, then uranium on that, then carbon, and so on, it was called an atomic pile. For a while all devices in which a self-sustaining nuclear reaction could take place were called piles. However, since later devices were formed in less makeshift fashion, they began to be called, much more properly, nuclear reactors.

The adjective nuclear is better than atomic anyway, if we are talking about nuclear reactions. Thus a bomb fired through the energy of uranium fission was called an atomic bomb, in the first announcement by President Truman that such a device had been dropped on Hiroshima in August 1945. The newspapers shortened this to atom bomb and to A-bomb, but all are really misnomers. An ordinary TNT bomb involves atomic reactions, and could be called an atomic bomb. The thing about the A-bomb is that it involves nuclear reactions; it should therefore be called a nuclear bomb.

Again, a submarine run by a nuclear reactor might be called an atomic submarine and said to be running on atomic power or atomic energy; but it should always be nuclear submarine, nuclear power and nuclear energy. The date December 2, 1942, mentioned above is even said to be the beginning of the atomic age, though we have been in the atomic age for thousands of years. It is the nuclear age we are now in.

But all this is useless. As in many other scientific names, the mistake has been made and it is probably already too late to correct it.

Nucleic Acid

IN A SERIES of experiments beginning in 1869 a Swiss investigator named Friedrich Miescher discovered an acidic substance in the nuclei (see PROTOPLASM) of cells. Because of its place of origin he named it nuclein. The -in ending seemed to imply a protein, which it wasn't, so in 1889 it was renamed nucleic acid.

But apparently one change was all the chemical profession could stand for. When, some time later, nucleic acid was discovered in the cytoplasm of the cell no effort was made to change it again. It remained nucleic acid, whether in or out of the nucleus.

There was one difference, though. The nucleic acid of the cytoplasm contained a kind of sugar called ribose in its molecule. Before its discovery in this nucleic acid by the American biochemist P. A. Levene in 1908 it had not been known to occur in nature. It had been synthesized by the German biochemist Emil Fischer in 1901, and he showed that it was very similar in molecular structure to another sugar called arabinose. Arabinose did occur in nature, in a kind of dried tree sap called gum arabic, whence its name. (Gum arabic was first imported from Arabia, and gum comes from the Latin *gummi,* which comes from the Greek *kummi,* meaning sap.) In naming the new sugar Fischer shuffled the letters in arabinose, left out a few, and came up with ribose.

The nucleic acid of the nucleus contained a sugar almost like ribose but with one oxygen atom missing. The usual way of naming such an oxygen-short compound was to add the prefix deoxy-, the Latin prefix *de-* having as one of its meanings 'something taken away'. So the name was deoxyribose. Americans added an s in order to improve the sound and called it desoxyribose, but in 1956 an international convention settled the matter in favour of the deoxy- prefix.

Because of the nature of the sugar the nucleic acid of the cytoplasm is ribonucleic acid, abbreviated commonly as RNA, while that of the nucleus is deoxyribonucleic acid, or DNA.

Nucleon

IN 1911 the New Zealand-born physicist Ernest Rutherford demonstrated that almost all the mass of the atom was concentrated in a tiny volume at its centre. Outside that volume were one or more very light electrons, nothing more. The tiny massive volume at the centre of the atom was called the atomic nucleus, from a Latin word meaning little nut.

By 1914 the existence of the proton was clearly established. It was 1836 times as massive as the electron, and every atomic nucleus contained one or more protons. All protons carry a positive electric charge, however, and repel each other. How could the protons all exist within the volume of the tiny nucleus, then? At first physicists decided that some of the negatively charged electrons were also present in the nucleus, and by their attraction for protons helped keep the nucleus together.

There were certain theoretical shortcomings to an atomic nucleus made up of protons and electrons only, however. In 1932 Sir James Chadwick discovered the neutron, and it was at once clear that the nucleus was composed of protons and neutrons. The neutrons possessed no electric charge at all, but what held the protons and neutrons together in the nucleus turned out to involve a new force field altogether, the strong nuclear interaction.

Except for electric charge, protons and neutrons are very similar. They have almost identical mass, and share many other properties as well. Some term was needed to include both. In 1941 the Danish physicist Christian Møller, noting that both particles occurred in the atomic nucleus, used the -on suffix present in the names of sub-atomic particles and proposed the term nucleon. It was quickly accepted. The word nucleonics then came to mean the study of the atomic nucleus and its particles generally.

Nucleus

THE WORD nucleus (Latin for a little nut from *nux* (genitive, *nucis*) meaning nut) has been used to describe a small body in the centre, more or less, of a plant or animal cell (see PROTOPLASM). It has been used for all sorts of objects at the centre of a larger mass, or for some small mass out of which a larger mass develops. The word has found its place in atomic physics too.

About 1906 the New Zealand-born British physicist Ernest Rutherford began to study the effect of firing alpha particles (see ALPHA RAYS) at thin sheets of metal. The vast majority of these particles went through as though nothing were there, but some were deflected, and a very few even bounced back. Rutherford decided, therefore, that most of the mass of an atom was concentrated in the centre, and this turned out to be the case.

The protons and neutrons (see PROTON), which are the heavy particles in an atom, are all concentrated in a very small central region, while all the outer regions are taken up by the very light electrons (see ELECTRON). Alpha particles plough through the electrons with no trouble, but every once in a while one will strike the tiny central portion and, as Rutherford observed, bounce back. This central portion is called the nucleus too. To distinguish it from the structure in the cell (the cell nucleus), it is called the atomic nucleus.

Ordinary chemical reactions—burning, rusting, all the activity that goes on in test-tubes and living tissue—involve only some of the outermost electrons of the atoms involved. Moderate amounts of energy are given off or taken up in the process, from a match flame, to body heat, to a dynamite explosion.

Radioactivity (see RADIOACTIVITY) involves an actual break-up of the atomic nucleus. This is a nuclear reaction and involves energy exchanges millions of times greater than those of the ordinary electron reactions. Enough energy is involved to make a hydrogen bomb explosion or, if the mass involved is large enough, a sun.

Occultation

THE VARIOUS stars are so far from us that they seem to keep the same relative positions year after year. Not so the sun, moon and planets, all of which appear to move in relation to the stars (see PLANET). It consequently happens that occasionally the moon, for instance, will pass in front of a star or planet, which will then remain hidden behind the moon for a period of time.

The Latin word for to hide is *occulare,* so anything that is hidden is occult. This term is frequently applied to supposedly mysterious arts hidden to all but a few experts. The star or planet that is hidden by the moon is also occult, and the hiding process is occultation.

Occultations involving the moon have been most useful to astronomers. For one thing, at the moment of occultation the precise position of the moon in the sky is known, so that occultations can be used to calculate the moon's orbit with considerable exactness. Secondly, the sudden disappearance of an occulted star, without preliminary dimming, is one of the proofs that the moon has no atmosphere to speak of. If there were an atmosphere the star would first dim, as it shone through that atmosphere prior to occultation.

The most dramatic use of occultation, however, involved the planet Jupiter. The four largest moons of Jupiter, as they circle their planet, pass behind it and are each occulted once a revolution. In 1676 the Danish astronomer Olaus Roemer had been timing those occultations, and discovered an odd fact. When the earth was moving away from Jupiter in the course of their orbital motions the time of occultation lagged, and was later than it should have been. When the earth was moving towards Jupiter the reverse happened; occultations grew more and more premature.

He deduced from this that it took light a finite time to travel through space. The distance between the earth and Jupiter could vary by over 300000000 kilometres and, apparently it took light about 16 seconds to travel that distance. From this he calculated the speed of light to be 224000 km/s. Present calculations show it to be 299793 km/s which is one of the fundamental constants of theoretical physics.

Oceanography

AS EARLY AS 3000 B.C., the island of Crete had developed a civilization based on sea trade. It was followed by the Phoenicians, who before the time of Julius Caesar had already explored the Atlantic Ocean as far north as the British Isles, and far enough south to have circumnavigated Africa.

Until modern times, however, the ocean was only a highway—only something to cross, and something (in case of storms) to be feared. Men knew virtually nothing about the ocean, except its size. They knew next to nothing about its currents, and nothing at all about its depths.

In 1769 the American scholar Benjamin Franklin had taken note of the warm current that moved steadily north-eastward along the eastern coast of central North America. Since it seemed to come from the Gulf of Mexico, it was called the Gulf Stream.

In the 1850s a retired American naval officer, Matthew F. Maury, subjected the Gulf Stream and other ocean currents to close study. He gathered data on ocean depths and other material of the sort, and in 1855 published a book on the subject. It was the first major book on what is now known as oceanography (short for ocean geography).

Also in the 1850s, the American businessman Cyrus W. Field was trying to lay a cable across the Atlantic to allow telegraphic communication between America and Europe. This led to interest in the ocean deeps, and to the discovery of life there. In 1860 a telegraph cable was brought up from the bottom of the Mediterranean, two kilometres deep, with the clearest evidence of deep-sea life, for it was found to be encrusted with corals.

Although oceanography is a century old, it is only now that it is becoming a major science. The study of the ocean bottom has helped us understand the past history of earth's crust, and it is to the oceans too that man may have to look for much of his food-supply in the future.

Octave

A MUSICAL tone is produced by the rapid compression and release of air by a vibrating object. This compression and release spreads out as longitudinal waves. The pitch of the tone depends upon the number of times per second (frequency) that the air is compressed and released. For instance, middle C (do) on the piano has a frequency of 264.

Combinations of musical tones sound pleasant to the ear when the individual frequencies are in simple proportions. For instance, the frequency of do is 264, of mi 330, and of sol 396. These frequencies are 4 × 66, 5 × 66 and 6 × 66, so that the three notes form a pleasant sequence and sound well if struck together as a chord (from the Greek *chorde,* meaning the string of a musical instrument).

The combination fa, la and do—and also sol, ti and re—form similar 4,5,6 ratios and sound pleasant chords. In fact the musical scale do, re, mi, fa, sol, la, ti, do was chosen just to allow a large number of pleasant combinations. The frequencies are: 264, 297, 330, 352, 396, 440, 495 and 528. There is nothing sacred about this; other note combinations are possible. However, Western ears are used to this one, and the use of other combinations by the Chinese or Arabs makes their music sound odd to our ears.

Notice that the second, or higher, do, has a frequency of 528, which is just twice that of the first do (264). By doubling all frequencies, we can build a new scale starting with the 528 do and ending with a still higher 1056 do. Or we can work downward from the 264 do to a still lower 128 do.

A new series is started with every eighth note therefore. The Latin word for eight is *octavus,* so an individual series from do to do is an octave. Furthermore, any note with a frequency just twice that of a second note is said to be an octave higher. If it is four times the frequency it is two octaves higher, and so on. (This notion has been extended to other forms of waves too, as for instance to light-waves.)

Oil

WHEN WE are not certain of the exact nature of substances we do what we can to differentiate them on the basis of outward appearance. For instance, tissues contain substances that are greasy to the touch and do not dissolve in water. These substances may be solid or liquid. The difference is not important really, since the solid can be melted and the liquid congealed.

Nevertheless, solid and liquid was all the difference men had to go on originally, so two names are generally given. In English the solid is fat, the liquid, oil. The word fat comes from the Anglo-Saxon, but oil is from the Latin *oleum,* meaning olive oil, which is related to the Greek *elaia,* meaning the olive-tree. (Olive oil was the chief oil for cooking and washing the body—no soap in those days!—in the ancient Mediterranean world.)

The word oil has passed on to anything that feels oily. For instance, there are materials that can be obtained from plants that seem to possess the very essence of the plant in concentrated form. Roses will yield a small quantity of fluid containing all the fragrance of roses. Similarly, fluids can be obtained that have the fragrance or flavour of jasmine, cloves, vanilla, wintergreen and so on. These fluids are oily to the touch and won't dissolve in water, so they are called essential oils (from essence).

Again, there is an oily liquid that is obtained from beneath the earth which is called petroleum. The petr- prefix comes from the Latin word *petra* (rock), so that the word means rock oil. One of the substances obtained from petroleum is called mineral oil.

Very strong sulphuric acid has an oily appearance (it may have an oily feel, but no one would dare touch it) and is called oleum, though anything less like olive oil in other ways would be hard to imagine.

In contrast to all this, the Greek word for a solid fat is *stear.* This enters chemistry at many point. For instance, one of the more common substances derived from solid fats is stearic acid, and the word for soapstone is steatite.

Omega-minus

BY 1960 the number of different sub-atomic particles that were known had passed the hundred mark, and was still climbing. Physicists could not help but wonder why so many particles existed. Could the particles exist in families? Perhaps within each family particles existed in different varieties, but were essentially the same in many respects. It would be easier to account for a small number of particle families than for a large number of individual particles.

In 1961 the American physicist Murray Gell-Mann and the

Israeli physicist Yuval Ne'eman each suggested a way of building such families by making use of eight different properties of the various particles.

For instance, Gell-Mann prepared a kind of triangle of particles: four delta particles at the bottom, three sigma particles above them, two xi particles above them, and at the apex a single particle. The properties all varied in a regular pattern. The four delta particles had electric charges of —1, 0, + 1 and + 2, in that order. The three sigma particles had charges of —1, 0 and + 1, and the two xi particles, charges of —1 and 0. The topmost particle would naturally have a charge of —1.

The catch in this triangle of particles was that the topmost particle was unknown. All the properties needed to make it fit the triangle could be worked out, though, and in particular it would have to have something called a STRANGENESS NUMBER (which see) of —3. Physicists were sceptical of this, for no particle with such a high strangeness number existed anywhere.

Since the undiscovered particle was the last one in the triangle, Gell-Mann called it an omega particle, since omega is the last letter in the Greek alphabet. Because it should have a negative electric charge, he called it omega-minus. Physicists began an intensive search for it in bubble-chamber tracks. Since they knew all the properties it was supposed to have, they knew what to look for, and in 1964 they found it!

Orbital

IN 1906 the New Zealand-born physicist Ernest Rutherford showed that atoms had a positively charged core, the atomic nucleus, with light electrons in its outskirts. The atoms seemed to be a very tiny solar system, with electron planets circling in orbits about a nucleus sun.

By nineteenth century views, though, electrons, moving rapidly in a curved path about the nucleus, should radiate energy away and spiral into the nucleus—yet they did not.

In 1913 the Danish physicist Niels Bohr suggested that electrons did *not* radiate energy while in orbit, but only when they changed orbits. When they moved into a larger orbit farther from the nucleus they absorbed a fixed quantity of energy; when they dropped to a smaller orbit they emitted that energy. This fitted the new quantum theory (see QUANTUM) advanced in 1900 by the German physicist Max Planck.

Further investigations into atomic structure showed that electrons could not be viewed as merely particles. In 1926 the Austrian physicist Erwin Schrödinger pictured sub-atomic particles as waves. The electron was a wave-form that spread all about the nucleus like a fuzzy shell. Under certain conditions the wave-form was concentrated in one section or another of the region outside the nucleus, and in 1932 the American chemist Linus Pauling showed how this sort of wave mechanics could be used to account for the atom's chemical behaviour.

With the electron as a wave-form, concentrated here and there, one could not consider it as moving in an orbit. One had to think of the various electrons in an atom as possessing certain energy levels instead. The old word orbit was not completely discarded, however, merely changed. Now physicists and chemists speak of electron orbitals to distinguish the different energy levels.

Organelle

WHEN CELLS were discovered in the 1830s they seemed to contain undifferentiated protoplasm. Microscopes were too poor to see much detail within the nearly transparent cell.

Microscopes improved, however, and biologists developed the technique of adding coloured chemicals which attached themselves to some parts of the cellular interior and not others—making parts of the interior clearly visible. The cell had already been found to possess a central nucleus (from Latin words meaning little nut). The part of the protoplasm inside the nucleus is the karyoplasm, the prefix coming from the Greek word for nut. The part outside is the cytoplasm, the prefix coming from the Greek word for cell.

Both the karyoplasm and the cytoplasm were found to contain smaller objects of definite shape. The karyoplasm contained the chromosomes, for instance, while the cytoplasm contained mitochondria, ribosomes and so on. These smaller objects had definite functions within the cell, so it seemed that just as the functions of an organism were divided among various organs, those of a cell were divided among these particles. The particles therefore came to be called organelles, from a Latin word meaning little organ.

The larger and more complex organelles, such as the mitochondria, have been found to possess small quantities of DNA, a substance characteristic of the nucleus. This has caused some biologists to wonder if perhaps the modern cell is really a complex combination of simpler, once independent objects. Perhaps in the beginnings of the development of life there were living sub-cellular fragments, whose chemistry was less complex than that of modern cells. Individual organelles may have combined to form a cell more

275

efficient than they themselves. What we see now as organelles may represent, therefore, the remains of pre-cellular life forms.

Organism

THE GREEK word *ergon* means work (see ENERGY), and from it the Greeks derived the word *organon* for any instrument that did work. This has come down to us as organ, which we apply chiefly to a type of musical instrument, but the more general meaning still shows up. A branch of the Government, such as the courts or the legislature, which does a particular type of work is an organ of government, and a newspaper is an organ of the Press.

The most important use of the term outside of music, however, is for any structure in a living creature that does a specific job of work. The heart, liver, lungs, skin are all organs. And since most living creatures are a collection of organs working together, they are, generally speaking, organisms. The word organism has lost its original significance and come to mean any living creature; thus a virus which may consist of a single large molecule only, and which cannot conceivably have organs in the usual sense, is still a micro-organism, from the Greek *mikros,* meaning small (see MIC-ROBE).

By 1800 it seemed quite obvious that there were considerable differences between chemicals found only in living tissue (or in material that had once been part of living tissue) and chemicals found in the non-living world all about. For one thing, chemists in the laboratory could not make any of the chemicals of living tissue unless they began with living tissue or the dead remnants thereof. In 1807 the Swedish chemist Jöns Berzelius divided all chemicals into two groups. Those from organisms, living or dead, he called organic; the rest he called inorganic (the Latin prefix *in-* means not).

But then in 1828 the German chemist Friedrich Wöhler upset things by preparing an organic chemical, urea, without any use of tissue, living or dead. Nevertheless the distinction remains useful and has been retained, but today organic refers to any compound with a molecule containing carbon atoms, while inorganic refers to any compound with a molecule that does not.

Ounce

IT IS NOT until modern times that the world has seen carefully standardized weights and measures on a universal scale. Until the nineteenth century every country and region had its own system of weighing and measuring things. Naturally, this led to confusion. Under such conditions merchants would want to use only those systems that were known to be reliable.

The city of Troyes, in north-eastern France, was a flourishing city in the Middle Ages, noted for its fairs. To keep its business going it regulated its weights and measures strictly; its unit of weight, the pound (from the Latin *pondus* meaning a weight), was widely adopted for valuable goods like gold, silver, jewels and drugs, where a small inaccuracy could make a large money difference.

The pound used even today in weighing such materials is still called the troy pound, or sometimes the apothecaries' pound (because it is used in weighing drugs). The troy pound is divided into twelve troy ounces. Ounce comes from the Latin *uncia,* meaning a twelfth. (Inch was a slightly more distorted version of the same word; and meant a twelfth of a foot.)

Potatoes, coal and other substances which are relatively cheap and bulky, and which are usually handled in considerable weights, were measured by the avoirdupois pound. Avoirdupois is just the French phrase *avoir du pois,* meaning to have weight.

The avoirdupois pound was made up of sixteen avoirdupois ounces, and in this case the word ounce is a misnomer, obviously. Nor are the troy ounce and the avoirdupois ounce equal. Both are divided into grains (a reminder of the time when grains of wheat or other cereals were used to weigh small quantities on hand scales, so that a tiny sliver of gold would weigh so many grains, you see). The troy ounce was equal to 480 grains, while the avoirdupois ounce was equal to only 437½ grains. The troy pound was equal to 5760 grains; the avoirdupois pound, to 7000 grains exactly.

Oxygen

THE GERMAN chemist G. E. Stahl, about 1700, developed a theory to explain why some substances burned or rusted when heated. He supposed such substances contained phlogiston (from a Greek word *phlogistos,* meaning inflammable).

When wood was heated, for instance, it lost its phlogiston to the air and changed to ash. If the air supply were limited, then the air after a while would have all the phlogiston it could hold, and burning would stop.

In 1774 the British clergyman and chemist Joseph Priestley studied a brick-red powder, today called mercuric oxide. He found this gave off a remarkable gas when heated. Substances burned in it more readily than in air itself. A glowing wooden splinter burst into brilliant flame if inserted into this new gas.

Well, if air that was full of phlogiston could not support combustion, then a substance that supported combustion so well should be completely empty of phlogiston. Priestley therefore called it dephlogisticated air.

The next year the French chemist Antoine-Laurent Lavoisier showed that burning was simply the result of combining chemically with this new gas, which occurred naturally in air. The phlogiston theory fell apart and Lavoisier earned his title 'Father of Modern Chemistry'.

But the great chemist was human and also made mistakes. He thought the new substance was found in all acids. (It isn't, but hydrogen is.) On the basis of this mistaken notion, he named the gas *oxygine* (oxygen in English), from the Greek word *oxys,* meaning sharp and the suffix *-genes,* meaning born. Oxygen is then that of which a sharp taste (or sourness) is born, and the process of combustion came to be called oxidation.

The Germans went along with this mistake but used their own language for the name, calling oxygen *Sauerstoff* (sour substance).

Pacemaker

THE HUMAN HEART beats steadily and rhythmically, hastening or slowing its beat when necessary, but by and large going about its work with calm perseverance for as long as a hundred years or even more.

This heart rhythm is a built-in property of heart muscle, and a heart taken out of the body, isolated from all nerve stimulation, but immersed in a solution containing the proper atoms and atom groups in the proper proportions, will continue beating if that solution is pumped through the heart's blood-vessels.

Not only will intact hearts beat in this fashion but even portions of hearts will. It was found, in fact, that different parts of the heart will beat at different rates if they are studied in isolation. In the intact heart, however, the part that beats most rapidly forces its rate on all the rest. The most rapidly beating part of the heart is therefore referred to as the pacemaker.

In the human being the pacemaker is located in a bundle of special cells in the right auricle. On occasion the pacemaker in the human heart ceases to function properly. This does not mean the heart stops beating altogether; it merely starts beating at the rate of the next most rapid natural rhythm. Naturally, the heartbeat is slowed and a person so afflicted is less able to indulge in hard work or hard exercise.

With the coming of miniaturized electronic devices after the Second World War, it became possible to manufacture small objects designed to yield a periodic electric discharge that would stimulate the heart to beat at something approaching its normal rhythm. This could then be implanted in the heart surgically as an artificial pacemaker. By the end of the 1960s many thousands of human beings were living normal lives, thanks to such devices.

Pachyderm

MOST ANIMALS, if they are common, have names stretching back to antiquity, with their ultimate derivations uncertain (cat, mouse). Or if they are first met with by explorers in foreign lands the local names are sometimes adopted (chimpanzee, opossum). Some animal names, however, have origins that are worthier of discussion.

The elephant, for instance, derives its name from the Greek *elephas*. This may in turn derive from the Phoenician *aleph* (ox). The elephant, after all, when brought to western Asia from India must have impressed the Asians with its sheer size. The largest animal with which they were acquainted was the ox, so this was called a kind of ox. (The Romans when they first came across elephants, while fighting the Greek King Pyrrhus, in Lucania in southern Italy, called them Lucanian oxen.)

The hippopotamus has a name that is almost pure Greek, coming from *hippos* (horse) and *potamos* (river). It is a river horse. The hippopotamus does frequent certain African rivers and it is a large animal, so the Greeks naturally named it after another and more familiar large animal, the horse (although the hippopotamus looks about as much like a horse as the elephant looks like an ox).

The elephant and hippopotamus were once included among the Pachydermata (from the Greek *pachys,* meaning thick, and *derma,* meaning skin; i.e., the thick-skinned animals), all of which were hoofed animals that did not chew the cud, so that horses and pigs were included. This grouping has been abandoned, but elephants are still sometimes called pachyderms.

The most conspicuous and unique feature of the elephant is its trunk, which is in Greek called *proboskis* (proboscis, in English) so that the elephant belongs to the order Proboscidea. The word proboscis comes from the Greek *pro-* (before) and *boskesthai* (to graze). An animal with a proboscis can graze a long way before itself, and that is a good description of what an elephant can do.

Pair Production

ACCORDING TO the special theory of relativity, evolved by Albert Einstein in 1905, mass and energy are different aspects of the same thing, and one can be converted into another. Mass is an extremely concentrated form of energy, and a tiny amount of mass can be converted into a large amount of energy, as in a nuclear bomb. The formula for this has become well known: it is $E = mc^2$, when E is energy, m mass, and c stands for the velocity of light. The reverse effect, the conversion of energy into matter, requires that a great deal of energy must be concentrated into a tiny volume. Even then only a small fragment of matter is formed.

What's more, matter can only be produced if certain of its properties are conserved. For instance, an electric charge cannot be created out of nothing. To create an electron, with its negative electric charge, out of energy a second particle with an equal positive electric charge must also be formed. The negative and positive charge, taken together, add up to zero, and therefore do not count as the creation of electric charge.

It is possible, then, to take an energetic gamma ray photon, and, under the proper conditions, convert it into an electron-positron pair. It is this pair production, first observed in 1933 by the British physicist Patrick Blackett, that is the clearest example of the conversion of energy into matter in practice.

Under the influence of a magnetic field the positron curves off in one direction and the electron in the other. As soon as a positron meets an electron (and our universe is full of them) it combines with it, and the matter of the two particles is converted into energy again. This is mutual annihilation. Occasionally the positron and electron glancingly approach and circle each other for something less than a millionth of a second before annihilation takes place. The combination of circling particles is called positronium.

Palaeobiochemistry

FOR CENTURIES NOW, men have been coming across stony objects in rock which resemble organisms. These fossils are the petrified bones, teeth, shells and sometimes traces of the soft tissues, too—of creatures that have been long dead. Some fossils are over 500 million years old.

The study of these fossils has made it possible to work out details of evolution. The evolution of the horse is well worked out, for instance, and the lines of descent of the ancient reptiles as well.

Fossils tell us only about the superficial appearance of past species. They don't ordinarily tell us much about the more intimate details: the colours of a dinosaur's scales; or whether a particular reptile might have been warm-blooded or had begun to develop hair.

It has become possible, however, to obtain remnants of organic compounds from some ancient bones. Amino acids have been extracted, and were found to be the same amino acids that occur in proteins today. That at least has not changed.

In the 1960s scientists went further in this field of palaeobiochemistry (from Greek words meaning the chemistry of ancient life). From rock formations of the type that would possess fossils if they were younger the American botanist Elso S. Barghoorn extracted organic compounds that might be remnants of very ancient life.

In rocks 2000 to 3000 million years old, for instance, hydrocarbons have been extracted that might have originated through the breakdown of chlorophyll. Other hydrocarbons have been found that seem related to certain coloured compounds, called terpenes, that are characteristic of plant life.

Simple life-forms, not possessing the kind of complex structures that will easily fossilize, thus make their presence known much more subtly, and life on this planet is now thought to be at least 3000 million years old.

Palaeomagnetism

EVER SINCE about 1600 scientists have been measuring changes in earth's magnetic field. Not only do the magnetic poles shift position with time, but the intensity of the field varies too. One might suppose, though, that there was no hope of determining the changes that took place prior to 1600.

Fortunately, certain minerals, when they crystallize out of the melted state, line up their crystals in the direction of the magnetic field and in a manner governed by the intensity of that field. Once the crystallization is complete, later changes in direction and intensity have no effect. It is possible, then, to measure the age of certain rocks through the radioactive breakdown of uranium, for instance, then study its crystalline formation carefully, and discover the direction and intensity of earth's magnetic field at some specific time in the long-distant past.

This new science of palaeomagnetism (the prefix is from a Greek word meaning ancient) has developed mightily after the mid-twentieth century. It was discovered, for instance, that every million years or so there is a long-term change in polarity. The

North Magnetic Pole becomes a South Magnetic Pole, and vice versa. This is not because the poles move bodily across the surface of earth from one polar region to the other. Apparently the magnetic field gradually decreases in intensity to zero and then begins increasing again in reverse.

During the period of zero field cosmic ray particles and other energetic, charged radiation from space are not trapped by the field and gradually decanted into the polar regions. Instead they strike earth's surface generally. What effect this would have on lifeforms is uncertain. Some think it may explain those periods in earth's history when many species seem to die out over a comparatively brief period of time.

Parabola

A CONE (from the Greek *konos*) is a solid figure that looks like a dunce's cap. If you were to imagine a cone cut by some kind of slicing machine in different ways the cut edges of the cone would make certain interesting mathematical figures. For instance, if a cone were to be sliced straight across, so that all points on the sliced edge were equally distant from the point of the cone, the edge would form a circle. If the slice were made obliquely, so that one side was farther from the point than the other, the edge would form an ellipse (see ELLIPSE). Circle and ellipse are examples of conic sections, the word section coming from the Latin *secare* (to cut).

In the case of both circle and ellipse the slice goes completely through the cone from one side to another. However, if you imagine the slice starting at one side of the cone in a direction that was parallel (see PARALLEL) to the other side the cut would continue for ever, without ever reaching the other side (if we assume the cone to be of infinite size).

The edge of the cone formed by such a section would be an open curve. Unlike the circle and ellipse, a point travelling along it would never return to its starting-place. The Greek geometrician Apollonius of Perga (about 220 B.C.) called such a curve a parabola, from the Greek *para* (beside) and *ballein* (to throw). He named it so from the mathematical properties of the curve, but we can see the reasoning most easily by supposing that the imaginary knife making the section was thrown exactly and evenly beside (i.e., parallel to) the far edge of the cone.

It is also possible to make a section of a cone by cutting it in such a way that the cut actually recedes from the far edge of the cone. The curve thus formed is also open, and Apollonius called it a hyperbola. The Greek prefix *hyper-* means over or beyond. It is as though the knife were thrown with the intention of making a parallel cut but had gone beyond the mark. In language we still call an extravagant exaggeration by the Greek form of the word, hyperbole.

Parallel

THE FIRST line was a textile cord, rather than a mark on paper. Line comes from the Latin *linea,* which probably comes from the old name for flax (hence our word linen for what may be man's oldest textile). Since the word straight is but another version of stretch (see ANGLE), the phrase straight line is related, in derivation and sound, to stretched linen, and it was a stretched linen cord that civilized man may have first used as a straight line in making land measurements.

Two straight lines running side by side indefinitely, neither approaching nor receding, are parallel, from the Greek words *para* (beside) and *allelon* (of one another). The Greek word for line is *gramme,* so a four-sided figure made up of two intersecting pairs of parallel lines is a parallelogram.

If the two pairs of parallel lines meet perpendicularly so that all four angles are right angles (see HYPOTENUSE), the figure is a rectangle. The prefix rect- comes from the Latin *rectus* (right); thus rectangle and right angle are different versions of the same word, even though the former refers to a closed figure, the latter to a single angle.

In general any four-sided figure, including the parallelogram and the rectangle, is a quadrilateral, from the Latin words *quattuor* (four) and *latus* (side). It is four-sided. The number four pops up again in the quadrilateral with all angles right angles and all sides equal. The Latin prefix *ex-* means out and *quadri-* is the common Latin combining form derived from *quattuor* (four). To draw a figure out of four lines (and making it, most naturally, the simplest and most regular possible) is to make it *ex quadre.* This becomes *esquarre* in Old French and square in English.

Paramecium

PROTOZOA WERE discovered by van Leeuwenhoek in 1675 (see MICROBE) but they did not receive that name until about 1818. At first the one-celled creatures were simply called animalcules, from the Latin diminutive *animalculum* (a little animal).

283

These animalcules could be easily obtained for study. It was only necessary to take some vegetable matter, steep it in water, and expose it to air. From those animalcules that might have been on the vegetable matter or in the water to begin with, or from those that were blown in, the numbers increased. Something soaking in water is called an infusion, from the Latin *in-* (in) and *fundere* (to pour); water is poured in to make an infusion, in other words. Because of their appearance in infusions, the animalcules were given the general name of Infusoria about 1763.

Today, though, Infusoria is a name restricted only to the most highly developed group of the Protozoa, to little creatures that almost ape the features of more complex organisms. There is a definite spot on the cell membrane where food is taken in and another where wastes are eliminated. The cells have definite shape and move about by the rapid and co-ordinated movement of tiny hair-like filaments all over the cell surface. These filaments are cilia, which is a Latin word meaning eyelashes. Even here, though, the name Ciliophora is replacing Infusoria. Since the Greek *pherein* means to bear, Ciliophora means cilia-bearing.

The most familiar of the ciliophores is the paramecium. The front of the paramecium is pointed and the back is rounded, while there is a constriction in the middle. In outline it has the shape of a slipper, so much so that a common name for it is slipper animalcule. Paramecium itself is a much less descriptive name, since it comes from the Greek *paramekes,* which means simply oblong.

Some non-ciliophores move by a similar method. They possess one or perhaps two long cilia-like processes that whip about and drive them along. These long processes are called flagella, a Latin word meaning whips.

Parapsychology

HUMAN BEINGS gain their awareness of the universe about them through certain senses, the chief ones being sight, hearing, smell, taste and touch. There are other senses, too. The nerve-endings in the skin do not merely react to touch; some react particularly to hot, cold or pain. There are also senses within ourselves that let us know the amount of tension on particular muscles, so that we know where every part of our body is at a given moment, or how it is oriented to the pull of gravity.

Do we know in detail all the different ways in which we might gain awareness of the universe? Are there undiscovered senses? Can we somehow detect a magnetic field, or radio waves, or cosmic ray particles, or properties of the universe we haven't yet studied, and don't know exist?

There are occasionally reports of people who become aware of something that is happening at a distance; or of something that has not yet happened. The former is clairvoyance (a French word meaning clear-sightedness), the latter precognition (from Latin words meaning to know before). There are also cases where one person seems to know what another is thinking. This is telepathy (from Greek words meaning to feel at a distance).

All such matters are lumped together as extra-sensory perception, commonly abbreviated ESP. The Latin prefix *extra-* means outside of, so that ESP is anything perceived outside the senses—at least, outside the senses as we know them. Presumably ESP is the result of senses we haven't properly analysed.

ESP came into prominence when the American psychologist Joseph B. Rhine began a scientific study of such phenomena. He wrote a book called *Extra-sensory Perception* in 1934. The scientific study of ESP is called parapsychology, where para- is a Greek prefix meaning beyond.

Parity

SUPPOSE WE GIVE each sub-atomic particle a label of either A or B. Suppose that when an A particle broke up to form two particles the products were always either both A or both B. We could then write $A = A + A$ or $A = B + B$. And if a B particle broke up to form two particles, suppose one were always A and the other B. Then $B = A + B$.

You might also find that if two particles collided and formed three, this would lead to cases of the following kind: $A + A = A + B + B$ or $A + B = B + B + B$, but never $A + B = A + A$ or $A + A + B = A + B + B$.

What does all this mean? Suppose you think of A as any even integer such as 2, 4 or 6; and B as any odd integer such as 3, 5 or 7. Two odd integers always add up to an even ($A = B + B$), and so do two even integers ($A = A + A$). An odd and an even integer always add together to form an odd one ($B = A + B$).

Parity is from a Latin word meaning equal. Originally even numbers were said to have parity, because they could be divided into two equal numbers ($14 = 7 + 7$). Later the word parity was applied to both even and odd. If two numbers were both even or both odd they were of the same parity; if one was even and one odd they were of opposite parity. Sub-atomic particles seem to act according to the rules that govern even and odd numbers, and they were said to have parity also.

The sub-atomic particles did not change parity in the course of interactions. If the parity of two or more particles added up to the equivalent of even (or odd) before interacting they added up to even (or odd) after interacting. This is called conservation of parity.

In 1956, however, two Chinese-American physicists, Tsung Dao Lee and Chen Ning Yang, showed that the even-odd arithmetic didn't hold for certain types of sub-atomic interaction governed by the weak nuclear field. For these the law of conservation of parity was broken.

Parsec

YOU WILL observe that if while looking at an object you move your head the object seems to move in the opposite direction as compared with the distant horizon. Furthermore, a distant object seems to move less than does a near-by object. For that reason a near-by tree might hide a particular house in the distance at first, and a different house after you have moved.

This apparent motion of near objects in comparison with far objects as the viewer moves is called parallax, from the Greek words *para* (beside) and *allassein* (make otherwise). In other words, with the motion of your head that which the near object was beside is made otherwise.

The most dramatic use made of the phenomenon of parallax is in the measurement of the distance of stars. As the earth moves about the sun the stars seem to move in tiny ellipses in the opposite direction. The stars are so far away that these ellipses are very small indeed.

The semi-major axis of this ellipse is the stellar parallax (the Latin word *stella* means star) and this is always less than a second. (The circuit of the sky is divided into 360 degrees, each degree into 60 minutes, and each minute into 60 seconds.) The nearest star, Alpha Centauri, for instance, has a parallax of about three-fourths of a second. Such a parallax, considering the position shift of the earth, corresponds to a distance of 40·25 billion kilometres.

This is an inconveniently large number to handle. A large unit is needed to measure astronomical distances. Light travels 299 793 kilometres per second, or 9 460 500 000 000 kilometres in a year, so the latter distance is known as a light-year. Alpha Centauri is therefore 4·25 light-years away from us.

Another way of measuring large distances is to take as a unit that distance at which a star will have a parallax of exactly one second. This distance is 3·26 light-years, and such a distance is said to be one parsec, a made-up word combining the first syllables of parallax and second.

Penicillin

PEOPLE AND animals who die of disease are buried underground, and yet the soil remains fairly free of disease germs. The bacteria and other microscopic organisms that live in soil destroy them.

In 1929 the English scientist Alexander Fleming became sharply aware of this when he noticed that spores of bread mould had got into a culture of disease germs he was growing, and that around each spore was a clear area where no germs grew. Fleming decided there was a chemical in the mould which stopped the germs. Since the scientific name of the mould was *Penicillium notatum,* he named the chemical penicillin. (He shared the 1945 Nobel Prize in medicine for this.)

When the Second World War began an Anglo-American research effort was put forth, penicillin was isolated, its structure determined and large-scale production begun. Since the War, penicillin and substances like it have largely replaced the sulpha drugs (see SULPHANILAMIDE) and a number of diseases and infections have suddenly been brought under control.

Beginning in 1940, the Russian-born American microbiologist Selman A. Waksman produced a number of bacteria-killing compounds. From moulds known as Streptomyces (from the Greek *streptos* (twisted) and *mykes* (fungus)—a fungus with twisted thread-like structures) he isolated one he called streptomycin. In 1942 he proposed the name antibiotic for such compounds, from the Greek *anti-* (against) and *bios* (life); that is, they acted against (germ) life. (Waksman received the Nobel Prize in medicine in 1952.)

Another group of moulds, the Actinomyces (from the Greek *aktis* (ray)—fungi with radiating thread-like structures) includes aureomycin (from the Latin *aurum* (gold) because of its golden colour), terramycin (from the Latin *terra* (earth) because it came from moulds that lived in the soil), and achromycin (from the Greek *achromos,* meaning without colour).

These last consist of molecules made up of four circles of carbon atoms joined together and are now called the tetracyclines, from the Greek *tettares* (four) and *kyklos* (circle).

Perfect Number

THE ANCIENT GREEKS enjoyed playing games with numbers, and one of them was to add up the factors of particular integers. For instance, the factors of 12 (not counting the number itself) are 1, 2, 3, 4 and 6. Each of these numbers, but no others, will go exactly into 12. The sum of these factors is 16, which is greater than the number 12 itself, so that 12 was called an abundant number. The factors of 10, on the other hand, are 1, 2, and 5, which add up to 8. This is less than the number itself, so 10 is a deficient number.

But consider 6. Its factors are 1, 2 and 3, and this adds up to the number itself. The Greeks considered 6, therefore, a perfect number. Throughout ancient and medieval times only four different perfect numbers were known. The second was 28, the factors of which are 1, 2, 4, 7 and 14. The third and fourth perfect numbers are 496 and 8,128. The fifth perfect number was not discovered till 1460 and the name of the discoverer is not known. It is 33 550 336.

There are no practical uses for the perfect numbers; they are merely a mathematical curiosity. Mathematicians, however, are curious, and have worked out formulas that will yield perfect numbers if certain conditions are met. If even the fifth perfect number is over thirty million, those still higher are, you can well imagine, terribly tedious to work out.

The break came with the development of computers during and after the Second World War. To demonstrate what computers could do, one could set them to work solving formulas for perfect numbers. By now twenty-one perfect numbers are known. The twenty-first perfect number, worked out in 1971, is a number with twelve thousand and three digits.

Such a number has no more practical use than any smaller perfect number, but wouldn't the Greeks have been astonished if they could have seen it!

Perihelion

THE ANCIENT Greeks always insisted that heavenly bodies moved in orbits that were exact circles, because the circle was the perfect curve and all things heavenly were perfect. The very word orbit comes from the Latin word *orbis,* meaning circle.

In 1609, however, the German astronomer Johann Kepler finally showed that planets moved about the sun in ellipses, not in circles, and that the sun was not in the centre but at one of the foci of the ellipse (see ELLIPSE).

Since the sun was out of centre, it meant that at some points in their orbits the planets would be closer to the sun than at other points. The point at which the planet is nearest the sun is the perihelion, from the Greek words *peri,* meaning around, and *helios,* meaning sun. It is a poor word unless you take around as meaning in the neighbourhood of, in which case perihelion is the point at which the planet is in the neighbourhood of the sun. Similarly, the point at which a planet is farthest from the sun is the aphelion, the prefix coming from the Greek word *apo,* meaning from. At aphelion a planet is away from the sun.

The ellipse of the earth's orbit is of low eccentricity, so that at perihelion the earth is only 3 per cent closer to the sun than at aphelion.

Similar prefixes are used when an orbit is about some body other than the sun. For instance, the moon travels about the earth in an ellipse with the earth at one focus. The point of closest approach of the moon is perigee, the suffix *-gee* coming from the Greek word *ge,* meaning earth. The point of the moon's greatest distance from the earth is apogee.

There are many known instances of two stars circling about their common centre of gravity. Each travels in an ellipse with the centre of gravity at one focus. The more massive star travels in the smaller ellipse. The point where the two stars approach most closely is the periastron (from the Greek word *astron,* meaning star) while the point of greatest distance is the apastron.

Perturbation

THERE IS something innately disorderly about a crowd. Even if they are perfectly peaceable, the mere fact that some are heading in one direction and some in another makes for disorder. Or if they are all standing still there are always some turning heads one way, some mopping their brows, some fidgeting.

The Latin word for crowd is *turba,* and a number of words implying disorder contain that as a stem. Water which has been stirred up so that mud from the bottom is mixed with it is turbid. Water being whipped up by waves (or emotions being whipped up by passion) are turbulent. A person who is upset and in disorder is disturbed or perturbed. A kind of flatworm that agitates the water as it moves is called a turbellarian.

Now, the law of gravity can only be applied exactly when no more than two bodies are involved. For instance, the earth travels about the sun in a course that would be exactly determined if the earth and sun were the only objects in the universe. However, the moon pulls at earth a bit and so do Mars and Venus and, in theory, all other bodies in the universe.

Fortunately, the effects of these other bodies on the earth is minor compared to that of the sun. So earth's orbit can be calculated as though the other bodies did not exist and then their minor effects can be worked out by noting the manner in which earth departs from its calculated orbit.

These minor gravitational effects of astronomical bodies upon each other, which cause a departure from the stately order of travel according to the two-body calculation, and set up, so to speak, a certain amount of disorder, are called perturbations. These had their most dramatic moment in astronomical history when perturbations in the orbit of Uranus which could not be explained by the effect of the other known planets led to the suggestion by John Adams of England and Urbain Leverrier of France that an unknown outer planet lay beyond Uranus, and could be found at that time in a certain spot of the heavens. Theory was verified by observation, and in 1846 Neptune was discovered.

Pesticide

ANIMALS COMPETE with each other for food, and man is included in this contest. Man gained an advantage when he invented agriculture. He grew the plants he desired, such as various grains, in greater quantities and more thickly than they occurred in nature. Other plants which grew in the area he found undesirable, and uprooted.

But the existence of solid banks of a particular plant served also to increase the numbers of those non-human creatures who fed on it. In a state of nature insects which feed on the occasional clumps of growing grain would be moderate in number. With acre after acre of nothing but grain, such insects would have an enormous food-supply for the taking, and they would multiply tremendously. Farmers had to fight hordes of insects that would not have existed in such numbers before the days of agriculture. They would also have to fight fungus infections, or small animals such as rabbits, moles and crows, who would want to feed on the plants he was growing. When he began to herd animals he had to fight off foxes, weasels, wolves and other predators.

All the living creatures—plant, animal, and microscopic—that competed with man were dangerous to his food-supply and sometimes to his own health. Rats not only preyed on his stores of grain but carried fleas and lice, which transmitted typhus, bubonic plague and other diseases.

In recent times man has developed chemicals that kill the competing creatures while leaving desirable organisms comparatively unharmed. There are herbicides, for instance, from Latin words meaning plant-killers, to remove weeds. There are also insecticides, fungicides and so on. Undesirable organisms can be lumped together as pests, from a Latin word meaning plagues. In consequence, chemicals that kill undesirable creatures of any sort have been lumped together now as pesticides.

Phalanges

PERHAPS THE greatest military genius among the ancient Greeks was Epaminondas of Thebes. Until his time Greeks had always lined up their infantry in straight lines several ranks deep and let them clash in battle. The Spartans, with the best-trained infantry, invariably won, if they were involved.

Epaminondas, however, set up the right flank of his army some fifty ranks deep. The centre and the left flank, each very light, was set back so that the heavy flank struck first. Its sheer weight smashed through the opposing line, like a large log of wood swinging against a door, and completely disorganized the enemy. This tactic was first used at the battle of Leuctra in 371 B.C., where Epaminondas defeated the Spartans completely, and ended their supremacy.

The Greek word for such a mass of infantry—*phalanx*—is of uncertain derivation, but it is also their word for log, and perhaps the infantry mass was likened to a swinging log, as I have done in the paragraph above.

In any case, Philip II of Macedon (a semi-Greek borderland) was a hostage at Thebes during the time of Epaminondas, and later as a monarch remembered what he had seen. He adopted and improved the phalanx, lightened it and increased its manoeuvrability, supported it well with cavalry, and equipped each man with a long spear so that the phalanx bristled like a porcupine. With it he conquered the Greeks, and his son, Alexander III (the Great), conquered the Persian Empire.

The most noteworthy thing about the phalanx was the close order maintained by the men; ranks and files kept together. This same close order is maintained by the bones in the fingers and toes. In each finger and toe are three small bones (two only in thumbs and big toes), one immediately behind the other. There are five sets of them, side by side, in each hand and foot (56 bones altogether) and because of their close order they are known to anatomists as phalanges, a plural form of phalanx.

Pheromone

HUMAN BEINGS can communicate by talking. Through sounds, gestures, written symbols, abstract ideas can be transmitted from one person to another. Human beings are unique in this respect.

Yet other creatures must be able to communicate in some fashion, if only so that there can be co-operation between two individuals of a species in order that they might reproduce. Within a body, the different parts are made to behave in some co-operative fashion by means of chemical messengers called hormones. Is it possible that chemical messages can be carried on, not only within an organism, but from one organism to another?

Such hormonal effects, carried through water or air from one member of a species to another, are called pheromones, the prefix coming from a Greek word meaning to carry. They are hormones carried over a distance.

Insect pheromones are the most dramatic. A female moth can liberate a compound which will act as a powerful sexual attractant on a male moth of the same species two kilometres away. Each species must have its own pheromone, for there is no point in affecting a male of another species. Each species must have receiving devices of tremendous delicacy because they must be able to react to just a few molecules in the air.

Pheromones are also used in inter-species conflict. Certain ants raid the nests of other ant species, kidnapping young, which they rear as slaves. The raiders use trails of pheromones which not only aid them in keeping together and co-ordinating their attacks, but also act to alarm and scatter the ant species they are attacking.

Biologists are endeavouring to use insect pheromones to lure members of troublesome species to destruction. In this way, they can be absolutely specific, doing no direct harm to any other species.

Phobos

IN THE LATTER half of the 1800s, as far as was known, the earth had one moon; Mars (next out from the sun) had none; Jupiter (next out) had four, and Saturn had eight. To make a perfect sequence, Mars should have had two moons; then it would be: 1, 2, 4, 8.

Of course, astronomers didn't take that sort of number-juggling seriously, though some people did. In *Gulliver's Travels,* written in 1726, the inhabitants of a mythical land named Laputa were supposed to have discovered two moons of Mars by means of superior telescopes, and Swift described the moons in considerable detail.

Then, in 1877, when Mars was at a point in its orbit close to earth, the American astronomer Asaph Hall decided to make a systematic search for any satellites Mars might have. He searched methodically, night after night, until it seemed certain there was nothing. He went home discouraged one day, having given up, when as the story has it Mrs Hall persuaded him to go back and try one more time.

And, as often happens in fiction, but practically never in real life, that one more time clinched it. Hall spotted something. He had to wait for the end of a spell of cloudy weather to check it, and when he did, there was not one satellite, but two. What's more their smallness, their closeness to the planet, their speed of revolution were all very unusual and yet all very close to what had been described by Swift. Certainly Swift's was probably the most inspired guess in all literature.

Naming the Martian satellites was easy. The Greeks had Ares (the Roman Mars) as god of war, attended by his two sons, Phobos (Greek for fear) and Deimos (Greek for terror). Hall called the inner satellite of Mars Phobos and the outer Deimos, so that Mars in the sky, as well as in myth and in reality, is attended perpetually by Fear and Terror.

But alas, the sequence of moons was spoiled anyway. In 1898 a ninth moon of Saturn was discovered, and in 1901 a fifth moon of Jupiter. As a matter of fact, twelve moons of Jupiter are now known.

Phosphor

SUBSTANCES WHICH ABSORB energy in some non-light form and then give it off as visible light are said to be luminescent, from Latin words meaning to become light (see ELECTROLUMINESCENCE). Sometimes there is a delay in giving off the light, and emission continues even after the energy source is shut off. This delayed luminescence is called phosphorescence, which also means to become light, the prefix in this case coming from the Greek word for light. Luminescence without delay is called

293

fluorescence, because it was first noticed in connection with a mineral called fluorite.

Substances which display phosphorescence are called phosphors. Examples are calcium tungstate and zinc sulphide, the phosphor qualities of which depend on their method of preparation, and also on the presence of certain impurities.

Fluorescence came into important commercial use through the researches of the French chemist Georges Claude, beginning in 1910. He showed that electric discharges through certain gases under low pressure could cause them to fluoresce. By putting them into glass tubes that could be twisted into various shapes, he developed neon light (neon was one of the gases used) in various colours.

Mercury vapour would also fluoresce under such conditions, producing radiation rich in ultra-violet light. In 1935, after methods for producing phosphors in quantity were developed, tubes coated on the inner surface were used. The ultra-violet light from the fluorescing mercury vapour excited a phosphorescent glow in the phosphor coating, with the result that a steady white light was given off. Such fluorescent lights consumed less electric energy for a given amount of light, and emitted light of greater whiteness and softness, accompanied by far less heat, than could ordinary incandescent bulbs.

Phosphorus

THE PLANET Venus is closer to the sun than we are. As a result, when viewed from earth it sometimes appears to the east of the sun, sometimes to the west; but in neither case does it get very far from the sun.

When Venus is east of the sun it reaches the western horizon later, so that for a time after the sun sets it remains glowing brightly in the western sky. It is then the evening star, which the Greeks called Hesperos (evening). In the morning Venus reaches the eastern horizon later than the sun; by the time Venus rises the sun is in the sky and Venus's light is drowned out.

When Venus is to the west of the sun it reaches the western horizon sooner than the sun, and by sunset it is gone. However, in the morning Venus reaches the eastern horizon sooner than does the sun, and for a time before the sun rises it glows brightly in the eastern sky. It is then the morning star, which the Greeks called Phosphoros. This word comes from *phos* (light) and *phoros* (bearing). The morning star was the bearer of light, for once it appeared the sun could not be far behind.

Since Venus cannot be both east and west of the sun simultaneously, the evening star and morning star can never appear on the same day. When the evening star is in the sky there is no morning star, and vice versa. This gradually dawned on the Greeks, and they came to realize that the two stars were one planet, which they then named the star of Aphrodite, the goddess of love, called by the Romans Venus.

The word Hesperos (Latin, Hesperus) does not appear in English except in poetry, but Phosphoros has taken on a new and permanent lease of life. In about 1669 a German alchemist named Hennig Brand isolated a waxy white substance from urine, which glowed in the dark. (It combined with oxygen and the energy of combination liberated light, but Brand did not know that.) Because it was a light-bearer, he called it phosphorus, and what was once the name of the morning star is now the name of a chemical element.

Photosynthesis

ALL ANIMALS eat food. They take large food molecules apart into the small units that compose them, absorb those small units, and put them together again into the large molecules that compose their own tissues. In doing so, however, 90 per cent of the original food is used to produce energy for life processes or else is wasted outright.

If animals ate only each other there would be constant attrition, and soon all would be dead. However, while some animals eat other animals, most animals eat plants. The wastage of animals eating animals is made up for by the fact that the animals that are eaten have usually built their own tissues at the expense of plants. Of course, the eating of plants is wasteful, too. Most of the plant tissue does not end up as animal tissue but is consumed for energy or ends as excreta.

How, then, do plants keep their own tissue volume up to the mark despite the day-in, day-out ravages of hundreds of thousands of species of animals?

Plants use as their raw materials water and carbon dioxide, which are among the excretion products of animals, and which are universally available. For the energy required to build up the large tissue molecules out of the small molecules of water and carbon dioxide (plus certain minerals), sunlight is used. Green plants contain a compound known as chlorophyll (from Greek words meaning green leaf), which absorbs sunlight and makes its energy available for certain chemical changes. This process is called photosynthesis (from Greek words meaning putting together by light).

Though the basic existence of photosynthesis has been known for two hundred years, it was only in the 1950s that the American biochemist Melvin Calvin, making use of radioactive isotopes, actually began to work out the detailed nature of the chemical changes involved.

Phylum

ONE WAY of dividing the animal kingdom is to include those animals which, like ourselves, have a bony skeleton under the heading vertebrates. Although these skeletons may differ in detail from animal to animal, they all include a backbone. The individual bone of a backbone is a vertebra (see VERTEBRA), and hence the name vertebrate.

All other animals may then be dismissed as invertebrates; that is, having no internal bony skeleton.

However, this is not a logical system of classification. Among the invertebrates there are groups of animals which differ from one another as much as do any of them from the vertebrates. Zoologists have therefore divided the animal kingdom into a number of groups, each called a phylum (from the Greek *phylon* meaning tribe). All the animals included within a phylum have the same general body plan, varying only in detail. Thus the vertebrates make up a single phylum, because all have a bony skeleton, a front and back end, no more than four limbs, and so on.

The invertebrates, however, are divided into more than twenty phyla, each with its own body plan. Some are small groups including a few odd animals known only to zoologists. Others are rich collections of animals more numerous and various than those of the vertebrates.

For instance, insects, spiders, crabs, lobsters and similar creatures are all assigned to the phylum Arthropoda, from the Greek *arthron* (joint) and *pous* (foot). They all have jointed limbs and bodies, too. They are covered by a shell which is also jointed.

Snails, slugs, oysters, clams, and squids, on the other hand, are members of the phylum Mollusca. These also have shells (or sometimes the remnants of one) but their shells have a different chemical composition from those of arthropods, and are not jointed. Within the hard shell is a soft, gooey creature (have you ever eaten oysters?) and Mollusca comes from the Latin *molluscus,* meaning soft.

Pi

ONE OF THE FIGURES in geometry easiest to construct is the circle. Since the distance from the centre to all points on the circle is equal, all you need for construction is a compass.

The Greeks discovered that the curve of the circle is a little over three times as long as the width of the circle. This fact is always true, regardless of how large or small the circle is. The Greeks tried to determine the exact ratio of the perimetron (meaning the measurement around) to the diametron (meaning the measurement through). In English these words become perimeter and diameter. The best the Greeks could do was the estimate by the mathematician Archimedes in the third century B.C. that the ratio was higher than $3\frac{10}{71}$ but lower than $3\frac{10}{70}$, or roughly about $3 \cdot 142$.

If the diameter is set equal to 1, then the perimeter of the circle is equal to whatever the ratio is ($3 \cdot 142$). About 1600 the English mathematician William Oughtred invented the symbol π to stand for that ratio. It is the first Greek letter in the word perimetron, and is called in English pi. Ever since pi has been universally used as the symbol for the ratio of perimeter (or circumference, the Latin equivalent, meaning to carry around) to the diameter.

The value of pi turned up in all kinds of mathematical equations, and not only in geometrical formulas. There were numerous attempts to get its exact value. By 1717 an English mathematician, Abraham Sharp, had worked it out to seventy-two decimal places, of which most people don't bother remembering more than $3 \cdot 14159$ . . . Actually, no exact value is possible. The decimal goes on for ever.

In 1955, however, a computer calculated the value of pi up to 10017 decimal places in thirty-three hours. Such a value has no practical use, but is of interest to mathematical theorists.

Piezoelectricity

IN 1880 two French physicists, the brothers Pierre and Paul-Jacques Curie, discovered that when certain crystals were compressed an electric potential was developed across them. The harder the pressure, the greater the potential. The Curies called this piezoelectricity. The prefix was from the Greek word meaning to press.

It was found to work the other way, too. If the crystals were placed under an electric potential, they pressed together slightly and became constricted. This is called electrostriction.

Later on, Pierre Curie married a Polish chemist, Marie

Sklodowska, who was to become famous as Madame Curie. She made use of her husband's discovery in early studies of radioactive substances. Under certain conditions, uranium, in breaking down, could produce ions that would cause a tiny electric flow. Madame Curie measured this flow by balancing it against an electric flow in the opposite direction produced by pressing a crystal. The amount of pressure required for balance gave the measure of the electric flow.

If a crystal is put under the influence of a very rapidly oscillating alternating current the crystal faces move in and out just as rapidly and produce sound waves. If the in-and-out movement is fast enough sound waves are produced of such high frequency and such short wave-length that they are ultrasonic. On the other hand, sound waves striking the crystal will produce an alternating current.

Piezoelectric crystals can be used, therefore, to connect electricity and sound. They can be used in microphones to pick up sound and convert it into an electric current, which can then be amplified and turned back into sound. The most common crystals used for this purpose are quartz. Others are ammonium dihydrogen phosphate, barium titanate and Rochelle salt.

Pinch Effect

ONCE THE HYDROGEN BOMB was invented men began to wonder if its energies could be controlled. The hydrogen bomb works by the fusion of four small hydrogen atoms into a large helium atom, so what was wanted was controlled fusion. In order to have that the hydrogen had to be kept together and made to undergo fusion at a slow and steady rate. If this could be done man would have an enormous energy source that would last millions of years.

A mass of hydrogen is kept together and allowed to undergo steady fusion in the sun and in other stars, but what keeps it together in those cases is an enormous gravitational field. Such a field cannot be produced in the laboratory, and other containing methods must be used.

An ordinary container won't work, for hydrogen will not undergo fusion until temperatures of more than a hundred million degrees are reached. At this stage no material body could possibly contain it. Either the material body would melt and the gas escape, or the gas would cool down upon touching the container. In either case fusion would stop.

At very high temperatures, however, hydrogen atoms break

down into a mixture of electrically charged electrons and protons. Charged particles respond to a magnetic field. If such a field surrounded a tube containing the hydrogen the particles would be pushed inward. They wouldn't touch the walls of the tube, but would be contained in an invisible magnetic bottle. Because the gas is pinched inward, the American physicist Lewi Tonks, who worked out the theory in 1937, called it the pinch effect.

A cylindrical magnetic field does not produce a stable pinch effect. The contained particles escape within a few millionths of a second. In order to produce the conditions for controlled fusion, physicists have therefore been working with magnetic fields of more complicated shape.

Pineal Gland

A CONICAL REDDISH-GREY BODY attached to the base of the brain is called the pineal gland because it somewhat resembles a pine-cone in shape. It is quite small, being only about 0·6 centimetres long in man.

In the early seventeenth century the pineal gland attracted considerable attention. The French scientist René Descartes was under the impression that the pineal gland was found only in humans and never in other animals. He maintained that this small scrap of tissue was the seat of the human soul. It was soon found, however, that the pineal occurs in all vertebrates and is far more prominent in some of them than in man.

It was recognized in more modern times as merely a gland, and in the late 1950s biochemists at the University of Oregon began work with 200 000 pineal glands obtained from slaughtered cattle. They finally isolated a tiny quantity of substance that on injection lightened the skin of a tadpole. It seemed that the pineal produced a hormone, which they named melatonin from Greek words meaning intensity of darkness. It is not known whether this hormone is of importance in the human body.

Still more exciting was the discovery in the 1960s that the pineal gland might be involved in the establishment of circadian rhythms, the way in which the activities of living tissue rise and fall with the alternation of day and night. The pineal was not always hidden deep in the head, as in most modern-day vertebrates. There was a time when the pineal was raised on a stalk, and reached the top of the skull. Then it had some of the structure of an eye. A primitive reptile on some small islands near New Zealand—the Sphenodon or tuatara—has such a pineal eye even today. It is particularly pronounced when the reptile is young. The pineal eye cannot really see, but it can detect the presence or absence of light, and could therefore initiate the circadian rhythm. How the pineal works in those forms of life where it is deep in the skull is still uncertain, however.

Pion

IN 1935 the Japanese physicist Hideki Yukawa advanced theoretical reasons for supposing that particles of intermediate size (with masses greater than that of an electron but less than that of a proton) ought to exist. In 1936 the American physicist Carl D. Anderson discovered such a particle, and it was eventually named a meson, from the Greek word meaning intermediate.

For a while it looked as though Yukawa's theory had been confirmed, but then some troubles arose. Yukawa's theory predicted the meson would be 270 times as massive as an electron, but Anderson's meson was appreciably less massive than that. Then again, according to Yukawa's theory, the meson would interact with atomic nuclei very rapidly, but Anderson's meson would not interact with nuclei at all.

In 1949 an English physicist, Cecil Powell, was studying cosmic ray particle tracks in the upper Andes. These struck atoms in the atmosphere, and produced whole showers of particles. Powell wanted to get as high as possible, hoping he would detect some cosmic-ray particles before they had slammed into atoms of the atmosphere. These original particles were called primary radiation.

In his studies, however, he uncovered tracks of particles of intermediate size. What's more, they weren't Anderson's mesons at all; they were distinctly more massive. Indeed, they were 273 times as massive as electrons, almost exactly what Yukawa had predicted. And they interacted strongly and rapidly with atomic nuclei. Powell's mesons were the particles that Yukawa had predicted.

To distinguish them from Anderson's mesons, the new particle was given the initial p (for primary radiation, which Powell had been studying). The Greek form of the letter (pi) was used, and the new particle was called the pi-meson, or pion for short.

Pithecanthropus

THE GREEK word for man is *anthropos,* and this shows up at once in the name of that science which concerns itself with the human species, anthropology. Again, those apes which most closely resemble man are called the anthropoid apes (the manlike apes).

However, the most dramatic use of the word occurs in connection with certain extinct creatures that are not exactly men and yet are closer to the human than is any living ape. On the basis of the different species of such ape men that have been found, anthropologists have tried to trace the possible route by which true man has developed.

These antique specimens are often called familiarly after the regions in which the remains were discovered; for instance, we have Peking man, Java man, Heidelberg man, Rhodesian man, and so on. However, anthropologists have tried to name them in Latin by genus and species, as they have done with other creatures, both living and extinct.

For instance, Peking man, an ancient form of 'ape-man', was formerly given the scientific name of *Sinanthropus pekinensis*. The prefix Sin- is a version of Chin- (as in Sino-Japanese war, for instance), so the name meant China man of Peking. Java man, one of the first of the ape-men to be found, was named as long ago as 1891, and given the title of *Pithecanthropus erectus*. Since the Greek *pithekos* means ape, the name actually meant ape-man, the ape-man who stands erect. More recently, however, it has been decided that Java man and Peking man are variants of the same creature, and that this was not sufficiently different from true man to be assigned to another genus. The generic name for man is the Latin word Homo, so these types have been termed *Homo erectus*. What appears to be a genuinely pre-human genus has, however, been discovered in South and East Africa, and to this has been given the name Australopithecus, or southern ape (Latin *australis*, southern, Greek *pithekos,* ape.

Pituitary

BEFORE THE nineteenth century no one really understood the function of the brain. The best the ancient Greeks could do was to suppose that it acted as an air-conditioner, cooling the overheated blood, or that it produced the mucus or phlegm which is the thickish fluid that coats the inner membranes of nose and throat, and is most noticeable when we have a cold. (*Mucus* is a Latin word and *phlegma* a Greek one. There is an Anglo-Saxon word too, which, however, is not used in polite society.)

Presumably, the Latins didn't like their own word, either, and used *pituita* instead, which they may have felt to be politer. It may perhaps be related to the Greek *ptyein* (to spit) and an unpleasant word always seems nicer if it is in a foreign language. In any case, the phlegm-producing function of the brain was centred in a small organ (they thought) attached to the bottom of the brain by a thin stalk, and that organ was therefore called the pituitary.

It amounts to a kind of Cinderella story, then, to point out that this small organ with the rather repellent name has turned out to be the master gland of the body. It produces a series of hormones which control the action of other glands. It has a less ludicrous name, the hypophysis (a Greek word meaning growing under; it grows under the brain), introduced about a hundred years ago by the American anatomist Burt Green Wilder, but it is almost always called the pituitary.

Many of the pituitary hormones have the stem troph in their names, from the Greek *trephein* (to cause to grow), since they seem to feed other glands and improve the yield of other hormones. This stem has been corrupted to trop, however. Thus one of the hormones is adrenocorticotropic hormone (which feeds the adrenal cortex and improves the yield of cortical hormones; see CORTISONE). This name has been abbreviated to ACTH, and in this guise the hormone has gained considerable newspaper fame as one of modern medicine's miracle drugs.

Planet

EVEN BEFORE the dawn of written history, men watching the stars at night must have noticed that they never changed their pattern. A group of stars forming a lopsided W kept that shape night after night, lifetime after lifetime. These were the fixed stars. As a group the stars moved about a point in the northern skies. Those far enough away from that central point rose and set, as did the sun and moon.

Each night the entire star-pattern shifted a little. Each star rose four minutes earlier and set four minutes earlier than the night before, so that stars in the west gradually disappeared behind the horizon while other stars appeared in the east. After a full year the circle was complete and the original pattern was again in the night sky.

There were, however, five star-like objects, as bright or brighter than the brightest stars, that did not stick to the pattern. One of these objects might lie midway between two stars on a certain night, for instance. The next night it would have shifted its position; the next night it would have shifted farther; and so on. Three of these star-like objects (we call them Mars, Jupiter and Saturn) made a complete circuit of the sky, by a rather complicated route. The other two (Mercury and Venus) did not move more than a certain distance from the sun.

These objects, in other words, were not fixed stars, but unfixed ones. They wandered among the stars. The Greek word for a wanderer is *planetes,* and so the Greeks called these bright objects by that name, and it has come down to us as planets.

In ancient and medieval times the sun and moon were numbered among the planets, because they wandered among the stars too. By the seventeenth century, however, the fact was accepted that the sun was the centre of the solar system, and a planet then became any astronomical body that revolved about the sun. The sun itself was no longer a planet, then, but the earth became one. The moon too is not a planet since it revolves about the earth primarily, and only secondarily about the sun.

Plankton

LIKE THE animal kingdom, the plant kingdom is systematically divided into groups and sub-groups. Some classifications give a grand division into two sub-kingdoms, the more primitive of which is Thallophyta. These include all the one-celled plants, plus related many-celled plants in which the individual cells are comparatively little specialized.

The largest of these, and an example sure to be known even to people without microscopes, are the seaweeds. These consist of mere undifferentiated shoots (in Greek, *thallos*) and lack roots, leaves or true stems. The Greek *phyton* means plant, so Thallophyta means shoot-plants.

The Thallophyta are divided into a number of phyla (see PHYLUM) which in turn fall into two groups, those which include plants with chlorophyll (see CHLOROPHYLL) and those which include plants without. The former are called algae (singular, alga), which is the Latin word for seaweed, an alga that can be seen with the naked eye. The latter are fungi (singular, fungus), which is the Latin word for mushroom, a fungus that can be seen with the naked eye. (Bacteria are often included among the fungi.)

The plant life of the oceans (which makes up about 85 per cent of all the flora on earth) belongs entirely to the Thallophyta. The algae of the oceans manufacture their food with the aid of sunlight and so must exist only in the topmost layers of the ocean where light can penetrate. There they float, drifting with the current and serving as food, directly or indirectly, for all the animal life of the sea. These algae are phytoplankton. The Greek *planktos* means wandering, so phytoplankton are plant-wanderers, because they must drift (wander) with the ocean currents. Small animal life that also drifts in the currents makes up the zooplankton. The Greek *zoon* means animal, so zooplankton are animal-wanderers. Together phytoplankton and zooplankton are referred to simply as plankton.

There are also animals in the surface waters which are independent of current and swim as they please. These are nekton, from the Greek *nektos* (swimming).

Plantigrade

THE LOWER surface of the foot is the sole, a word which comes from the Latin *solea* (sandal); the sole is the part of the body covered by a sandal. Man walks on the entire sole of his foot, putting the foot down on toes and heels with each step. We are therefore plantigrade, from the Latin *planta* (sole) and *gradior* (to walk). We are sole-walkers. Actually, though, only a minority of mammals walk in this way; we, the apes, bears and raccoons are the chief examples.

Most mammals walk on their toes only. The heels are lifted high, and in normal walking never touch the ground. Walking on toes has the advantage of adding the length of the foot to that of the leg. The animal is higher and can see farther; with longer legs, it can also run faster. These mammals are digitigrade. The Latin *digitus* means toe, so they are toe-walkers.

The extreme occurs in those mammals that lift up to the very toe-tips, which are sheathed in enlarged nails called hoofs (see UNGULATE). These are mostly animals which depend on running for safety (though the fastest land-runner, as it happens, is the cheetah, a large cat, and certainly without hoofs).

The hoofed mammals are divided into two orders, depending on the number of hoofs. Those with two hoofs on each leg (cattle, sheep, goats, swine, deer, etc.) belong to the Artiodactyla, from the Greek *artios* (even) and *daktylos* (toe); they are the even-toes. Those with an odd number of toes (horses, donkeys, zebras, with one hoof per leg; rhinoceroses and tapirs with three) belong to the Perissodactyla, from the Greek *perissos* (odd): the odd-toes.

The Greek words for odd and even show signs of having originated from the necessity of sharing things. An even number of anything can be shared exactly, and the Greek *arti* means exactly; hence *artios* is an even number. An odd number when shared always result in one item left over. The Greek *peri* means over, hence perissos for an odd number.

Plasma Physics

ORDINARILY three states of matter are recognized: solid, liquid and gas. In solids, atoms or molecules are in virtual contact and maintain their position. In liquids the atoms or molecules are still virtually in contact, but they slip and slide past each other freely and do not maintain any given position. In gases the atoms or molecules are virtually free of each other and move independently, with large spaces between. In all three states the atoms (whether singly or in molecular groupings) are whole and complete.

As the temperature rises the atoms themselves begin to break down, losing electrons, which carry a negative electric charge, and leaving behind atom residues that carry a positive electric charge.

Once a gas is largely or entirely composed of electrically charged particles, its properties are different in many ways from ordinary gases. It can be manipulated by magnetic fields, for instance, as ordinary gases cannot. It is in fact a fourth state of matter.

The American chemist Irving Langmuir chose a name for this fourth state in the early 1930s, and chose a rather poor one. In Greek the term plasm was used for something that had a form, a definite shape. In the nineteenth century the German anatomist Max Schultze used the term for the liquid part of blood, which had no shape but within which objects with a shape were to be found. This liquid part has been called blood plasma ever since, directly reversing the meaning of the word. Langmuir compounded the error by applying plasma to the formless mass of charged particles.

In trying to control hydrogen fusion physicists must work with gases at extremely high temperatures of a hundred million degrees or more. The plasma that results must be held in place by magnetic fields. Plasma physics has thus become an extremely important branch of science, one on which mankind's future energy needs will depend.

Pleasure Centre

SINCE THE BRAIN is the controlling organ of the body, it was a natural belief that different parts of the brain might control different parts of the body. Some thought that by studying the bumps on the skull one might determine over-developed portions of the brain, and deduce specially prominent character traits. This enticing but worthless thought gave rise to the folly of phrenology (from

Greek words meaning study of the mind) in the nineteenth century.

A closer study came in 1870, when two German physiologists, Gustav Fritsch and Eduard Hitzig, stimulated different portions of the cerebral cortex of a dog in order to note what muscular activity (if any) resulted. It was also possible to destroy a patch of the cortex and take note of what paralysis might or might not result. In consequence parts of the brain were indeed associated with particular muscle systems. Other parts of the brain were associated with sensations received from particular nerve-endings.

Something a little more subtle had come about in 1861, when a French surgeon, Pierre Broca, had noted that a particular area of the brain controlled not only the ability to speak but also the ability to understand speech. Damage there caused aphasia (from Greek words meaning no speech).

In 1954 the American physiologist James Olds discovered something still more astonishing, a specific region in the brain which on stimulation apparently gave rise to a strongly pleasurable sensation. An electrode fixed to the pleasure centre of a rat, which the animal could stimulate itself, was stimulated up to 8000 times an hour for days at a time, to the exclusion of food, sex and sleep. All the desirable things in life may be desirable only in so far as they stimulate the pleasure centre. Direct stimulation would make all else unnecessary.

Pleistocene

THE MAIN eras of the earth's history (see FOSSIL) were divided into sub-groups with names often derived from the places where the rock strata belonging to a particular period of time were first studied. For instance, rock strata belonging to the Palaeozoic era were mainly studied in Wales, so the first three sub-eras were named the Cambrian, Ordovician and Silurian periods after Cambria, the old Latin name for Wales, and after two of the Celtic tribes that had lived there in pre-Roman times, the Ordovices and the Silures. The Cambrian period is the earliest from which definite fossils are obtained. Though there are dim evidences of life before that, people who are interested in fossils often lump all the epochs before that as Pre-Cambrian and let it go at that.

The next division of the Palaeozoic era is the Devonian, from Devon. This is followed by the Carboniferous period. This breaks the place-name habit, since it is derived from the Latin *carbo* (coal) and *ferre* (to carry). It was the era during which our modern coal-beds were laid down, so it carries the coal. The final period of the Palaeozoic is named after the region of Perm in the Ural Mountains of the Soviet Union.

The second great geological era, the Mesozoic, is customarily divided into the Triassic (from Greek *treis* or *tria,* three, as being in three divisions); the Jurassic, from the Jura Mountains of France and Switzerland, and the Cretaceous, from the Latin *creta,* chalk.

The most recent of the eras in time is the Cainozoic, which is divided on a different plan into the Palaeocine, Eocene, Oligocene, Miocene, Pliocene, Pleistocene and Holocene periods. The suffix -cene comes from the Greek *kainos* (new) while the prefixes are derived from the Greek *palaios,* (ancient), *eos* (dawn), *oligos* (few), *meion* (less), *pleion* (more), *pleistos* (most), and *holos* (whole). The periods, in other words, are: Palaeocene, the ancient new; (2) Eocene, the dawn of the new; (3) Oligocene, a few bits of the new; (4) Miocene, less than half of the new; (5) Pliocene, more than half of the new; (6) Pleistocene, most of the new; and (7) Holocene, the whole of the new. The first five periods are commonly referred to as the Tertiary (Latin *tertius,* third), and the last two as the Quaternary (Latin *quater,* fourth). These are the two great divisions of the Cainozoic: the Quaternary was formerly considered an era in its own right.

Polarized Light

LIGHT CONSISTS of waves which undulate in all directions. When light passes through certain transparent crystals the regular rank and file of atoms in the crystal make it possible for light to undulate in only two directions, at right angles to each other.

These two wave forms behave differently as they pass through the crystal. Both are bent, or refracted (from Latin words meaning to break back), but by different amounts. One beam goes into the crystal but two come out, in a phenomenon called double refraction.

Sir Isaac Newton knew of the phenomenon of double refraction, but thought that light consisted of small particles rather than of waves. In an attempt to explain double refraction, he speculated that the particles consisted of two types which were separated in doube refraction. If they were, this might be analogous to the north and south poles of magnets.

In 1808 a French physicist, Etienne Malus, studied double refraction and called each separate beam, consisting presumably of light of a single pole, polarized light. This was a poor name, for Newton's notion was soon shown to be wrong, but it stuck.

Reflection often polarizes light, and it would be handy if one could find a substance which would let through light polarized in one direction but not light polarized in the other. That would eliminate much of the reflected glare without too much general darkening. Some organic crystals could do this but they were too fragile to use as spectacles, for instance. In the 1930s, though, a Harvard undergraduate, Edwin H. Land, conceived the idea of getting the organic crystals properly lined up and then embedding them in clear plastic. The plastic once it hardened would do the job and be strong enough to be used as spectacles. The product was known by the trade name Polaroid (not to be confused with the Polaroid camera, also invented by Land).

Pollution

EVERY ORGANISM produces waste products no longer useful to itself and which if allowed to accumulate would in fact be harmful. In every case, though, the waste products are of use to other forms of life, which often restore them to a form useful to the original waste-producer.

Thus all animals make use of oxygen in air, or dissolved in water, combining it with the carbon in foodstuffs, and excreting carbon dioxide as a waste product. No animal can live in an atmosphere with too great a carbon dioxide content. Plant life, however, can utilize carbon dioxide and, using the energy of sunlight, build it up to foodstuffs again, excreting oxygen as a waste product. Of course, animals can make use of the oxygen again.

In this way there is an oxygen-carbon dioxide cycle, and through the activities of both plants and animals both oxygen and carbon dioxide remain in a constant concentration in the atmosphere. There is also a nitrogen cycle, a water cycle and so on.

For millions of years such cycles have remained more or less in balance, but man's coming has made a difference. Human agriculture upset the balance in some ways, but with the coming of man's industrial civilization the upset became truly dangerous. Wastes are produced in quantities so great that other forms of life can't handle them quickly enough. Some wastes, such as non-rusting metals, plastics and so on, can't be restored to circulation at all. Some wastes are actively poisonous.

Wastes which cannot be comfortably cycled or which are actively poisonous are pollution, a well-known English word from a Latin one meaning to render thoroughly filthy. As the 1970s opened the new application of this old word came to stand for a growing nightmare to mankind.

Polymer

LIVING TISSUE builds up its giant molecules by joining together strings of smaller molecules (see NEOPRENE and MONOSACCHARIDE). Sometimes the giant molecules have properties that are startlingly different from the units of which they are composed. For instance, sugars are soft, powdery solids, but they can join together to form the fibrous compounds in wood, which are strong enough to build houses.

The chemist has tried to design chemicals with new and useful properties by imitating this process, beginning with some simple compound and encouraging it to combine into long-string giant molecules. The original unit molecule is the monomer, from the Greek *monos* (one) and *meros* (part); it is the one part with which you begin. The final giant molecule is a polymer, from the Greek *polys* (many); it is made up of many parts. The process of joining up the original units is polymerization.

The best-known products of such laboratory polymerizations are the various plastics. The first product of a polymerization is a soft solid that can be moulded into any shape desired. The Greek *plassein* means to mould, and *plastikos* means fit for moulding. Sometimes the plastic after cooling can be heated and moulded a second time, or any number of times, into a new shape. This is a thermoplastic, from the Greek *thermos* (heat). Other plastics once cooled set permanently into shape and cannot be moulded thereafter. These are the thermosetting plastics.

Some familiar plastics are named after the monomer. For instance, one well-known plastic is built up out of molecules of a gas called ethylene (chemically related to the carbon chains in ether—see ETHER—whence its name) and is therefore called polyethylene (many ethylenes). Another monomer is styrene (a liquid obtained from the resin of a tree called Styrax or Storax, whence its name) and a plastic built up out of that is polystyrene.

There are also a number of well-known polymeric fibres. Of these nylon and Dacron or Terylene are probably the best known. Nylon is a polyamide, and Terylene a polyester.

Polywater

IN 1965 a Soviet scientist, B. V. Deryagin, studying liquid water present in very thin glass tubes, was astonished to find it had unusual properties. It was 1·4 times as dense as water ought to be; it could be heated to 500 °C before it boiled (instead of 100 °C), and it could be cooled to —40 °C before it froze (instead of 0 °C).

American chemists were dubious at the news, but when they repeated Deryagin's work they came out with the same results. It didn't seem possible that so common a substance as water could present surprises, but it did.

Explanations were at once advanced. The molecules in ordinary water form weak attachments to each other, with the hydrogen atom of one molecule lined up with the oxygen atom of a neighbour to form a hydrogen bond. This represents a weak attachment in which two atoms are farther apart than is usual for atoms held together by chemical bonds.

In the narrow space within a hair-like glass tube, the water molecules seemed to be forced closer together. The hydrogen bond attraction, exerted over a shorter distance, held neighbouring molecules together strongly. Indeed, still more water molecules joined and clung, till what amounted to a large molecule made up of units of thousands of ordinary water molecules were thought to have been built up.

When a large molecule is made up of numerous identical small ones bound together the small molecules are said to polymerize (from Greek words meaning many parts). The new form of water was therefore called polymerized water or polywater, for short.

Some chemists still doubted. They suspected that water might dissolve some of the chemicals in glass and that it was this solution, not pure water, that had the unusual properties. Evidence accumulated that this was so, and after creating excitement for several years, polywater faded away.

Porphyrin

THE MOST obvious thing about blood, to the naked eye, is its colour and this crops up in the names of things that have nothing to do with blood. For instance, the most common iron ore is a brownish red mineral which, because of its colour, received the name haematite (from the Greek *haima,* meaning blood and *haimatites,* blood-like stone) as early as Greek times.

As it happens, there is iron in the compound that gives blood its colour, though it isn't the iron that's responsible. The red compound in blood contains a complicated ring of carbon and nitrogen atoms which is red-purple in colour all by itself. It is called the porphyrin ring, from the Greek *porphyra,* the name of the mollusc from which purple dye was obtained. (There are also red-purple minerals which have nothing to do with the porphyrin ring which are called porphyry because of the colour.)

The particular porphyrin compound found in the red substance of blood is protoporphyrin IX. The proto- prefix comes from the Greek *protos* (first) because this particular variety of porphyrin is first in importance in the body. The reason for the IX is more complicated.

By making minor adjustments in the outer atoms of the protoporphyrin molecule, it is possible to imagine fifteen different compounds, all protoporphyrin varieties, each of which might qualify as the red compound of blood. Only one could really be that compound. The German physician and biochemist Hans Fischer numbered these compounds from one to fifteen and he and his students set about synthesizing each one in the laboratory. Each as it was prepared was compared with the natural product obtained from blood. It turned out that the particular protoporphyrin which Fischer had happened to number nine was the natural one, hence protoporphyrin IX. In 1930 Hans Fischer received the Nobel Prize for chemistry for his work on blood compounds.

The protoporphyrin IX exists in blood in combination with iron and the combination was first named haematin, but is now generally called haem, so the Greek word for blood ends by being given to the compound responsible for the colour.

Potassium

SOAP WAS not known to ancient peoples. The Greeks and Romans used oil as a cleaning agent. This sounds odd to us now, but oil will dissolve grease and help remove grime. Naturally, any substance that would help the oil do this would be much in demand. The ancients showed their desperation in that they sometimes added sand or other gritty material to the oil, for the sake of the scouring action, but that has obvious disadvantages.

A more suitable additive was eventually found in the ash of certain woods. This ash would be stirred in water, which would dissolve out some of the substances in the ash. The water, with the substances dissolved in it, would be poured off into a large pot. The water would then be boiled off and the dry residue heated strongly. The powdery material that resulted was called in English potash. The two syllables run together in pronunciation, so that most people don't notice that it simply means pot ash. The Arabs, who were the great chemists of the Middle Ages, called the same material *alquili,* meaning the plant ash.

When oil was heated with potash a kind of soap was formed, so that a new and better cleaning material was developed.

The British chemist Sir Humphry Davy isolated a hitherto unknown metal in 1807, and because it occurred in potash he gave it the Latin-sounding name potassium. The Germans, oddly enough, gave it a Latin-sounding name derived from the same substance, but they used the Arabic word and called the metal Kalium. For that reason the chemical symbol for the metal is K, even in those countries that call it potassium.

Potash is one of a group of substances that share properties opposite to those possessed by acids. Such opposite-to-acid compounds are named alkalis, a word which obviously descends from the Arabic word for potash.

Power

UNTIL THE middle of the eighteenth century, work was done by the muscles of humans and animals or by such natural forces as the wind, which mankind found ready to hand. Periodically men would try to do something with the force generated by boiling water as it expands into steam. The first really useful steam-engine (see ENGINE) was constructed by the Scottish inventor James Watt in 1765 and patented in 1769.

Watt's steam-engine was first used to work the pumps draining water out of mines. Previously they had been worked by men, or more usually by horses. The dryness of the mine depended on how fast the water could be removed, in other words, on the rate at which the work of lifting the water could be performed. The rate of doing work is termed power (from the Latin word *posse,* meaning to be able).

Watt measured the power that could be exerted by a horse. By means of a rope and pulley he found that a particular horse could lift a weight of 150 pounds through a height of 221 feet in one minute. Power is measured by multiplying the weight lifted by the height through which it is lifted and dividing the product by the time. One horsepower was, therefore, $150 \times 221 \div 1$ or, in the round numbers which have come to be accepted, 33 000 foot-pounds per minute.

By measuring the power of a car or aeroplane engine in horse-power we are still harking back to the days when Watt was curious to know how many horses could be replaced by his steam engine.

In electrical measurements, the power of a current is measured in watts, in honour of James Watt. The power can be obtained by multiplying the volts of electromotive force (see VOLT) by the amperes of current strength. Thus an electric bulb with a current strength of half an ampere under an electromotive force of 120 volts has a power of 60 watts. One horsepower, by the way, is equal to 745·7 watts.

Prime Number

GREEK MATHEMATICIANS were fond of playing with numbers, and began games with them that still exercise the brains of our own mathematicians.

For instance some numbers (from the Latin *numerus*, meaning number) can be divided evenly by smaller numbers. For instance, the number 12 can be divided by 2, 3, 4 or 6. Twelve divided by two is six; twelve divided by three is four; twelve divided by four is three; and twelve divided by six is two. Each of these numbers is a factor of twelve (from the Latin *facere*, meaning to make, since out of these smaller numbers you can make the larger one).

Of course, all numbers can be divided evenly by the number 1 (twelve divided by one is twelve), or by themselves (twelve divided by twelve is one). Such universal factors are of no interest and can be ignored.

If we concentrate on the other factors, though, it turns out that some numbers possess none. The numbers 2, 3, 5, 7, 11, 13, 17, 19, 23, 29, 31 and 37, for instance, have no factors other than 1 and themselves. (And numbers of any size can lack factors. There is no upper limit.) Such factorless numbers are prime numbers, or just primes, from the Latin *primus* (first).

The reason for that is that other numbers, those *not* prime, can be broken up into prime factors. The number 12, for instance, can be expressed as $3 \times 2 \times 2$. (The number 2 is the only even prime. All other even numbers can be divided by 2, and so have that as a factor and are not prime.) The number 15 is 5×3; 143 is 11×13; 370 is $37 \times 5 \times 2$, and so on. Such numbers, with factors, are composite numbers, from the Latin *com-* (together) and *ponere* (past participle *positus*), to place; they are built up by placing together smaller numbers. So the primes can be viewed in a poetic way as having existed first, while other numbers came into existence afterward only by the building up of primes.

Probe, Space

IN ORDER TO BE LAUNCHED into orbit a satellite must be sent into space at a velocity of at least eight kilometres per second. At lower velocities it would fall back to earth before having circled the planet even once. The greater the speed it attained the farther away from earth it would loop as it circled.

At velocities of just over 11 kilometres per second it would move so rapidly away from earth that it would outstrip the ability of the planet to pull it back. The earth's gravitational field grows less intense with distance, and the speeding satellite would be subjected to less and less pull. Though its speed would decrease, it would never decrease to zero; therefore it would never return. The satellite would have exceeded the escape velocity for earth.

Such a satellite would not, however, be free of the sun's much greater gravity. Though not circling earth, it would move out into an orbit about the sun and would become an artificial planet. The first such artificial planet was launched on January 2, 1959, by the Soviet Union.

The term artificial planet is not used very often. A satellite circling earth sends back information about every point of its rapid orbit. An artificial planet circling the sun usually can send back information for only part of its mighty orbit, which takes a year or more for each revolution. Usually such a device is sent towards some other world with hopes that it will send back information about that world or its vicinity. It is a space probe (from a Latin word meaning examination). It examines the body towards which it is launched.

The first artificial planet was launched towards the moon. It was the first lunar probe. Since then the Soviet Union and the United States have launched space probes towards Venus and Mars, and a good deal of information has been gathered about both planets, and brilliant photographs returned of Mars.

Propellant

THE WORD propel is from Latin words meaning to drive forward. Something which drives a vehicle forward is a propeller, and the term is used for a mechanical device that turns rapidly and forces water or air backward. The backward motion of water or air forces a ship or plane forward.

It is also possible to drive an object in one direction if fuel is burned and if the gases formed are expelled in the opposite direction. Suppose you have a hollow cylinder with a pointed end (for streamlining) and a stick trailing off behind to give it stability. (This is shaped vaguely like a distaff, which was a stick for holding the fibre, from the days when women spun their own yarn. The cylinder is therefore called a rocket, from an Italian word for distaff.)

If the cylinder is filled with gunpowder which is ignited the gases that form push out the rear opening and the rocket flies with increasing rapidity in the other direction. The gunpowder is therefore a propeller, though in this case another form of the word, propellant, is employed.

Gunpowder was the first rocket propellant used, but in the twentieth century much more powerful and efficient propellants were developed. If a rocket is to penetrate beyond the atmosphere it must carry its own oxygen (or the equivalent) with it. In 1926, when Robert H. Goddard fired the first rocket potentially capable of going beyond the atmosphere, he used a mixture of petrol and liquid oxygen as the propellant. This was a bipropellant (two propellants), because two substances were involved. It is possible to use a single substance which will break down into gases on heating. Acetylene is an example. This would be a monopropellant (one propellant).

Eventually propellants may involve nuclear reactions.

Prostaglandin

AROUND THE URETHRA in the human male, between the bladder and the penis, is a small piece of glandular tissue, which was observed by the Greek anatomist Herophilus in the third century B.C. He thought of it as standing before the bladder, which is true if you are considering a person lying on his back and look at him from the direction of his feet. Herophilus called it *prostatai adenoidis*, meaning the gland that stands before. In English that becomes prostrate gland.

The fluid that surrounds the sperm cells produced by the male is called seminal fluid or semen. The word seminal is from the Latin word for seed, since the sperm cells are considered a kind of male seed. The seminal fluid is produced partly by the seminal vesicles, two little pouches behind the bladder. Vesicle is from a Latin word meaning a little hollow. The prostrate gland also contributes to the seminal fluid.

In the early 1930s various scientists noticed that something in seminal fluid stimulated smooth muscles and lowered blood-pressure. This was what one would expect of a hormone. Since it was thought that such a hormone would be formed by the prostrate gland, the Swedish physiologist Ulf Von Euler named it prostaglandin in 1935.

In 1960 the hormone was actually isolated, with several tons of sheep vesicular glands used as starting material. Prostaglandin was found to exist in numerous varieties (at least six occur naturally in the human body), and has a structure resembling certain fatty acids. They are present in tiny quantities, and are manufactured elsewhere than in the seminal vesicles, for they are found in women too.

Prostaglandins have a variety of actions in the body and are involved in reproduction, in the regulation of blood-pressure, in nerve transmission and so on. In 1971 it was reported that aspirin may exert its pain-relieving action through its effect on prostaglandins.

Protein

THE FIRST person to distinguish the three main groups of substances in food was William Prout (see PROTON), in 1827. He named them the saccharine, oily and albuminous substances.

The word saccharine comes from the Greek *sakchar* (sugar) and includes the various sugars (sugar, one can see, is a distant relative of *sakchar*) and the starches, which can be converted to sugar by acid. Starch comes from the Anglo-Saxon *stearc* (strong), since when added to a limp textile it will make it stiff. Sugars, starches and related compounds are today lumped together as carbohydrates. This name originated through the mistaken notion on the part of early chemists that carbohydrate molecules consisted of a string af carbon atoms to which water molecules were attached (the Greek *hydor* means water). As usual, mere wrongness did not prevent the name from sticking.

Prout's oily substances included both oils and fats (see OIL). The Greek word for fat is *lipos,* and to the modern chemist oils and fats are lumped together as lipids. (Actually, this is a name settled on only recently, and one will still find lipids spoken of as lipoids or lipins.)

The third group of substances in foods contained nitrogen atoms, which lipids and carbohydrates did not. The white of egg was the best example of a food containing this nitrogenous matter. (It contains that and water and little else.) The nitrogenous matter was therefore named after it and called albumin, from the Latin *albus* (white).

Early experiments on diet proved that albuminous substances were the most essential of the three. Dogs fed on carbohydrates and lipids alone died in about a month. The German biochemist Gerardus Johannes Mulder suggested in 1839 that albuminous substances be therefore called proteins, from the Greek *proteion,* meaning in first place. To this day, though, certain comparatively simple proteins, including the one in white of egg, are called albumins. The one in egg white is egg albumin.

Proteinoid

IN 1936 a Russian biochemist, Alexander Oparin, published a book, *The Origin of Life,* in which he speculated on the chemical steps that might have led to the formation of the first living organism.

Eventually biochemists, following Oparin's lead, began to experiment with mixtures of simple chemicals such as those which might have existed in earth's primitive atmosphere and ocean, and studied the reactions which took place when energy in the form of electrical discharges or ultra-violet light was added. In 1953 an American biochemist, Stanley L. Miller, found that a mixture of simple chemicals, in the presence of electric sparks, formed certain organic acids plus a few amino acids, the building blocks of proteins.

Others joined in the search, and more and more complicated molecules were built up in this fashion. It seemed quite certain that, little by little, given a whole ocean of chemicals under blazing sunlight and millions of years to work in, complicated proteins and nucleic acids, like those of living tissue, might be built up.

The American biochemist Sidney W. Fox wondered if the process need necessarily be step by step. Once amino acids were formed might they not join together all at once to form protein? In 1958 he heated amino acids under conditions that might have existed on a hot, volcanic earth and did find that they formed long chain molecules resembling proteins. He called them proteinoids (protein-like).

Fox dissolved the proteinoids in hot water and let the solution cool. He found that the proteinoids collected in tiny spheres about the size of small bacteria. Fox called these microspheres.

The microspheres seemed to have some of the properties of cells. They had a membrane, could swell or shrink, and even produce buds. They were not alive, but perhaps they represented a step towards life.

Proton

IN 1815, during the very early days of the modern atomic theory, the British chemist and physician William Prout suggested that all atoms were built up out of hydrogen atoms. Carbon atoms, for instance, weighed exactly twelve times as much as hydrogen atoms, and were therefore made up of twelve hydrogen atoms apiece. Oxygen atoms weighed sixteen times as much as hydrogen atoms, and so on. Prout suggested that hydrogen, as the primary substance out of which all else was built, be called protyle, from Greek words *protos,* meaning first, and *hyle,* meaning matter.

As more information was gathered it came to seem obvious, however, that Prout's notions were wrong. The chlorine atom, for instance, was 35½ times as heavy as the hydrogen atom, and at the time chemists were certain there was no such thing as half a hydrogen atom.

Nevertheless, in 1896 and thereafter, it was discovered that atoms were made up of still smaller particles. Over 99·9 per cent of the mass of a hydrogen atom, it turned out, consisted of a single tiny particle located at the very centre of the atom. The centres of atoms heavier than hydrogen were found to contain varying numbers of this particle, so that they were made up of hydrogen atoms (in a way) after all. An atom of chlorine turned out to be 35½ times as heavy as hydrogen, because chlorine was made up of two varieties of atoms. One kind weighed 35 times as much as hydrogen atoms; the other kind 37 times as much. Since the first type was three times as numerous as the second, the average weight was 35½.

It should be noted, moreover, that the atomic weight of carbon is not exactly twelve times that of hydrogen, nor is oxygen exactly sixteen times. Until 1961 oxygen was the standard for the whole table, but now the place has been given to carbon 12, but the same observation holds good.

In 1920 the British physicist Ernest Rutherford suggested that this centrally located sub-atomic particle be called a proton. This was a deliberate tribute to Prout's protyle, with the substitution of an -on suffix since that suffix had become conventional for particles within the atom.

Protoplasm

IN 1665 the English physicist Robert Hooke described the tiny holes in cork as cells. This is a perfectly good name for holes, since it comes from the Latin *cella,* meaning a small room, or more generally any small, hollow place. However, when later investigators, using microscopes, found that other plant and animal tissues were made up of small units marked off from one another, they persisted in calling them cells, too, though the units in living tissues, as opposed to dead tissues such as cork, were not empty.

In 1839 the German physiologist Theodor Schwann and the botanist Matthias Schleiden established the cell doctrine: that is, the thesis that all living tissue was made up of cells, and that the individual cell was the unit of life.

Plant cells (but not animal cells) are surrounded by cell walls that contain a fibrous material that was named cellulose, because of its place of origin. The -ose ending is used for sugars and related compounds and fits here, since the cellulose molecule can be broken down by acid into simple sugar molecules (see GLUCOSE).

Inside all true cells is a little spherical body called the nucleus. This is a diminutive form of the Latin *nux* (nut); that is, a little nut.

The material that fills the cell is protoplasm. The word was first used by the Czech physiologist Johannes Purkinje, in 1840, and applied to the material composing young animal embryos. It comes from the Greek *protos* (first) and *plasma* (a moulded form); it was the first form into which the animal was moulded, in other words. The German botanist Hugo von Mohl first applied it, in 1846, to the material within the cell (which is really the first form into which an animal is moulded, since we all start as a single cell).

The word is, however, falling out of use now, since protoplasm is not a single substance but rather a complex mixture of various different things, and it is these constituents of protoplasm that interest biologists and biochemists, rather than protoplasm as a whole.

Psychology

PSYCHE IS the subject of one of the most beautiful of the Greek myths. She is a young girl with whom Eros (the god of love) falls in love. Eros marries her without allowing her to see him. Goaded on by her jealous sisters, Psyche tries to see Eros by candlelight and he forthwith leaves her. To win him back Psyche is forced to undergo many trials and dangers, but in the end succeeds. She is converted into a goddess and joins Eros—Cupid—in endless heavenly bliss.

Like most Greek myths, this is really an allegory—that is, a story in which the characters and events are symbols of something else. For instance, Psyche can represent anything that undergoes trials before winning a victory that consists partly of being changed into a new and more glorious form. The ugly duckling becomes a swan; the caterpillar becomes a butterfly. The latter case was probably in the minds of the Greeks, for Psyche is usually represented in their art as having butterfly wings. (This custom persists to this day, for the fairies in storybooks of today are usually represented with butterfly wings; angels, on the other hand, have bird wings.)

However, Psyche really symbolized the human soul which through the days of life wearies in hardship and labour but then at death breaks out like a butterfly from a cocoon into a new and heavenly existence. In fact, the Greek word for soul is *psyche* (without the capital).

The psyche or soul includes that part of the human being that is not flesh and blood. Modern scientists have applied the word psyche to a man's intellect, emotions, temperament and personality. The study of these things is therefore psychology.

A person who studies the mental and emotional processes is, generally speaking, a psychologist, but one who studies it from the medical points of view particularly, and is mainly interested in mental disease, is a psychiatrist (the Greek *iatros* meaning physician).

Pterodactyl

IN WALT DISNEY'S cartoon feature *Fantasia,* one of the most dramatic sequences showed a fight between a *Tyrannosaurus rex* (see DINOSAUR) and another giant reptile. The other was a tiny-headed plant-eater some 9 metres long with spikes at the end of his tail and a series of triangular bony outgrowths standing on end in a double row down his backbone. When the bones of this creature were first found the large triangular bones didn't seem to fit anywhere in the skeleton, and at first people thought they covered the beast flatwise like the slates on a roof. The creature was named Stegosaurus, from the Greek *stegos* (roof) and *sauros* (lizard). It was the roofed lizard.

The Triceratops was a reptile that resembled a giant rhinoceros. Its huge skull was extended into a frill covering the nape of his neck, while two horns grew over his eyes and a third on his nose. The name comes from the Greek *trikeratos* (three-horned).

The most curious of the ancient reptiles were perhaps those that had developed the ability to fly. Some of them were the largest flying creatures ever to live. They had long, rather narrow, leathery membranes stretched out to the end of an enormously enlarged fourth finger, the remaining fingers being small and free. (Bats, in contrast, have their flying membranes stretched across four fingers, with only the thumb free.) Such an animal is a pterosaur, from the Greek *pteron* (wing). It was a winged lizard.

A small variety was called a pterodactyl. The Greek *daktylos* means finger, so it was a winged finger, which is just about right. A larger variety is the pteranodon, which had a wingspread of as much as 4·5 metres. Its skull was stretched backward into a long, narrow keel which served perhaps to keep it steady in flight, and it had no teeth. Since the Greek word for without teeth is *anodon,* the creature's name is the curtly descriptive wings, no teeth, though that would apply to modern birds as well.

Pulsar

IN THE 1960s astronomers noted that radio waves from particular sources changed in intensity. In some cases the intensity was changing very quickly; almost as though it were a radio-wave twinkle. Special radio telescopes were designed to catch the twinkle.

In 1964 the English astronomer Antony Hewish was using such a telescope. He had hardly begun when he detected very brief and very regular bursts of radio energy from a particular place in the sky. Each burst lasted only $1/30$ of a second and came every $1\,1/3$ seconds. The intervals were so equally and accurately spaced that the value could be worked out to a hundred-millionth of a second. The period of the twinkle he had discovered was $1\cdot33730109$ seconds.

Hewish at once searched for other such sources, and by February 1968 he was able to announce he had located four. Other astronomers began to search avidly, and more sources were quickly discovered. In two more years almost another forty radio-wave twinkles in the sky were discovered. At first the astronomers hadn't the faintest notion as to what caused the twinkle. The radio waves came in pulses, so they called the phenomenon pulsating stars. By combining the first four and last three letters of the phrase, this became pulsars.

The search was on to see if anything could be seen in the spots where the pulsars were located—that is, if they emitted visible light as well. One of the pulsars was located in the Crab Nebula; it pulsed more quickly than any other which had been discovered, and there was reason to think this might mean it was the youngest, and therefore might be the brightest. Astronomers zeroed in on it, and sure enough, in January 1969, it was discovered that a dim star within the Crab Nebula did flash on and off in precise time with the radio pulses. It was the first optical pulsar to be discovered.

Pyrite

BEFORE THE days of matches one of the ways of starting a fire was to strike iron against rock. The resulting friction heated the metal to glowing heat (helped along by chemical combination of iron and oxygen), while the force of the rock broke off small pieces of the heated metal, sending sparks flying. If the sparks could be made to land in a nest of dry, tindery wood, then, with luck, the fire would be started. (We still use this principle in cigarette lighters.)

Naturally, not every rock would do in this early fire-making device. It is not surprising that a rock that turned out to be appropriate would be named in accordance. Such a rock is iron pyrites or, simply, pyrite. In Greek *pyrites* means 'of fire', from *pyr,* meaning fire. Pyrite was a compound of iron and sulphur that had a yellowish metallic cast to it. The name is now applied to any sulphur-containing ore with a metallic cast.

There are copper pyrites, for instance, and tin pyrites. The former is called chalcopyrites, from the Greek *khalkos* (copper), and the latter stannite from the Latin *stannum* (tin).

But iron pyrites remains the most famous for a rather dramatic reason. Its yellowish metallic cast invariably deluded a certain percentage of the inexperienced prospectors who rushed to the California and Alaskan gold fields. Gold is hard to find, but iron pyrites with its golden glint is easy to find, and many a luckless amateur went rushing in with a sack of the pyrite to stake a claim. He received scant sympathy from the old-timers whose point of view is shown in the popular name given to iron pyrites: fool's gold.

Minerals can fool even experts, though, and some have received their scientific names with that in mind. There is a rock called apatite, from the Greek *apate* (deceit) because it is so easily mistaken for other minerals. (Bones and teeth, by the way, are made up of a kind of mineral related to apatite. Oddly enough, false teeth, which are intended to deceive, are not.)

Quantum

IN THE nineteenth century physicists were quite interested in the manner in which a heated body radiates energy, if only because the stars (which are heated bodies) radiate energy, and it is that energy which is our only source of knowledge of the outer universe. To simplify matters they imagined a so-called black body, an object which would absorb all the radiation falling upon it and reflect none, therefore appearing black. If such a body were heated there was reason to think that some relatively simple rule ought to govern the manner in which it emitted energy.

Actual experiments with objects that were nearly black bodies showed that energy was radiated mostly in a certain narrow range of frequencies. (Radiant energy is in wave form, and frequency refers to the number of wave vibrations per second.) Less energy is radiated at frequencies both higher and lower. If the temperature of the body is raised more energy of every frequency is emitted, but the point of maximum emission shifts to higher frequency.

Unfortunately, despite expectations, no reasonable theory seemed to account for the exact manner in which energy was radiated, and physicists were nonplussed. Finally, in 1900, the German physicist Max Planck advanced a completely new idea. He supposed that energy, like matter, was made up of tiny particles. He showed that such energy particles would vary in size according to the frequency of a particular energy wave. The size of the energy particle, divided by its frequency, always gave the same figure, called Planck's constant (constant because it never varied). It is interesting that this was in some ways a return to Newton's corpuscular theory of light, which had been superseded by the wave theory.

The energy particle was called a quantum (plural, quanta), from the Latin word *quantus,* meaning how much? and Planck's theory is called the quantum theory. Because quanta do behave like particles, they are also called photons, from the Greek word *phos,* meaning light (which gives rise to the common prefix photo- in scientific words), and the usual -on suffix used for sub-atomic particles.

Quark

IN 1961 the American physicist Murray Gell-Mann worked out a system for grouping sub-atomic particles into families. It simplified the jungle of particles that physicists had discovered after the Second World War. Gell-Mann, however, looked for still further simplification.

Was it possible to imagine a very few particles still simpler than the protons, neutrons and other ordinary sub-atomic particles? Could different combinations of these few sub-sub-atomic particles make up all the different sub-atomic particles there were? If that were the case physicists might reach a more basic understanding of the structure of matter.

Gell-Mann did come up with three possible particles (and three corresponding anti-particles) which were particularly unusual because they had fractional electric charges. All charges known to this time were equal to that on an electron (-1) or on a proton ($+1$), or were exact multiples of those charges. On the other hand, one of Gell-Mann's suggested new particles had a charge of $+\frac{2}{3}$ and the other two, $-\frac{1}{3}$. (The equivalent anti-particles had charges of $-\frac{2}{3}$ and $+\frac{1}{3}$, respectively.) A proton might be built up of two

323

+⅔ particles and one −⅓ to give +1 altogether, while a neutron might consist of one +⅔ and two −⅓ to end up with 0 charge.

To name these particles Gell-Mann turned to *Finnegans Wake,* a book that contained a riotous series of complicated puns and verbal distortions. One sentence went "Three quarks for Musther Mark" (a distortion, perhaps, of "Three quarts for Mister Mark").

Since it would take three of the Gell-Mann particles to make up each of the more prominent sub-nuclear particles, Gell-Mann decided to call them quarks. The queer name caught on, something that might have astonished James Joyce, had he not died in 1941, nearly a quarter-century before his word had entered the scientific vocabulary.

Quasar

AFTER THE SECOND WORLD WAR astronomers searched the sky to find regions which were emitting radio-waves. Beginning in 1960, they found radio-waves associated with what seemed like certain dim and undistinguished stars. Yet why should these emit strong radio-wave beams while ordinary stars did not? The spectra were studied, and dark lines were discovered which represented certain absent wave-lengths in the light the stars emitted. Usually much information could be gained from these dark lines, but in this case their nature was a puzzle.

In 1963 the Dutch-American astronomer Maarten Schmidt wondered if the dark lines were those which would ordinarily be located in the ultra-violet region, but were shifted so far towards the long wave-length as to appear in the visible light region. There was usually such a shift towards long wave-length (red shift) in distant objects. The greater the distance, the greater the shift. No one, however, had ever before observed a shift as great as the one Schmidt now suspected. Other radio-emitting stars were studied, and they too showed this enormous red shift. (See RED SHIFT.)

The conclusion was that these radio-emitting stars were extremely distant, over a thousand million light-years away, farther than any other objects in the universe. Clearly, they could not be ordinary stars, for no individual stars could possibly be seen at that distance. Since their nature was unknown, they were called quasi-stellar objects (quasi-stellar being a Latin term for star-like). In 1964 the Chinese-American astronomer Chiu Hong-Yee used the first and last groups of letters of quasi-stellar and called them quasars for short.

The name stuck. As astronomers now think, a quasar is rather small, much smaller than a galaxy, yet it shines with ten times the brilliance of a galaxy. Its exact nature is still in dispute, and some astronomers even deny that the enormous red shift can really be interpreted as representing great distance.

Quasi-Mammal

SHORTLY BEFORE 1800 certain primitive mammals discovered in Australia surprised zoologists. Although they had hair, the hallmark of the mammal, they laid eggs, something no other mammals did.

These egg-laying mammals are called monotremes (from Greek words meaning one opening, since in place of the separate openings for faeces and urine that all other mammals have, they have one opening for both wastes, and for eggs too—a situation characteristic of birds and reptiles). The best-known monotreme is the duck-billed platypus.

Mammals are descended from reptiles. In particular, they are descended from a now extinct group of reptiles called therapsids (from Greek words which mean beast openings, because they possess openings through skull bones similar in types to those in the skull of mammals).

The only remains we have of the therapsids are bones and teeth. We don't know anything about their soft tissues and skin. We don't know whether they might not have developed hair and warm-bloodedness even while they were still technically reptiles. Perhaps the duckbill and the other monotremes are even today closer to the ancestral therapsid line than to other mammals.

An American zoologist, Giles T. MacIntyre, studied the problem in the late 1960s. He considered the trigeminal nerve which leads from the jaw muscles to the brain. This nerve follows a slightly different path in reptiles to that taken in mammals. In the adult duckbills the nerve follows the mammalian path, or seems to. In young duckbills the path can be more easily followed, and in them MacIntyre saw that the nerve travelled the reptilian path. It seems possible, then, that the monotremes might be the last surviving therapsids rather than being the most primitive mammals. MacIntyre speaks of the monotremes as quasi-mammals, where quasi is a Latin word meaning having some resemblance to.

Radar

UNDER THE proper conditions sound waves will be reflected from a hillside or other such obstruction. Sound travels only at the rate of about one-third of a kilometre per second in air, so it usually takes a perceptible interval of time for sound to travel to the hillside and back. If the obstruction is 333 metres away there is a two-second interval between the sound and its reflection (which is called the echo, from the Greek *echos*, meaning sound). In fact by timing the interval between sound and echo you can estimate the distance to the hillside.

You can do the same with any other form of radiation that can be emitted, reflected and detected. For instance, you could send out a beam of light to a distant mirror, detect it on its return and determine the mirror's distance that way. (Of course, light travels 299792·5 kilometres a second, so to go 150 kilometres and back would take barely a thousandth of a second: it takes only eight minutes from the sun.)

During the Second World War, the British worked out a practical application of this long-known principle. They used short radio waves instead of light, since radio could penetrate fog, clouds and other obstructions that ordinary light could not. The radio waves travelled at the speed of light, but there are electronic devices that will determine the interval between emission and return.

Using this radio-echo principle, the British were able to detect German planes on their way to bomb London, long before the enemy were near their target. The outnumbered RAF always seemed to the puzzled Germans to be lying in wait at the right time, and never to be surprised. It was radio echoes more than anything else that won the Battle of Britain.

Since the radio-waves were used to tell the direction in which to send the RAF planes and the distance to send them (their range of flight, in other words), the device was called *ra*dio *d*irecting *a*nd *r*anging, and from the initials the word radar was coined.

Radical

MATHEMATICIANS speak of roots of a number. The number 2, for instance, is the square root of 4 and the cube root of 8 (see SQUARE ROOT). The Latin word for root is *radix* (preserved by us today in the word radish, which is an edible root, eradicate and deracinate, which mean to root out). Any mathematical expression which contains roots is said to contain a radical. The sign of the radical is $\sqrt{\ }$.

To find the root of a number by mathematical computation is spoken of as extracting a root. The word extract comes from the Latin *ex-* (out) and *trahere* (to draw), and that, after all, is what is done to roots. To be made use of they must first be drawn out of the ground.

The root is the basic part of the plant, or of anything—the part from which plants draw their sustenance, or ideas their significance. Any person, therefore, who wants to change the basic principles of something is a radical.

The word is used, then, in both mathematics and in political science, with meanings that have apparently no connection until the derivation is considered. In chemistry the word exists in still a third completely different meaning.

In the late 1700s the French chemist Guyton de Morveau talked about that portion of an acid other than the oxygen (in those days it was wrongly thought that oxygen was the element characteristic of acids; see OXYGEN) as the radical. It was the root of the acid, in other words, out of which the acid itself was built by the addition of oxygen.

The notion didn't last, but meanwhile it meant that the word radical had been applied to a group of atoms in a molecule. In the early 1800s the French chemist Joseph Louis Gay-Lussac took to using the term to mean any group of atoms that could be passed from molecule to molecule during chemical reactions as a unit. That is the meaning of radical in chemistry to this day.

Radioactivity

IN LATIN a spoke of a wheel is *radius*. The same word in both Latin and English also applies to any line connecting the centre of a circle with a point on its circumference. Radii (the plural form) can be drawn from the centre to all points on the circumference, emerging from the central point in every direction. Such a family of lines are said to radiate from a point.

Light behaves in this fashion. It emerges from a candle flame or an electric bulb, travelling in straight lines in all directions. Light, therefore, is an example of a radiation. There are other kinds of radiations, some of which can't be detected by our senses.

For instance, in 1894 one set of radiations, imperceptible to us and much less energetic than light, was made use of by the Italian Guglielmo Marconi to transmit code messages. At the time the usual instrument for transmitting messages was the telegraph, which used wires. Marconi's instrument used radiations instead, so it was eventually referred to as a wireless telegraph or a radio-telegraph. These names were too long. One was shortened to wireless and the other to radio.

In 1896 uranium was found to give off new types of radiation much more energetic than light. The French scientist Pierre Curie and his Polish-born wife, Marie Sklodowska Curie, suggested in 1898 that this phenomenon be named radioactivity.

The Curies discovered a new element that same year which gave off radiations must more strongly than did uranium. They therefore named the new element radium. In 1900 the German physicist Friedrich Ernest Dorn discovered a gas released by radium during its breakdown. That gas is now called radon.

The Greek word for ray is *aktis* (genitive *aktinos*) and that was used too. When the French chemist Andre Debierne discovered a new radioactive element in 1899 he called it actinium. Then when the German physicists Otto Hahn and Lise Meitner discovered still another, which broke down and changed into actinium, they called it protactinium (that is, first actinium, from the Greek word *protos*, meaning first). See also FALLOUT, RADIOACTIVE.

Radiocarbon Dating

IN RECENT DECADES uranium breakdown to lead has been used to date extremely old rocks. Uranium breaks down at a known, very slow, rate, and from the relative amount of uranium and lead in certain rocks it can be shown that particular rocks might be a thousand million years old or more.

Uranium breaks down so slowly, however, that in a rock only a few thousand years old the amount of lead formed is too small to be detectable. What was needed for shorter date determinations was a substance that broke down more quickly. In 1939 two American biochemists, Martin D. Kamen and Samuel Ruben, discovered a form of carbon, carbon 14, which broke down at such a rate that half was gone in only 5770 years. In breaking down it gave off radiations of speeding electrons, so it was called radiocarbon.

Although carbon 14 breaks down quickly, cosmic-ray particles, smashing into atmospheric atoms, are continually producing new carbon 14 atoms. The carbon 14 content of the atmosphere, while very small, therefore remains steady. As long as an organism is alive, it keeps incorporating carbon 14 into its tissues as fast as those atoms break down, so there is a steady small content. Once the organism dies, however, incorporation ceases and the carbon 14 present slowly disappears at a steady rate.

In 1946 the American chemist Willard F. Libby developed methods for determining the carbon 14 content in old materials that had once been part of a living organism. Charred wood from ancient campfires, textiles that had been used to wrap mummies, anything of that sort was used. The carbon 14 content was determined, and from that one could tell how long it had been since the material had ceased to be part of a living organism. This procedure was called radiocarbon dating, and it had helped archaeologists define more clearly certain events in the last thirty thousand years, notably the exact times of the coming and going of the great glaciers. There are also other means of chemical dating. Bones take up fluorine from the soil at a given rate, and by this means in 1949 Piltdown Man was exposed as a forgery. Then there is a potassium-argon method, which depends upon a radioactive isotope of potassium.

Rayon

NOWADAYS, with artificial fibres numerous and varied, we tend to think of them as something new, but the first such thing was prepared over a century ago. Chemists' first attempts to imitate the fibres of nature began with the natural cellulose fibre, cotton. If cotton is treated with a mixture of nitric acid and sulphuric acid (so called because they are acids that contain a nitrogen and a sulphur atom in their molecules respectively) nitrogen-containing groups attach to the glucose units of the cellulose, as many as three per unit. The resulting nitrocellulose still looks like cotton but is explosive, and is called guncotton therefore.

If cellulose is treated less thoroughly (only two nitrogen-containing groups being attached per glucose unit), what results is not explosive, though still dangerously inflammable. It is called pyroxylin, from the Greek *pyr* (fire) and *xylon* (wood). It is a variety of wood (i.e., cellulose), in other words, that is easily set on fire.

Pyroxylin, unlike cellulose, will dissolve in certain liquids—in a mixture of alcohol and ether, for instance—to form a thick, sticky solution called collodion, from the Greek *kollodes* (glue-like). If collodion solution is forced out of small holes jets of liquid spray out. The alcohol and ether evaporate almost at once leaving a fine fibre of pyroxylin. This first artificial fibre was patented in 1855.

Since then other less inflammable varieties of chemically treated cellulose have been used. An important fibre resulting was rayon. The name was a made-up one, implying a ray of light, since the fibres were glossy and shiny. Such cellulose can also be forced through narrow slits to make transparent, flexible films which may be used in photography (the word film has become a slang term for motion pictures) or, if thin enough, Cellophane, from the Greek *phanein* (to appear). In other words, Cellophane is made from a substance that has the appearance of cellulose but is not cellulose.

Rayon and Cellophane both date back to the 1890s. In more recent years, many polymeric fibres, of which nylon and Terylene are probably the best known, have made their appearance (see POLYMER).

Red Shift

IN 1842 an Austrian physicist, Christian Doppler, demonstrated that when sound of a particular pitch was emitted by a source that was approaching the observer the sound was higher than it would have been if the source were motionless. The approaching source squeezed the sound waves together, and made them shorter. On the other hand, if the source were moving away, the sound waves were pulled apart and made longer, so that the pitch became deeper.

This Doppler effect could apply to any wave phenomenon—to light, for instance. If a light source were approaching us the waves it gave off would squeeze together and become shorter. If a spectrum were being studied in which the light was separated into different wave-lengths, all the wave-lengths would shift toward the short-wavelength end of the spectrum. That end was occupied by light seen as violet in colour, so that the Doppler effect in an approaching light source was called the violet shift. If the light source were receding the wave-lengths would lengthen and shift towards the other end, occupied by red light. That would be the red shift.

The violet shift and the red shift were easily seen when the spectrum included dark lines because of the wave-lengths that were absorbed. The dark lines shifted, and since their proper position was known, the exact extent of the shift could be easily measured.

In the 1920s it became clear to the American astronomer Edwin P. Hubble that the distant galaxies were all receding from us. In fact, the more distant the galaxy the more rapid the recession, and therefore the greater the red shift. With that deduction the measurement of the red shift became an important way of learning some fundamental things about the universe as a whole. It was by the detection of an unexpectedly enormous red shift that the puzzling quasars were discovered.

Relativity

IF YOU'RE sitting down now you're not really at rest. You are whirling about the earth's axis, and the earth itself is moving about the sun, and the sun is moving about the centre of the galaxy, and the galaxy itself is moving in some direction. But in the nineteenth century physicists considered space to be filled with something called ether which they thought did not move. So they tried to determine the earth's motion through the ether to get an idea as to what its real motion was.

In 1887 the American physicist A. A. Michelson tried to do this by a most ingenious experiment that should certainly have worked if there were an ether. It did not work!

In 1905 a 26-year-old Swiss patents clerk, Albert Einstein, worked out a theory to account for this. There was no ether, he said; there was nothing in the universe that was motionless. There was no such thing as rest or motion all by itself. Rest or motion depended on comparing one object with another.

The motion of the moon took it about in an ellipse if it were measured relative to the earth, something else if it were measured relative to the sun, and it was motionless if its motion were measured relative to itself. (The word relative comes from the Latin *re-* (back) and *latus* (carried). If motion is relative you can only measure it by carrying back your observations to another body and making a comparison of motions. Human relatives are such because their blood line can be carried back to a common ancestor.)

But does the moon really move, and if so, how? Einstein insisted you could not tell. From this insistence that only relative motion existed, plus the further assumption that the speed of light in vacuum was always the same, Einstein worked out a new system of the universe that explained many things older systems did not. Because relative motion was what Einstein began with, his system is referred to as the theory of relativity.

REM Sleep

SLEEP IS A NECESSITY. Lack of sleep will kill more quickly than will lack of food. Yet why do we sleep? It can't be merely to rest, for we might just lie quietly in bed wide awake, and that would not substitute for sleep. Indeed, sleep is sometimes full of activity, what with twisting and turning in the bed, so that there is little rest indeed, and yet it will accomplish its purpose.

When wakefulness is enforced no bodily functions go seriously awry except those of the brain. Extended wakefulness affects the co-ordination of various parts of the nervous system, and there is the onset of hallucinations and other symptoms of mental disturbance. So sleep must do something for the brain that mere wakeful rest does not.

What about dreams? Few people seem to think of dreams as being of physical importance, and indeed a dreamless sleep is sometimes spoken of as particularly restful. Yet dreamlessness merely means not remembering dreams, not the lack of them.

The American physiologist W. Dement, studying sleeping subjects in 1952, noticed periods of rapid eye movements that sometimes persisted for minutes. He called this rapid-eye-movement sleep or, abbreviated, REM sleep. During this period breathing, heartbeat and blood-pressure rose to waking levels. This occupied a quarter of the sleeping time. If a sleeper was awakened during these periods he generally reported being in the midst of a dream. Furthermore, if a sleeper was continually disturbed during these periods he began to suffer psychological distress and periods of REM sleep multiplied during succeeding nights, as though to make up for the lost dreaming.

Apparently it is not just sleep the brain needs if it is to recover from the wear and tear of the day, but REM sleep. Dreams somehow restore the proper functioning of the nervous system.

Resonance Particles

AFTER THE DISCOVERY of radioactivity in the 1890s it was found that some types of radioactive atoms only existed very briefly before breaking down. Polonium 212 breaks down in less than a millionth of a second. This could be measured by the energy of the particle it emits (the greater the energy, the faster the breakdown).

Sub-atomic particles themselves could break down. For instance, the muon breaks down after about two-millionths of a second; the pion after only two hundred-millionths of a second; some of the hyperons, in less than a thousand-millionth of a second.

These ultra-short intervals can be measured by noting how far a particle travels in a bubble chamber before breaking down. If a particle is formed, travels three centimetres, then breaks down, and if it travels at nearly the speed of light, it would have taken about 1/10000 million seconds to go that distance.

Actually, even 1/10000 million seconds is a long time on a sub-atomic scale. In the 1950s it seemed that pions and protons interacted very readily at some energies and not at others. When something happens at some energies and not at others it is referred to as a resonance in an analogy with what happens in sound when some sound-waves but not others cause a receiver to resonate (from Latin words meaning to sound again) in response.

In the 1960s the American physicist Luis W. Alvarez maintained that actual particles were being formed in these resonance events and that they broke down in a few quadrillionths of a second. Even at the speed of light particles could only travel a sub-microscopic distance in this time. The existence of these resonance particles could only be demonstrated indirectly, but that evidence is quite convincing, and physicists now accept their existence.

Retinene

PLASTERED UP AGAINST the internal surface of the rear of the eyeball is a coating about the size and thickness of a postage stamp. It is called the retina, which may come from a Latin word meaning net (readers will remember the retiarius), but may not. No one knows the real origin of the name.

The cells of the retina are sensitive to light and enable us to see. They come in two varieties, cones and rods (so called from their shapes). The cones are stimulated by comparatively bright light, and react to the different colours. The rods are stimulated by dim light, but do not distinguish between colours.

In 1877 a biologist working in Rome, Franz Boll, reported that the frog retina had a rose colour that bleached on exposure to light. The German biologist Fritz Kuhne extracted the coloured compound the following year. It seemed purplish to him, and he called it by a German phrase translated into English as visual purple. The more formal chemical name rhodopsin is from Greek words meaning rose eye. Rhodopsin occurs in the rods and makes vision in dim light possible.

In the 1940s the American biochemist George Wald studied rhodopsin and found that it could be broken into two parts, a large protein portion and a smaller coloured compound, or pigment. The protein portion he called opsin, the pigment, retinene, from retina.

Retinene, it turned out, is closely related in chemical structure to vitamin A, but is less stable. Retinene undergoes delicate changes in structure which make vision in dim light possible, and as some of it is destroyed more is formed out of vitamin A.

It is for this reason that a vitamin A deficiency in the diet affects vision. The supply of retinene fails, and the result is night blindness, an inability to see in dim light. This, however, is not the only result of vitamin A deficiency.

Rh Negative

THERE IS a common Indian monkey, given the name of rhesus by the French naturalist Jean Baptiste Audebert in 1797. Audebert insisted that the name was simply made up and meant nothing, and yet. . . .

In 1900 the Austrian physiologist Karl Landsteiner discovered that human blood might contain one of two substances, or neither (or, as was discovered two years later, both). The substances were called simply A and B, so that four blood types—A, B, O and AB—were possible. Blood also contained antibodies (see ANTI-BODY) for the substance or substances it did not possess, so that B blood, for instance, could not be added to A blood or vice versa without causing the blood corpuscles to stick together and grow useless. It was only after Landsteiner's discovery, therefore, that blood transfusion (from the Latin *trans,* meaning over, and *fundere,* meaning to pour) became practical; and physicians knew enough to pour over blood from a well person to a patient who needed blood without killing the patient.

Blood substances not interfering with transfusion also exist. One was discovered in 1940 by Landsteiner and the American physician Alexander S. Weiner in the blood of a rhesus monkey. The new substance was therefore called Rh from the first letters. Some eight varieties of this factor are known today. No natural antibodies can exist against Rh, but in the case of all but one variety antibodies can be developed artificially. The exceptional variety is called Rh negative, the others Rh positive.

It sometimes happens that a mother with Rh negative blood is carrying an unborn child who has inherited Rh positive from the father. Some of the child's Rh positive may filter across to the mother's blood, which may then develop antibodies against it. If these antibodies filter back into the child's blood they may ruin enough red corpuscles to allow a very sick baby to be born. Physicians then have to replace the baby's blood with fresh blood quickly, and, to be prepared for that possibility, expectant mothers are routinely typed for Rh these days, so that at least one-third of the made-up name Rhesus has become very significant indeed.

Ribosome

WHEN THE EXISTENCE of tissue cells was first recognized in the 1830s little could be discovered concerning their intricate internal structure. With time, however, new techniques, both chemical and physical, were evolved, and these made the tiny structures within the cell visible. Eventually, for instance, small structures called mitochondria were located in the cytoplasm and were found to be the power-houses of the cells, the structures within which energy-releasing reactions took place.

Then in the 1930s the electron microscope was invented, which magnified much more than ordinary microscopes. The electron microscope was continually improved, and by the 1950s the tiny mitochondria were in turn found to have a complex structure. In addition, still smaller objects were observed. These smaller bodies were about one ten-thousandth the size of mitochondria and little could be made out concerning them. They were called microsomes, from Greek words meaning small bodies.

About the only thing that seemed remarkable in the microsomes was the fact that they contained quantities of ribonucleic acid, or RNA. This was coming to be considered an important substance in connection with the chemistry of inheritance, and interest in the microsomes increased. In 1953, the Rumanian-American biochemist George E. Parlade found tiny particles densely distributed on the microsome membranes. By 1956 he had isolated these tiny particles (each about a thousandth the size of a microsome) and found that they contained nearly all the RNA of the microsomes. These still smaller particles he therefore called ribosomes.

In 1960 it was discovered that it was on the ribosomes that specific protein molecules within the cell were formed, and that these tiny objects were therefore a key component of the system by which inherited characteristics were passed from cell to cell and from generation to generation.

Robot

THROUGH HISTORY, men have dreamed of manufacturing mechanical men who would perform all tasks without ever getting tired or rebellious. In the Greek myths there was a bronze man who protected the shores of Crete, and golden girls who helped the smith-god Hephaistos. In the Middle Ages there were legends of the construction of mechanical men too.

This view entered the public consciousness most sharply in 1818, when a novel *Frankenstein,* by Mary Shelley (the wife of the poet) was published. Frankenstein was a student of anatomy who produced a living body out of scraps of corpses. The being so created eventually killed Frankenstein.

In 1921, after the First World War with its aeroplanes and its poison gas had shown how deadly science could become, a more elaborate version of this view was published. The Czech dramatist Karel Capek wrote a play in which a scientist produced large numbers of mechanical men designed to be the world's workers, freeing mankind for other, higher tasks. Unfortunately, the scheme went awry and in the end the mechanical men wiped out all of mankind.

Capek's scientist was named Rossum, and his industrial organization was R.U.R., which was also the name of the play. That stood for Rossum's Universal Robots, where *robot* was the Czech word for worker. The play was popular enough to cause the word to enter many languages, including English, to represent a mechanical man designed to simulate human activity and do human work.

Over the last forty years robots have played a part in many science-fiction stories, though the tendency has been to drop the view that they are inevitable enemies of mankind and to stress the safeguards that will be built into them to make them generally useful. And meanwhile science gets ever closer to developing at least simple robots.

Schizophrenia

AS MEDICAL SCIENCE has succeeded in controlling various physical ailments, mental ailments have come to make up an increasing percentage of those illnesses requiring hospital treatment. It is estimated that one American out of ten suffers from some form of mental illness, and of the severe forms the most common is schizophrenia.

This name was coined in 1911 by a Swiss psychiatrist, Eugen Bleuler. It comes from Greek words meaning split mind. Bleuler used it because persons suffering from this disease seemed to be dominated by one set of ideas (or one complex) to the exclusion of others, as though the mind's harmonious working had been disrupted, and as though that one portion had seized control of the rest.

At least half of all patients in mental hospitals are schizophrenics of one type or another, and it is estimated that 1 per cent of mankind (perhaps 35 million people altogether) is affected.

Some way of preventing the disease, or curing it, or at least mitigating its effects, would be very desirable. Could it be caused by some vitamin deficiency, for instance? There is a disease, pellagra, brought about by a lack of the vitamin niacin, and it produces mental disorders. In the body, niacin forms part of a more complicated molecule called niacin adenine dinucleotide

(NAD), and there were indeed reports in the middle 1960s that massive treatment with NAD brought about improvement in some schizophrenics.

An older name for the disease was dementia praecox, from Latin words meaning early-ripening madness. This was intended to differentiate it from mental illness affecting the old through the deterioration of the brain with age (senile dementia), since schizophrenia makes itself manifest at a comparatively early age, generally between the years of eighteen and twenty-eight.

Schmidt Camera

FROM 1609, when the Italian scientist Galileo Galilei first used a telescope to observe the heavens, astronomers have been using better and better instruments for the purpose. Now telescopes with huge mirrors, 500 centimetres across or more, are in use or are being constructed, and with them objects millions of light-years away can be studied.

These huge telescopes have mirrors whose shapes are paraboloids (like that of the rear reflecting surface of a car headlight). Only such a surface can exactly focus the light of tiny, distant stars and give a sharp image. However, the larger and more precise a telescope, the smaller the area of the sky that can be focused on.

For the eye this is all right, since the eye can only look at a small area at one time anyway. From the mid-nineteenth century on, however, astronomers used photography more and more, and it was very tedious to photograph tiny bits of the sky and try to fit all the photographs together. Yet if one were to try to photograph a large section of the sky it turned out that objects became more and more distorted, the farther the light hit from the centre of the paraboloid mirror.

If a spherical mirror was used the stars wouldn't come to a sharp image anywhere, but it didn't matter where the light hit. The images were equally fuzzy everywhere.

In 1931 an Estonian-born German astronomer, Bernhard Schmidt, designed a correcting lens that could be placed in front of the spherical mirror. This made the stars equally sharpe everywhere. With such a Schmidt telescope photographs could be made over large parts of the heavens at once. In fact, such instruments are used so exclusively for photography that they are usually called Schmidt cameras.

The largest Schmidt camera now in use has a 135-centimetre mirror and began operations in East Germany, in 1960.

Science

THE GREEKS were the first who searched for understanding of themselves and the world apart from religion. They called this search *philosophia* (in English, philosophy), from the Greek *philos* (lover of) and *sophia* (wisdom). They searched, because they loved wisdom.

The most famous philosophers of all time, Socrates and Plato, concerned themselves with questions such as 'What is virtue?' and 'What is justice?' This is moral philosophy. The Latin *mores* means manners, so that moral philosophy deals with man's manner of living.

But the search for understanding leads everywhere and may well concern itself with any object in nature, from the universe to a blade of grass. Nature, by the way, comes from the Latin *nasci,* past participle *natus* (to be born). It includes, therefore, all that has been born, all that has come into being—all creation, in short.

The type of philosophy that deals with nature, rather than with the inner soul, is natural philosophy. To Plato this seemed an inferior or second-class branch of philosophy, and such was his intellectual influence that for many centuries that attitude stuck.

Now, one way of avoiding a stigma is to choose a new name, or, if one must placate the gods, to use an auspicious word for an inauspicious one—a process known as euphemism, from the Greek *eu-* (well) and *phanai* (to speak); a word, in other words, that speaks well of its subject. The euphemism that came into use for natural philosophy was natural science or just science, from the Latin *scientia* (knowledge).

This, by the way, explains the very famous quotation from *Hamlet:* 'There are more things in heaven and earth, Horatio/Than are dreamt of in your philosophy'. Now, Horatio was not a philosopher in the modern sense. He was a very rational-minded student of natural philosophy at the University of Wittenberg, and he didn't believe in ghosts even when he saw one with his own eyes. Change philosophy in the quotation to science and it will make more sense.

Even today scholars who qualify for the doctor's degree in one of our modern sciences can be awarded the degree of Doctor of Philosophy, as well as that of Doctor of Science.

Scintillation Counter

WHEN SUB-ATOMIC PARTICLES were first discovered in the 1890s physicists were at a loss for methods to study them.

Since they were so tiny, and moved so quickly, it seemed a forlorn hope to be able to detect the effects of a single particle. Yet it proved surprisingly easy to do so. In 1908 the New Zealand-born physicist Ernest Rutherford and his German assistant Hans Geiger allowed sub-atomic particles to impinge upon a screen of a chemical called zinc sulphide. Every time a particle smashed into the screen it struck an atom, and the energy of its impact was translated into a tiny spark of light, or scintillation, from the Latin word *scintilla,* meaning spark.

By counting the scintillations under a lens important information could be gathered. This was tedious work, however, and the invention of the cloud chamber, which detected not only single particles but even detected the paths they took, made scintillations obsolete.

In the 1940s, however, both British and German physicists used electronic devices to amplify the scintillations to the point where the amplified light could trigger a counter. In this way scintillations were automatically counted, and the device was naturally called a scintillation counter.

Scintillation counters have their special uses. For one thing, they can easily discern types of particles that are detected with difficulty, if at all, by cloud chambers and bubble chambers (which work well only for electrically charged particles). Scintillation counters are very effective in picking up gamma ray photons, for instance, which are uncharged.

Then, too, scintillation counters are very quick in their responses. They can pick up light flashes that last for only a few thousand-millionths of a second, and are therefore useful in studying events that are very brief.

Sco XR-1

IN THE 1960s rockets scanned space for X-ray sources and found the Crab Nebula to be one. The Crab Nebula is the remains of a huge stellar explosion, and it is still heaving with energy. Astronomers were anxious to give the source a name according to some methodical system. The Crab Nebula is in the constellation of Taurus and so the X-ray source in it was called Tau XR-1; that is, the first X-ray source to be detected in Taurus.

The Crab Nebula is by no means the strongest X-ray source in the night sky, however. From the direction of the constellation Scorpio, there comes a beam of X-rays eight times more intense than those emitted by the Crab Nebula. This strongest of the X-ray stars was named Sco XR-1. For some years, despite the probings of astronomers, the sky in the direction of Sco XR-1 seemed blank. But then in 1966 a seemingly ordinary thirteenth-magnitude star was identified as the source. It produced a thousand times as much energy in the form of X-rays as of visible light. Why? Nobody knows.

There are other mysteries connected with these X-ray sources. There is a strong source in the constellation of Cygnus, Cyg XR-1, which, however, in the course of a single year, lost much of its X-ray intensity. It is a variable X-ray star. A second source in that constellation, Cyg XR-2, is weaker because it is very far away, perhaps as much as 2000 light-years away. (This was the first X-ray source to have its distance determined—in 1968.) Allowing for its distance, its X-ray emission is a thousand times as energetic as all the radiation, of all kinds, coming from the sun.

A radio source in Centaurus, Cent XR-2, came into being in late 1965, got stronger till April 1967, then faded. It was the first X-ray nova to be discovered.

Scotophobin

ARE MEMORIES represented by special molecules in the brain? What about learning in general? This is an important question. If the brain learns and remembers by reorganizing its connections and setting up special pathways, then what is learned and remembered is peculiar to each particular brain and can never be physically transferred. If special molecules are involved these might be transferable.

In the early 1960s experimenters found that a very primitive creature, a flatworm, could be trained to perform some simple activity, like responding to light. If these trained flatworms were chopped up and fed to untrained ones the latter would possess the ability, or develop it more easily than otherwise. Some molecule in the trained flatworms seemed to have been incorporated into the untrained, and it meant the same to both.

In 1965 the Danish physiologist Ejnar Fjerdingstad found he could do the same with a much higher animal, the rat. He trained rats to go to light, then injected brain material from these into untrained rats, and found that the latter could learn to respond to light more easily than could untrained rats which had not been injected.

The Hungarian-American physiologist Georges Ungar went further. In 1970 he subjected rats to an electric shock in the dark, so that they finally developed strong fear of the dark. Brain extracts when injected into unshocked animals caused them to show fear too. From several kilogrammes of brains from animals trained to show fear he isolated a chemical compound which would induce fear in an untrained animal. This he called scotophobin, from Greek words meaning fear of the dark.

The compound is a simple one, made up of a chain of nine amino acids, and it induces fear of the dark not only in rats but in goldfish as well. It seems clear that this simple compound is the closest approach yet to an actual memory molecule.

Semiconductor

MOST COMMON SUBSTANCES either conduct an electric current very well, like metals, or like glass, porcelain and sulphur, hardly at all. The former are called conductors, and the latter non-conductors.

There do exist, however, certain substances which are rather intermediate in this respect. The elements silicon and germanium conduct a current far more poorly than metals do, but with far more ease than glass does. Such substances are half-conductors, so to speak, or to use the proper Latin prefix, semiconductors.

These semiconductors did not rouse great interest until the 1940s, when several physicists—William B. Shockley, John Bardeen and Walter H. Brattain—began to examine them. They found that germanium and silicon served best as semiconductors when they had small traces of certain impurities. The germanium and silicon had four outer electrons per atom, but a trace of arsenic added an occasional atom with five outer electrons. This fifth electron did not really fit in the orderly array of atoms that made up the silicon or germanium crystal lattice. It drifted from atom to atom under an electric voltage, and it was this which carried an electric current from the negative pole to the positive pole. The mobility of the electron increased with temperature, so that the conductivity of a semiconductor rose with temperature instead of falling, as was the case with an ordinary conductor.

If a small trace of boron was added to silicon or germanium something similar happened. The boron had only three outer electrons, so that there was a hole where the fourth electron ought to have been. A neighbouring electron would fill it, forming a new hole which would be filled, forming a new hole again, and so on. Under an electric voltage this hole travelled from the positive pole to the negative pole, and again the semiconductor effect was produced.

Sequenator

PROTEINS, giant molecules characteristic of life, are made up of long chains of amino acids of about twenty different varieties. In determining the structure of a protein molecule not only must all the amino acids be identified, but their exact order in the chain must also be determined.

During the 1950s methods were devised for breaking up the protein chain into short fragments. The particular amino acids making up the short chains were identified, and their order worked out. (It is much easier to do this for short chains than for long.) Once this was done one could reason out the exact structure of the original long chain.

In 1964 the exact structure of trypsin, a protein made up of a chain of 223 amino acids, was deciphered. By 1967 the technique was actually automated. The Swedish-Australian biochemist P. Edman devised a system which could work on as little as five milligrammes of pure protein. From the protein chain amino acids could be peeled off and identified, one by one. In one test sixty amino acids were peeled off and identified in four days. Edman called the device a sequenator, because it determined the sequence of the amino acids in the chain.

Once the sequence of amino acids was known biochemists had begun to try to put amino acids together in the right sequence to form a synthetic version of the protein molecule. This was difficult to do, since each time the chain had to be dissolved, a particular amino acid added—and the new chain then separated out and dissolved afresh.

In 1959 the American biochemist Robert B. Merrifield used a technique in which an amino acid was bound to beads of a resin. Additional amino acids were added one by one, but at each step, the growing chain was easily separated by filtering out the beads. Synthesizing protein chains became much more efficient. In 1970 the Chinese-American biochemist Choh Hao Li synthesized a 188-amino-acid chain of a growth hormone.

Seyfert Galaxy

ONCE THE VERY DISTANT QUASARS were discovered astronomers had to determine their nature. To be visible over such great distances they had to be up to a hundred times as luminous as an ordinary galaxy, yet they were quite small.

This was discovered when in 1963 they were found to vary in their light-emission. The variation was so rapid that the quasars had to be small, since nothing could vary in brightness as a whole unless some sort of impulse could travel from end to end in the time of variation—and nothing could travel faster than light.

It began to seem that a quasar might have the mass of a galaxy, but that the mass was concentrated into a ball only one light-year across or less, instead of the 100 000 light-year stretch across an ordinary galaxy.

It was as though an ordinary galaxy could compress itself into a quasar as an ordinary star could compress itself into a white dwarf. Was a quasar a white-dwarf galaxy?

If so might there be signs of some galaxies that were part-way along in the process? Back in 1943 an astronomer, Carl Seyfert, had studied a rather unusual galaxy. Like many other galaxies, it had a nucleus and spiral arms, but, unlike most others, the nucleus was very bright, as though something very energetic were going on there. Others were found, and by 1968 a dozen of these Seyfert galaxies were known, though probably they are much more numerous.

Seyfert galaxies are not very distant; but as the process of central compression continues the centre must get smaller and brighter. In the case of advanced Seyfert galaxies that are very distant the tiny, enormously bright centre may be the only thing left visible, and it is that centre that we detect (perhaps) as a quasar.

Sickle Cell Anaemia

IN 1910 an American physician, James H. Herrick, examined a West Indian Negro with anaemia (from Greek words meaning no blood), a condition in which the red corpuscles fail to supply adequate oxygen. (The term anaemia was first used in this way in 1849 by an English physician, Thomas Addison.)

Herrick studied the red blood corpuscles of his patient and found that instead of being normally round and flat, with a shallow depression in the middle, they were distorted into a kind of crescent shape, rather like the blade of a sickle. Herrick called them sickle cells, and named the condition sickle cell anaemia. Other people, almost invariably Negroes, were also found to suffer in this way, and it was recognized as an inherited condition.

In 1949 the American chemist Linus Pauling showed that red corpuscles sickled because they contained an abnormal form of haemoglobin. The abnormal form was named haemoglobin S, with the S standing for sickle cell, of course.

Haemoglobin S is less soluble than ordinary haemoglobin, and it has a tendency to form solid crystals within the red blood corpuscle. It is these crystals that distort the shape of the corpuscle so that it becomes less efficient in transporting oxygen. Sickle cell anaemia was the first molecular disease to be recognized, the first disease to result from the inheritance of a gene that produced a particular abnormal molecule.

In the 1950s a German-born biochemist, Vernon Ingram, working at Cambridge University, analysed both normal haemoglobin and haemoglobin S and found that each consisted of almost exactly the same chains of over six hundred amino acids. There was a difference in only two amino acids out of all those hundreds, but that was enough to produce a serious disease which few sufferers could survive into adulthood.

Silicone

THE ELEMENT which is most similar to carbon in its electron structure is silicon. Since carbon atoms easily form long chains and complex rings, it might be expected that silicon atoms would do so too. To a certain extent, indeed, they do, but silicon atoms are larger than carbon atoms and don't form nearly such firm bonds. This means that chains of silicon atoms tend to be unstable, and it is difficult to hook more than a very few of them together.

Silicon atoms, however, form a very tight bond with oxygen atoms, and, indeed, the rocky structure of earth consists very largely of minerals in which silicon and oxygen atoms are bound tightly together, along with smaller quantities of other atoms, to form silicates.

In 1899 the English chemist Frederic Kipping began a systematic study of chains of silicon and oxygen atoms in alternation. Such chains could be made of any length, and were even more stable than carbon chains. Because carbon and oxygen are bonded together in certain well-known compounds called ketones, Kipping used the same ending for his silicon-oxygen chains, and called them silicones. Each silicon atom in a silicone chain could attach itself to two different atoms or atom groupings, and the result was that a vast variety of silicones were possible, depending on the length of the chain and on the nature of the added groupings.

During and after the Second World War a variety of silicones were prepared that had useful properties. There were silicone liquids that were used as wetting agents; silicone greases that were used as lubricants; silicone resins that were used as electrical insulators; and soft silicone solids that were used as artificial rubber. In all cases they had the useful properties of being stable,

344

resistant to heat and unchanging in properties as temperatures altered.

Skeleton

TO CHILDREN (and some adults) a set of human bones is something so frightening that it is used as one of the symbols of Halloween. It is such a caricature of a human being, with its arms and legs that are too thin and its fingers and toes that are too long; with its slatted chest, its grin, its hollow eyes. It is like a human being that has completely shrivelled and dried up. The Greek word *skeletos* means dried up, and hence skeleton.

The word skull, despite its similarity in sound, has no relationship to skeleton. It comes from the same Anglo-Saxon word from which shell is derived. Just as a soft egg or nut is surrounded by a hard shell, so is the soft brain surrounded by a hard skull. The Greek word for skull is *kranion,* from which we get our word cranium.

In general various bones have both common Anglo-Saxon names, like skull, and fancier classical names, like cranium.

The classical name may be derived from the bone's appearance. The collarbone, for instance, is the clavicle, from the Latin *clavicula* (little key), because it is a long bone which bends at the end, so that it looks like an old-fashioned key. The shoulder-blade is a flat bone which is called the scapula, a Latin word which comes from the Greek *skaptein* (to dig) because it resembles the business end of a spade.

The name of a bone may also come from the part of the body in which it is found. The thighbone is the femur, which is the Latin word for thigh, and the breastbone is the sternum, from the Greek *sternon* (chest).

Again, the large curving bone that makes up the main portion of the hipbone is the ilium, which is the Latin word for flank. It is connected to the sacral vertebrae (SEE VERTEBRA) and the combined bones are referred to commonly as the sacroiliac. This is the Latin name most familiar to the general public, as far as the bones are concerned, because (alas) of the back pains that sometimes occur in that region.

Smog

WE USUALLY THINK of the atmosphere as consisting of gases such as oxygen and nitrogen. Actually it contains tiny fragments of

345

liquids and solids too. The liquids are the drops of water formed as clouds, usually high in the air, when water vapour condenses. Sometimes, however, the clouds are at ground level, and then we speak of fog.

Tiny solid particles enter the air when something burns. This happens even in the absence of any human activity. Lightning, for instance, might set a forest fire that would dump huge amounts of tiny particles of unburnt or partly burnt solids into the air. Volcanic eruptions may send cubic kilometres of fine, solid material into the stratosphere. After the great Krakatoa explosion of 1883 the dust did not settle for years. In general, solid particles suspended in air are spoken of as smoke.

In the last century men have burned coal and oil at a steadily increasing rate. If the coal and oil contained nothing but carbon and hydrogen atoms (as they would if they were absolutely pure), then only carbon dioxide and water would be formed if they were well burnt. Combustion usually isn't entirely complete, and carbon monoxide and hydrocarbon fragments are also formed. In addition, sulphates and nitrates are formed from impurities that are present.

The sulphates, particularly, tend to dissolve in any water droplets that might be present in the air, forming an acidic, irritating substance. If the air over a city is stagnant this combination of smoke and fog (called smog) may linger and accumulate, damaging lungs and eyes, making respiratory diseases worse and sometimes killing those who, through age or illness, cannot tolerate the added stress on their lungs.

The situation grew much worse after the Second World War. Killing smogs took place in Donora, Pennsylvania, in 1948, and in London in 1952. Smog is one of the growing problems with which science must deal; although clean air legislation, enforcing smokeless zones, is proving very effective.

Solar Wind

IN 1859 the English astronomer Richard Carrington, who was studying sunspots, noted a sudden brightening on the face of the sun. Almost immediately after that observation disorders were noted in the behaviour of the magnetic compass, and the Northern Lights were particularly brilliant.

Since then it has been discovered that there are periodic violent eruptions of incandescent matter on the sun, eruptions even hotter than the sun's surface generally. These, which usually take place near sunspots, are called solar flares.

The constant heaving of the sun's surface, particularly in flares, sends matter thousands of kilometres upward, and some escapes even the sun's giant gravity. As a result there are quantities of sub-atomic particles shot out into space. Chiefly because of the flares, the sun is surrounded by matter moving away from it in all directions at speeds of hundreds of kilometres an hour. About a million tonnes of matter leave the sun's surface each second, and this steady outward movement of matter has been called the solar wind.

It is the particles in this solar wind that are trapped in the magnetic lines of force of earth, making up the magnetosphere. The force of the wind streamlines the magnetosphere, making it spherical and blunt on the side towards the sun and drawn out, like a teardrop, into a long tail on the night side. The solar wind causes the magnetosphere to have a rather sharp boundary called the magnetopause.

When a flare happens to shoot upward in earth's direction the solar wind toward us is intensified. Unusual numbers of charged particles flood the magnetosphere and enter the upper atmosphere in the polar regions, creating brilliant Northern Lights, upsetting the compass and interfering with radio and television reception. This is a magnetic storm.

Solstice

THE EARTH'S axis is not perpendicular to its plane of rotation, but is tilted at about 23½ degrees, and retains that orientation as the earth revolves about the sun. As a result the axial tilt moves first towards the sun, then away from it, and repeats the process each year.

To people on earth it seems as though the noonday sun creeps higher in the sky each day (as one can tell by the shrinkage of the shadow of some vertical structure) until it reaches a maximum height. After that, from day to day, the nonday sun sinks lower and lower (and the shadow lengthens) until it reaches a minimum height. Then it repeats.

Primitive man naturally watched the sun's sinking with apprehension. (As it sank winter came on and life grew hard.) Naturally, he had no assurance that the sun would ever start going up again. When the day arrived that the sun stopped sinking and turned to rise there was general rejoicing and holiday.

The day of the year on which this happens is December 21, and this is the winter solstice. The word solstice comes from the Latin *sol* (sun) and *sistere* (to stand still). It is the day the sun stands still.

June 21 is the summer solstice, when the sun, having reached its highest point, stands still and begins to sink again.

On December 21 the sun is at zenith 23½ degrees south of the equator. That latitude is the Tropic of Capricorn, because the sun is then in the constellation Capricorn (see ZODIAC). The word tropic comes from the Greek *tropikos,* which in turn comes from *trepein* (to turn). It is at that point, you see, where the sun turns and reverses its movement.

On June 21 the sun is at zenith 23½ degrees north of the equator. This is the Tropic of Cancer, the sun being then in the constellation of Cancer. The region of the earth's surface that lies between Cancer and Capricorn is called the tropic zone, or just the tropics.

Sonar

BATS FIND their way around in the dark by producing high-pitched squeaks that bounce off obstacles in an echo. By listening for the direction of the echo, and noting the time lapse between squeak and echo, the bat knows the direction and distance of an obstacle, and can thus avoid it.

It seemed reasonable to suppose that men might develop instruments that could take advantage of this principle. What was needed were sounds sufficiently high-pitched. The higher the pitch (particularly if it were ultra-sonic—that is, too high-pitched for the human ear to detect) the further it would penetrate in a particular direction, and the more easily it would be reflected.

In 1880 the French physicist Pierre Curie and his brother Paul-Jacques devised a method of producing high-pitched sound-waves easily. In the early twentieth century the system they used was improved to the point where sound waves far in the ultra-sonic range could be produced. During the First World War, a French physicist, Paul Langevin, applied these ultra-sonic sound-waves to the detection of submarines. (Submarines are vulnerable vessels, actually, and can be put out of action easily if one only knows where they are. It is only the fact that they are hidden that makes them dangerous.)

Suppose a surface ship emitted a series of ultra-sonic waves. There would be echoes from the sea-bottom, or even from schools of fish, but one might identify an echo sent back by a submarine, and thus learn its direction and distance.

The First World War ended before Langevin had perfected his device, but by the Second World War such systems for echolocation were working. The system was called sound navigation and ranging, where ranging is used in the nautical sense to mean distance determination. Using the initial letters of the words, the system was called sonar.

Sonic Boom

AN AEROPLANE travelling through air must push air molecules out of the way. This is possible because air molecules move so quickly that when they strike the surface of the plane they can almost always bounce away more rapidly than the plane is moving.

Sound travels through the air at a rate that depends on the natural motion of the molecules. This meant that if an aeroplane travels at less than the speed of sound the air molecules can bounce away. (The speed of sound is about 1207 kilometres an hour. Smaller speeds are subsonic, from Latin words meaning less than sound, and higher speeds are supersonic, or more than sound.)

When the plane approaches the speed of sound it is travelling as fast as the air molecules, which can't get away. More and more molecules are collected in front of the plane, and a region of high pressure is set up. At first planes weren't properly designed to withstand the high pressure, so that it seemed this might be an upper limit to their speed. With the proper design, however, the American pilot Charles E. Yeager managed to fly faster than sound on October 14, 1947.

If a plane breaks the sound barrier and flies at supersonic speed it leaves the high-pressure region behind. If the plane slows up the region races on ahead. In either case, the region manages to spread out and become air of ordinary pressures in time. If, however, the plane is close to the surface and its nose is aimed somewhat downward the region of high pressure, like an enormous sound-wave, will continue moving downward and will reach the surface before it is spread out. There will then be a sonic boom, which will sound like a loud bang. The air vibrations will rattle houses, break windows, and do considerable damage if strong enough. This was one of the American objections to building the SST, or giant supersonic transport plane. Some people objected to the sonic booms it would be continually producing. Concorde and the Russian TU-144, of course, went ahead despite these objections.

Space Station

SIR ISAAC NEWTON supplied the theory that explained how an object might be placed in orbit about earth, but it was not until '1957 that sufficiently powerful rockets were built for the purpose.

It was only then that adequately advanced computers for use in plotting the rocket orbits were developed, for that matter.

The initial satellites were relatively small, only large enough for some instruments. Eventually rocket engines were made with adequate power to lift into orbit a capsule large enough to hold men. In 1969 a vehicle holding three men was fired to the moon. It retained sufficient fuel to land two of the men on the moon in a smaller vessel, and bring all three back to earth.

Even so, these larger vessels could only serve as temporary homes for human beings. Both the United States and the Soviet Union have expressed interest in something more ambitious.

It would be possible to place a very large vehicle in orbit if, instead of attempting to launch it all at once, it were sent into space in parts. Men could be sent out also to put the individual parts together and, in the end a large structure, capable of housing eighteen to thirty-six men, might be circling earth at a height of three to four hundred kilometres above its surface.

Within the structure men could carry out astronomical observations of earth, the moon, the sun, and, indeed, of the universe, being unhampered by obscuring atmosphere. An observatory established on the moon might be larger and more comfortable, but it would also be more distant and harder to reach. The structure nearer to earth is called a Manned Orbital Laboratory, abbreviated MOL, but the popular term for it is space station. The Russian space station *Salyut* was launched in April 1971, and then in May 1973 America put up her Skylab. The Russian project was initially successful, but three cosmonauts perished during re-entry after their mission; Skylab had very serious difficulties at the outset, but in the event was a brilliant success.

Spectrum

IF A BEAM of light passes from air into glass at an acute angle it is bent or refracted, from the Latin *re* (back) and *frangere* (break); the beam is broken (or bent) back. If the glass is in the form of a triangular prism the light on emerging is refracted farther in the same direction.

Sunlight consists actually of a mixture of light of varying wavelengths. These affect our eyes differently, so that we see the components of the mixture as colours. The different colours are refracted by different amounts. Red light is refracted least; orange, yellow, green and blue are refracted in increasing amounts; violet is refracted the most.

The result (as was first noted by Sir Isaac Newton in 1672) is that a beam of light that has passed through a prism and been allowed to fall on a white surface becomes a rainbow of varying colours: red at one end, through orange, yellow, green and blue to violet at the other.

The coloured strip remains pure light, and is called a spectrum, which is the Latin word for image or apparition.

Particular substances when heated to a white heat give off only certain colours. If the radiating light is passed through a slit each colour will form a sharp image of the slit, falling in a certain position in the spectrum, leaving the rest black. On the other hand, sunlight passing through a cool gas will have certain of its colours absorbed, and dark lines will appear against a coloured background. The outer layers of the sun are cool enough to do this, so that the solar spectrum is actually crossed by dark Fraunhofer lines, so called after the German optician Joseph von Fraunhofer, who first observed them in 1814.

An instrument through which one can view a spectrum against a marked scale so that the position of each bright or dark line can be located exactly is a spectroscope. From the position of the lines mankind can learn and has learned the composition of the sun and the stars.

Spiral

CURVES THAT lie in a plane and are two-dimensional have rather familiar names: circle, ellipse, oval, and so on. Partly, this is because such curves are easy to show on a blackboard or on the page of a book, and are thus more frequently talked about. Curves which extend through all three dimensions are a little less familiar.

Imagine trying to draw a circle in the air with your finger and simultaneously moving your finger away from you. The path made by your finger is like that of the groove on a screw. Such a screw-like curve is called a helix in both Greek and English.

The helix is important in physics because wires wrapped around an iron core to form an electromagnet follow such a path. A magnet can be moved through a helix of wires to set currents going. The emphasis in such a wire helix is usually on the channel within the coil through which the magnet can be moved. Such a helix is often called, for that reason, a solenoid, from the Greek *solen,* meaning channel.

The Greeks had another word for coil or twist, and that was *speira,* which may have been derived from *sparton,* meaning rope. The common rope, after all, is formed by coiling or twisting a number of fibres about a common axis. The fibres, when so coiled, automatically take up a helical shape, so the word spiral is often used as a synonym for helix. For instance, a stairway that curves upward in a helix is called a spiral stairway rather than a helical stairway.

But spiral has an additional meaning. A common example of a helix or spiral in nature is the snail shell, which coils round and round as the snail grows. (Its viscera has to be spiralled too!) But as the snail grows the shell grows, in such a way that in coiling about it makes larger circles with each coil. For this reason, spiral has come to mean a two-dimensional curve which recedes from a centre as it curves (like a snail-shell seen from the side).

Square Root

THE ANCIENT Greeks loved to arrange dots in geometrical shapes and count up the number of dots it took to make triangles, squares and so on. A perfect square could be made with four dots (two on each side), or with nine dots (three on each side), sixteen dots (four on each side) and so on. For this reason numbers like 4, 9, 16, 25, 36, 49, etc. were termed squares (Latin *quadra*). For each square there was a side (*latus* in Latin). The side of 4 was 2; the side of 9 was 3; the side of 16 was 4 and so on.

If you forget about sides and squares you might concentrate on the fact that 2×2 is 4; 3×3 is 9; 4×4 is 16 and so forth. You can therefore define a square as any number that results from the multiplication of a digit by itself. And the side of a square is the number which when multiplied by itself would give the square.

The medieval Arabs, who studied and helped to preserve Greek mathematics during the Dark Ages, were less interested in geometrical figures than were the Greeks and more interested in arithmetical relationships. To them if 2×2 was 4, then 4 was built up out of 2's as a plant is built up out of a root. They didn't think of 2 as the side of a square but as the root of something that was larger. They called it that, and we call it that to this day.

Now, dots can be arranged in cubes also. (Cube is from Greek *cubos*, a cube or die.) Eight dots can be arranged in a cube, two on each side. Twenty-seven dots can be arranged in a cube, three on each side, and so on. The Greeks called numbers like 8, 27, 64 and 125 cubes—and we do also. Arithmetically cubes are obtained by multiplying a number by itself twice: $2 \times 2 \times 2$ is 8; $3 \times 3 \times 3$ is 27; and so on. So 2 is a root of 8 as well as of 4, and 3 is the root of 27 as well as of 9. To distinguish between these roots, it is only logical to speak of square roots and cube roots. The number 2 is the square root of 4 and the cube root of 8.

You can extend this even further, so that 2 is the fourth root of 16, the fifth root of 32, and so on.

Stellarator

IN THE 1950s physicists began to attempt to confine hydrogen gas at temperatures over a hundred million degrees, in order to initiate controlled nuclear fusion and liberate vast quantities of energy.

At first they tried to confine the gas in a straight cylindrical tube, holding it away from the walls (which would cool the hot gas and end the chance of fusion) by a magnetic field. This could be done, but only for a millionth of a second or so. Some arrangement had to be worked out which would produce a more stable confinement.

One possibility was to make the straight cylinder into a round hollow tube like a doughnut. This would eliminate the ends, which were the weak points of the earlier arrangement. Unfortunately, other problems showed up. The inner edge of the doughnut was shorter than the outer edge, so that the magnetic field was more concentrated and therefore stronger along the inner edge. With an uneven magnetic field, the gas was forced to the outside of the hollow and made contact with the walls.

In 1951 the American astronomer Lyman Spitzer suggested that the doughnut be twisted into a figure eight. In that way the inner edge in one half of the figure becomes the outer edge in the other. As a result of the twist one part of the tube crosses over another part, so that the whole design carries the plasma through three dimensions.

With such a tube the hydrogen atoms could be accelerated to very high energies, and therefore temperatures. The temperatures that were obtained were similar to those at the centre of stars. The Latin words for star is *stella,* and so the figure-eight tube was named a stellarator.

Although stellarators and other devices have not yet helped scientists achieve controlled fusions, the goal is coming nearer and nearer each year.

Strangeness Number

THERE ARE two force fields involved in nuclear reactions: the strong nuclear field and the weak nuclear field. Reactions that involve the former take place in a few billionths of a billionth of a second. Reactions that involve the latter take place much more slowly, taking as much as 1/100 million seconds or even longer.

In the 1950s new particles were discovered, such as the kaons and the hyperons. These formed very rapidly, so that it was clear that they were affected by the strong nuclear field. It seemed that when they broke down they should break down by way of the strong field as well, so that after being formed they would exist only for a few billionths of a billionth of a second, as is true of resonance particles.

The kaons and hyperons, however, broke down much more slowly through the weak nuclear field. Strange, thought the physicists, and so the particles came to be called strange particles.

In 1953 the American physicist Murray Gell-Mann suggested that in strong interactions, a certain quantity had to be conserved—that is, had to be kept unchanged. It was this quantity that involved the strange particles. In the breakdown, the quantity was *not* conserved, and hence they could not break down by way of strong interactions, only by way of the weak field, where the conservation was not necessary. This conserved quantity, in strong but not weak fields, came to be known as the strangeness number.

The strangeness number for ordinary particles, such as protons, neutrons and electrons, is zero, but for strange particles it is some positive or negative integer.

Later it turned out the strangeness number could be added to something called the baryon number to give a quantity that was easier to manipulate. This new quantity, since it applied chiefly to hyperons with their various electric charges, was called hypercharge.

Streptococci

THE TERMS microbe and germ, often applied to bacteria, are really too broad for this use (see MICROBE), and, strangely enough, the term bacteria is too narrow. It is derived from the Greek *bakterion,* meaning a little rod, but not all bacteria have the shape of little rods. In fact, a bacterium that does have that shape is now referred to as a bacillus, from the Latin meaning a little rod, a diminutive of *baculus,* meaning staff.

The question of names has always plagued bacteriologists anyway, simply because they deal with creatures that are so small, mere blobs of life.

Nevertheless, names have been applied to certain broad groups on the basis of differences in appearance. For instance, many bacteria are not rod-like but are simply tiny spheres. Such a bacterium is a coccus (plural, cocci, pronounced cock's eye), from the

Greek *kokkos,* meaning grain or seed. Sometimes they are called micrococci; the Greek *mikros* means small.

Cocci sometimes form pairs with a common cell wall about the two, and these are diplococci (from the Greek *diploos,* meaning twofold). One of these causes some types of pneumonia, and is sometimes referred to as pneumococci for that reason.

Some varieties of cocci divide in such a way that the daughter cells remain attached, each dividing in its turn so that eventually long strings of them are formed which, as strings will, may curve or twist. These are streptococci, from the Greek *streptos,* meaning twisted. One variety of these can give rise to throat infections.

If cocci divide in such a way that descendants stick together in a bunch rather than simply in a string they are called staphylococci, from the Greek *staphyle,* meaning a bunch of grapes. Ordinary boils are staphylococcal infections.

Rod-like bacteria that twist into coils are spirilla (a Latin word meaning a little coil, from *spira,* a coil, see SPIRAL). The word spirilla was actually made up after Latin ceased to be a living language.

Sulphanilamide

AN ATOM combination made up of sulphur atom, three oxygen atoms and a hydrogen atom is known as a sulphonic acid group.

Now, an amine group (see AMMONIA) which replaces an oxygen and a hydrogen atom of an acid group is given the special name of amide. The special name is convenient because an amine group attached to an acid group has properties that are different from those of an amine group attached elsewhere.

A sulphonic acid to which an amine group has attached itself is therefore called a sulphonamide group. If a sulphonamide group is attached to a molecule of aniline (see ANILINE) the resulting compound is given the composite name sulphanilamide (sulph-anil-amide).

Sulphanilamide was first synthesized in 1908, but for nearly thirty years afterward was important only as a component of dye compounds. One such compound was Prontosil. In 1934 the German chemist Gerhard Domagk found that Prontosil was amazingly effective in clearing up certain types of infections. (In 1939 he was awarded the Nobel Prize in medicine for this, but was forced by Germany's Nazi Government to decline it.) In 1935 French chemists showed that it was the sulphanilamide portion of Prontosil that did the trick.

At once there was a mad scramble to prepare other drugs related

355

to sulphanilamide. These new drugs, it was thought, might be even more effective than sulphanilamide, or might attack some kinds of germ that suphanilamide did not. Over five thousand such drugs were prepared in the next ten years. For the most part changes were made by attaching various types of atom groupings to the amide group of sulphanilamide. When the chemicals pyridine, thiazole or diazine were attached, for instance, the resulting compounds were named sulphapyridine, sulphathiazole and sulphadiazine, by analogy with sulphanilamide.

This was false analogy because sulphanilamide is a name made up of sulph-anil and not sulpha-nil. Nevertheless, this has persisted, and the whole group of drugs is popularly known as the sulpha drugs.

Superconductivity

AN ELECTRIC CURRENT will flow through a metal wire; it will not flow through glass, sulphur, rubber or many other substances. Metals in general conduct an electric current from one place to another. They are electrical conductors, and have the property of conductivity.

The conductivity is never perfect. Every metal displays a certain resistance to the flow of the electric current. This resistance, acting like friction, turns some of the current to heat. If there weren't a battery or some other source of current continually working the electric current would die out almost at once, with all its energy converted to heat. This would be true even in silver wires, though silver has the highest electrical conductivity known for any ordinary substance, and the lowest resistance.

In general, vibrating atoms get in the way of the current. As temperature goes up—so that atoms vibrate more rapidly—resistance goes up. Conversely, as temperature goes down, resistance goes down.

Throughout the nineteenth century scientists reached lower and lower temperatures. In 1908 the Dutch physicist Heike Kammerlingh Onnes produced a temperature of less than $4 \cdot 2°$ above absolute zero ($4 \cdot 2°K$) and thus liquified helium, the last substance to resist liquefaction.

Using liquid helium, he measured the electrical resistance of metals at very low temperatures. He expected this resistance to become low indeed but in working with mercury he found to his surprise that at $4 \cdot 12°$ above zero ($4 \cdot 12°K$) the resistance dropped to zero. An electric current going through mercury below that temperature went on for ever, even with the source of electricity shut off. This was superconductivity. Other substances were found to be superconductive too, and in 1968 an alloy of three metals was found to be superconductive at a record temperature high of 21 °K.

Superfluidity

ONE OF THE MOST remarkable substances in the universe is helium. It is an absolutely inert gas and will not combine with anything. It is lighter than any gas but hydrogen, and it stays a gas at lower temperatures than any other known gas. It does not liquefy until a temperature of 4·2° above absolute zero (4·2°K) is reached.

Liquid helium is even more remarkable. It does not freeze even at absolute zero, unless it is placed under pressure.

It is also remarkable for its ability to conduct heat. Heat is conducted with varying ease by different substances. Metals conduct heat more rapidly than non-metals, and copper conducts it faster than any other metal. In 1935, though, the Dutch physicist William Keesom and his sister A. P. Keesom found that at temperatures below 2·2 °K liquid helium conducted heat with the speed of sound. Nothing else on earth conducted heat so rapidly.

The Russian physicist Peter Kapitza found that the reason liquid helium conducted heat so well was that it flowed with unprecedented ease, carrying heat from one part of itself to another at least two hundred times as rapidly as heat would travel through copper.

All gases and liquids have the ability to flow, and they are therefore called fluids. The rate of flow is limited by the internal friction of molecules against molecules (viscosity), but in liquid helium there seems to be almost no viscosity at all. Helium not only has fluidity, it has superfluidity.

As a result of its superfluidity, liquid helium can leak through holes too small to allow gases to leak through. Something can be gas-tight but not liquid-helium tight. The remarkable properties of helium below 2·2 °K are such that it is called helium 2, to distinguish it from the more ordinary helium 1 above that temperature.

Synchrotron

IN THE 1930s physicists developed methods for accelerating sub-atomic particles in order to give them high energies and send them smashing into atomic nuclei. The most successful of these was invented by the American physicist Ernest O. Lawrence in 1931. It whirled particles around and around (thanks to the driving force of a magnetic field), and it was therefore called a cyclotron.

By making larger and larger magnets one could whirl the particles to greater and greater energies. The device only works well, however, if the mass of the particles doesn't change. As the particles go faster their mass increases considerably (as Albert Einstein predicted they would in his special theory of relativity). This lowers the efficiency of the cyclotron, and limits the energies it can produce.

In 1945 the Soviet physicist Vladimir Veksler and the American physicist Edwin M. McMillan, each independently, worked out a way to alter the strength of a magnetic field so as just to match the increase in mass. The two effects were synchronized (from Greek words meaning at the same time) and the efficiency remained high. Such a modified cyclotron was called a synchrocyclotron.

In cyclotrons the whirling particles spiral outward, and eventually pass beyond the limits of the magnet. If the particles could be held in a tight circle they could be whirled many more times before being released, and still higher energies would be attained.

The English physicist Marcus Oliphant worked out a design for such a device in 1947, and in 1952 the first of the kind was built in Brookhaven National Laboratory on Long Island. It still made use of a synchronized increase in the strength of the field, but the spiralling of the particles, as in a cyclotron, was gone. The new device was therefore called simply a synchrotron.

Tachyon

THE SPECIAL THEORY of relativity, presented by Albert Einstein in 1905, says it is impossible for any particle with mass to reach or surpass the velocity of light in a vacuum (299792·5 kilometres per second). On the other hand, particles of zero mass (like the photons that make up light) must travel through a vacuum at exactly 299792·5 kilometres per second at all times.

According to relativity, any particle moving at faster-than-light velocities would have to have an imaginary mass. The mass would be some quantity multiplied by the square root of −1.

In 1962 a Russian-American physicist, Olexa-Myron Bilaniuk, and an Indian-American co-worker, George Shidarshan, pointed out that particles with imaginary mass might exist without violating relativity. They would merely be required to move *always* at a speed faster than light. Such particles would slow down as they gained energy, but no matter how much energy they gained, they could never slow down to the velocity of light. That was the limit for these particles, as for ordinary particles, but from the other direction.

In 1967 the American physicist Gerald Feinberg popularized the notion and gave the faster-than-light particles the name tachyons, from a Greek word meaning fast. Bilaniuk and Shidarshan then suggested tardyon (from a Latin word for slow) as the name for all particles that travel more slowly than light. As for particles without mass that travel at exactly the speed of light (such as photons, neutrinos and gravitons), Bilaniuk and Shidarshan suggested the name luxons, from the Latin word for light.

Tachyons ought to leave a trail of light as they travel through a vacuum and therefore might be detected. However, they would move so fast that such flashes would last for unimaginably brief periods of time, and so far tachyons have not been detected.

Tantalum

IN GREEK mythology Tantalus was a king of Lydia who had seriously offended the gods. He was therefore condemned to suffer harsh punishment in Hades. He stood in water up to his neck, but when driven by thirst he stooped to drink the water swirled down and out of sight. Branches laden with fruit dangled within inches of his face, but when, driven by hunger, he reached out to eat they swayed just out of his grasp. The king has left his name in our language in the word tantalize.

His name enters chemistry too, as an element. In 1802 the Swedish chemist Anders Gustaf Eckberg discovered element number 73, and thereafter for a number of years there was some dispute as to whether it was really a new and distinct element, and if so what its name should be. In 1814 the Swedish chemist Jöns Jakob Berzelius, who was the great chemical authority of the day, decided it was indeed a new element, and also decided in favour of the name tantalum.

He reasoned that the new element was unusual for a metal in resisting the action of acids, even of aqua regia (see AQUA REGIA). Though it stood in acid, in other words, it was not affected by it, any more than was Tantalus by the water in which he stood.

Some people have suggested that tantalum received its name because the discoverer had been tantalized by near misses before finding it. Though this is apparently a false derivation, there are other elements that do bear witness to their discoverers' difficulties.

For instance, the Swedish chemist Carl Gustav Mosander discovered element 57, and called it lanthanum, from the Greek *lanthanein* (to escape notice) because he found it so difficult to isolate. This element is the first of the rare earth elements (see YTTRIUM) and in consequence this rather peevish name is applied to the whole series of rare earth elements, which are termed the lanthanides. Another of these elements (number 66) was discovered in 1886 by the French chemist Lecoq de Boisbaudran, who named it dysprosium, from the Greek *dysprositos* (hard to get at).

Technetium

BY 1925 all but four elements in the periodic table (see ISOTOPE) had been discovered. Two of these holes in the table were among the heavy radioactive elements and were expected to be rare and hard to find. The other two, in positions 43 and 61, were surrounded by stable elements, and it seemed there ought to be no great problem in locating them.

Chemists concentrated on the search and there were a number of reports that one or the other had been detected. For instance, in 1925 three German chemists reported the detection of element 43 and named it masurium after Masuria, a district in East Prussia (now part of Poland). For fifteen years periodic tables listed masurium (but with a question mark).

The next year American chemists at the University of Illinois and Italian chemists at the University of Florence reported the detection of element 61. The first group called it illinium, the second florentium, after their respective universities. There was quite an argument about it, but American periodic tables listed illinium (again with a question mark).

It has since turned out that everyone was wrong. Elements 43 and 61 are radioactive, and do not exist naturally on earth. They can, however, be formed by nuclear reactions which, since 1919, mankind has known how to bring about. In 1936, for instance, the American physicist Ernest Lawrence bombarded the element molybdenum (element 42) with sub-atomic particles, and investigation showed that small quantities of element 43 had been formed. Eventually element 43 was named technetium (now the official name), from the Greek word *technetos,* meaning artificial, since technetium was the first element to be discovered through formation by artificial means.

In 1945 element 61 was found by American chemists at Oak Ridge among the uranium fission fragments. It was named promethium after the Greek demigod Prometheus, who had brought down fire to mankind from the sun. The new element, after all, had come out of the man-made sun of uranium fission.

Teflon

ATOMS OF CARBON have the unusual property of being able to join together in long chains, straight or branched, and in complicated systems of rings. Carbon atoms, so hooked up, can attach other atoms to themselves. In particular they can be attached to hydrogen atoms, which are the smallest of all atoms, and most easily fit into the angles where the carbon atoms form branches and rings. For this reason there are many hundreds of hydrocarbons that have been found or have been synthesized, and countless millions more can exist.

The only element with atoms small enough to substitute for hydrogen in this respect is fluorine. Fluorine is so hard to work with that fluorocarbons remained virtually unknown right down to modern times. The first example, and the simplest, carbon tetrafluoride (CF_4), with one carbon atom and four fluorines, was prepared in 1926.

During the Second World War uranium hexafluoride (UF_6) was used in connection with nuclear bomb research. Chemists began to study fluorocarbons in earnest. Fluorine atoms, it turned out, hold on to other atoms so strongly that they simply don't let go. Because they hang on as they do they don't engage in chemical reactions, so that fluorocarbons are much more inert than are hydrocarbons. They don't even stick physically to other substances.

One of the fluorocarbons is tetrafluoroethylene ($CF_2 = CF_2$). Carbon is tetravalent (Greek *tetra-* combining form for four, Latin *valere,* to be worth), fluorine monovalent, (Greek *monos,* one), so that each carbon atom is linked to two fluorine atoms, and to the two spare 'places' on a fellow carbon atom, forming a double bond. Molecules of this compound can thus be made to hook together to form a long chain of carbon atoms, each with two fluorine atoms attached. The proper name for the chain would be polytetrafluoroethylene, the prefix coming from a Greek word for many. The chemists who prepared the chain called it Teflon for short (a brand name), taking the letters from tetrafluoro.

A thin coating of Teflon on frying-pans will resist heat, and won't stick to food. For this reason Teflon pans are easier to clean than ordinary ones, and require little or no fat to be used when frying meals.

Tektite

BACK IN THE eighteenth century, numbers of small, rounded pieces of greenish glass were discovered in what is now Czechoslovakia. In the last century such pieces of glass have been found in other places: the Philippines, Australia, Texas and the Ivory Coast. Some have been found on the sea bottom near those regions.

About 1900 an Austrian geologist, Eduard Suess, suggested they might be meteorites. Their composition was something like that of stony meteorites but they might have melted in their passage through the atmosphere and then cooled into a glassy substance. In fact, their shape was what would be expected of something moving rapidly through the air against atmospheric resistance.

This theory has come to be accepted. These objects are now thought to be meteorites that have partly melted in passage, and so are called tektites (from a Greek word meaning to melt). The youngest ones, from the Far East, seem to have fallen only 700 000 years ago.

One suggestion is that the tektites originated on the moon. Large meteor strikes, like that which formed the crater Tycho, might have splashed lunar material upward fast enough to allow it to escape the moon's small gravitational field. Some of the material then moved far enough to be captured by earth.

In 1969 scientists working on ways of keeping men alive under conditions in space decided it would be helpful to work on similar problems involving the undersea world. Four men spent two months 15 metres under the surface of the sea in small quarters designed to supply them with the necessities of life. This was called the Tektite Project, because the work had been concerned with space to begin with and had ended under the sea, rather like the tektites themselves.

Teleostei

ALTHOUGH vertebrate is a common term used to include all animals that have bony skeletons (see PHYLUM), there are actually some creatures that are quite similar to the vertebrates, similar enough to be classified with them, which yet do not have a bony skeleton. A skeleton, yes, but not of bone.

There is a time when all vertebrates have a skeleton not of bone but of cartilage (from the Latin *cartilago,* of uncertain derivation, which has replaced the Anglo-Saxon gristle). Cartilage is a tough, flexible tissue which gives support but is not particularly hard. It develops first and bone follows. A baby's skeleton is largely cartilage, and hardens only slowly over the years as it turns to bone, or ossifies (from the Latin *os* for bone and *ficare,* meaning to make). Adults retain cartilage in many places, notably in the ear and in the tip of the nose.

In some animals the skeleton remains cartilage throughout life. The best examples are found among the animals in the class Pisces (the Latin plural word for fish, which is the type of animal mainly contained in the class). Fish itself comes from the Anglo-Saxon *fisc,* which is obviously related to the Latin word.

Ordinary fish—cod, mackerel, salmon, herring—have skeletons of bone, and are sometimes called bony fish to emphasize that fact. Zoologists put them all in the sub-class Teleostei, from the Greek *teleos* (complete) and *osteon* (bone). The skeletons of these fish are completely bone.

However, included in Pisces is another sub-class consisting of creatures which in some cases look very much like bony fish (with some minor but characteristic differences) but which have skeletons made up entirely of cartilage through life. These include the sharks and rays. These are the Elasmobranchii, from the Greek *elasma* (a metal plate) and *branchia* (gills) because their gills have a flat plate-like structure.

Telescope

A COMMON plaything today is the glass lens which will magnify print and focus sun's rays sufficiently to scorch paper. (Lens is a Latin word meaning lentil. The most common form of lens is shaped like a lentil seed.)

But lenses are much more than toys. In 1608 a Dutch spectacle-maker named Jan Lippershey placed two lenses in a tube and made distant things appear close. He applied for a patent, but the Dutch Government scented a secret weapon, refused the patent, but bought all rights and told Lippershey to keep experimenting.

Secrecy, as usual, did no good. Rumour of the discovery spread, and in 1609 the Italian physicist Galileo Galilei reinvented a similar instrument and began exploring the heavens. The instrument was called a telescope, from the Greek *tele* (distant) and *skopein* (to watch). It gave man the means to watch the distance.

In rapid succession Galileo discovered mountains on the moon, spots on the sun, the phases of Venus and the four largest moons of Jupiter, still called the Galilean satellites in his honour.

The use of -scope as a suffix in the names of scientific instruments that help man to 'see the unseen' is widespread (see SPECTRUM, MICROBE, COSMIC RAY). However, man can watch his environment by senses other than sight, and in the case of one very familiar instrument, it is the sense of hearing that counts.

For centuries, doctors have tried to gain knowledge of what is going on within the chest by placing the ear against the chest wall to catch the noise of the heartbeat and of the pumping air. This is known as auscultation, from the Latin *auscultare* (to listen). In 1819 the French physician René Laennec devised a tube, one end of which could be placed against the chest to sharpen and make clearer the noises within to an ear placed at the other end. That was the birth of the indispensable stethoscope of the modern physician, named from the Greek *stethos* (breast).

Terametre

THE METRIC SYSTEM, as first worked out by its French originators in 1795, had kilo- (from a Greek word meaning thousand) as a prefix for its largest measurement. Thus a kilometre is a thousand metres and a kilogramme is a thousand grammes. (The prefix is used so familiarly now even outside the metric system that a kiloton is a thousand tons.)

Scientists began to find it convenient to use prefixes for still larger measurements. They began to use the prefixes myria- and mega- for the purpose. These are from Greek words meaning ten thousand and great, respectively. Thus a myriametre is equal to ten thousand metres and a megametre to a million metres.

In 1958 there was international agreement to make the prefix mega- official and to add two more prefixes that would signify even larger measurements still. A thousand megametres would be a gigametre, from a Greek word meaning giant, and a thousand gigametres would be a terametre, from a Greek word meaning monster.

Such units could be convenient astronomically. For instance, the circumference of earth is about 40 megametres and the moon is about 380 megametres from earth.

Venus is about 42 gigametres from earth at its closest, and the sun is 145 gigametres away. Jupiter, when at its farthest from earth, is not quite a terametre distant. Pluto's orbit from side to side is not quite 12 terametres wide.

Even the largest prefix yet used, when applied to the metre, takes in only the solar system, which is a mere speck in the universe. In one year light travels a little less than ten billion kilometres or nearly 10 000 terametres. This is one light-year, and even the nearest star is 4·3 light-years away.

Terpene

THERE IS A SMALL tree growing on the shores of the Mediterranean which the Greeks called *terebinthinos*. If the bark of the tree is cut a yellow, sticky fluid wells out, which hardens after a while on exposure to air. This substance is called turpentine, which is merely a corruption of *terebinthinos*.

Terebinth turpentine is now called Chian turpentine because it was originally collected on the island of Chios in the Aegean Sea. A more important modern source is a number of cone-bearing trees (or conifers, the Latin *ferre* meaning to bear) such as pines and firs.

If crude turpentine is heated with boiling water some of it is driven off with the steam. If this portion is trapped and cooled an oily liquid called spirits of turpentine (for the meaning of spirit, see ETHER) results. What is left behind is a yellow-brown brittle solid called rosin. This word is a corruption of resin, which is the name given to the general class of gummy fluids that flow from the bark of trees and harden in air. The Greek word for it was *rhetine,* a close relative of *rheein,* meaning to flow.

Spirits of turpentine contains a number of organic molecules, each of which contains ten carbon atoms arranged in such a way that they can be divided into five-carbon halves, with each half consisting of four carbon atoms in a line and a fifth carbon attached to the second. Such ten-carbon compounds are terpenes, from turpentine.

The five-carbon unit out of which terpenes are built was called isoprene in 1860 by a chemist named C. G. Williams. The iso-prefix is often used in naming carbon chains with one carbon atom branching off a second carbon in line, but the -prene part seems to have been improvised and to have no special meaning. Because isoprene is half a terpene molecule, it is sometimes called hemi-terpene, the Greek prefix *hemi-* meaning half.

Thermodynamics

THE STEAM-ENGINE, developed and made practical in the eighteenth century, was quite inefficient. Very little of the fuel consumed in heating water and making steam was converted into useful work. Scientists grew interested in studying the way in which heat flowed from one point to another and the manner in which it was converted to work. They hoped to learn to make the steam-engine more efficient and, perhaps, to understand the universe better. In this way a new science, thermodynamics (from Greek words meaning heat movement) came into being.

Anything which could be converted into work was called energy. Heat, therefore, as well as certain other phenomena—such as light, sound, motion, and so on—came under that heading. By the 1840s three different physicists, the Englishman James Joule and two Germans, Julius von Mayer and Hermann von Helmholtz, were convinced that if all the forms of energy were lumped together the total remained constant in any closed system. Energy might be changed from one form to another but it could neither be created out of nothing, nor totally destroyed. This is called the law of conservation of energy.

It is also called the first law of thermodynamics because the new science could not be properly understood unless to begin with it was accepted that energy was conserved. Conservation of energy may be the most basic rule that scientists have yet discovered concerning the universe. No case has ever been found in which it doesn't hold.

On one occasion there was serious doubt. In the 1890s radioactivity was discovered and large quantities of energy seemed to be appearing out of nowhere within the atoms. But then in 1905 the German-born physicist Albert Einstein showed that mass was a form of energy, and that the huge energies produced by radioactivity were compensated by the disappearance of tiny amounts of mass.

Thermonuclear Reaction

ALTHOUGH THE first practical use of energy derived from nuclear reaction involved the splitting of large nuclei (see FISSION), scientists had speculated about the possibility of energy from the joining of small nuclei long before anyone had even conceived of splitting large ones. To combine four hydrogen nuclei, for example, into one helium nucleus would liberate tremendous energy. This process is called nuclear fusion, from the Latin *fundere* (past participle *fusus*) meaning to melt. Four lead pellets, for instance, if melted would run together and form one larger pellet, if cooled again. So fuse came to mean the joining of many small pieces into one large one.

The trouble in achieving nuclear fusion was that the hydrogen electrons got in the way. The hydrogen nuclei just couldn't be forced close enough to fuse by anything scientists could do until the A-bomb was invented (see NUCLEAR REACTOR). The A-bomb developed a heat of about a hundred million degrees and under certain circumstances this was enough to strip away the electrons and bang the hydrogen nuclei together hard enough to cause them to fuse. An explosion much more powerful than an ordinary atomic explosion resulted.

This new type of bomb was as much an atomic bomb as the old A-bomb, or as little, since both are really nuclear bombs (see NUCLEAR REACTOR), but to distinguish between them the new bomb, which fused hydrogen nuclei, was called a hydrogen bomb, or an H-bomb. There was some attempt to call the older bomb a uranium bomb or a U-bomb, but that didn't catch on. A more logical division, which is becoming popular, is to call the H-bomb a fusion bomb and the ordinary A-bomb a fission bomb.

Ordinary nuclear reactions, including fission, are brought about in the laboratory by bombarding atomic nuclei with sub-atomic particles. Hydrogen fusion, however, is brought about by sheer heat. The Greek word for heat is *therme,* so fusion is a thermonuclear reaction, and the H-bomb is sometimes called a thermonuclear device.

Thermosphere

IT IS CLEAR that temperature drops steadily as one climbs mountains. Very high mountains have their peaks covered with snow even in the summer—even on the equator. In 1646 the French mathematician Blaise Pascal sent his brother-in-law up a mountain-side with a barometer, and found that air pressure decreased with height. That meant that air grew less dense with height.

It is not surprising, then, that temperature drops with altitude, since the thinner air is, the less heat it can hold. In the 1890s the French meteorologist Léon Teisserenc de Bort sent thermometers up in unmanned balloons and recorded a continuing drop of temperature. At a height of sixteen kilometres it was down to —55°C.

The thinner the air the less heat it can hold altogether. But as the air grows thinner the atoms and molecules within it become fewer, and the heat striking it from the sun need be distributed among only those fewer atoms and molecules. Each atom or molecule has a larger share of heat, and the temperature (heat content per atom) tends to rise.

This latter effect becomes more marked the higher one goes. Up to a height of from eight to sixteen kilometres, depending on latitude, the temperature does drop steadily. As one goes still higher the second effect gains and for a while the temperature remains steady.

Eventually, at heights of beyond 80 kilometres the temperature actually starts to go up, as revealed by rocket measurements in the 1950s. There continues to be less and less total heat, since there are so few atoms or molecules that even at high temperatures very little heat can be stored altogether. Nevertheless, by a height of 480 kilometres the temperature has reached 1000°C. This region of very thin air with high temperatures is sometimes called the thermosphere, from Greek words meaning heat sphere.

Thiophene

THE COMPOUND benzene is usually obtained from petroleum or from coal tar. When it is so obtained, unless special precautions are taken, small quantities of another substance accompany it. The benzene molecule, you see, is made up of six carbon atoms in a ring. The molecule of the other substance is made up of four carbon atoms and a sulphur atom in a ring. The sulphur atom apparently takes up just the space of two carbon atoms, so that the molecules are sufficiently alike in shape to respond to similar treatment. Whatever course of events isolates one, isolates the other.

Chemists were so unaware of the presence of the impurity that they used to test for the presence of benzene in a liquid by adding some concentrated sulphuric acid and a crystal of a substance called isatin. The result was a beautiful blue colour which did not appear unless benzene was present in the liquid. What the chemists didn't know for quite a while was that the impurity gave the colour and not the benzene.

The German chemist Victor Meyer used to demonstrate this test to his class until one day in 1883 Meyer's laboratory assistant supplied him with some benzene prepared in a new way—from chemically pure benzoic acid. This new sample had only benzene in it and no impurity. Naturally, it did not give the test. One can imagine the scholarly Meyer staring dumbfounded at the test-tube and shaking it helplessly while his class longed to laugh and dared not.

Meyer did not let the matter drop. He investigated the peculiar happening, and did not rest until he had located the impurity and worked out its structure. (The moral to this story is that the good research scientist pays strict attention to any peculiar event, and does not shrug off a mystery as just one of those things.)

The benzene twin was named thiophene. The thio- prefix in

chemical names invariably implies the presence of a sulphur atom in the molecule, from the Greek *theion* (sulphur), while -phene is from *pheno,* one of the early names of benzene (see BENZENE). Thiophene therefore means sulphur-benzene, a good and descriptive name.

Thyroid

WHEN THE GREEKS of the heroic age (as in the *Iliad*) went into battle, they took care to take along a large shield behind which they could remain whenever possible. Ajax (the Grecian fighter second only to Achilles in valour) was noted for his curved, oblong shield that reached from neck to ankles. Now, the Greek *thyra* means door, so such a large shield was called a *thyreos* (carrying such a shield was like carrying a door).

Such a really large shield had a notch on top over which the hero's head could take a quick look when necessary. If you put your fingers on your larynx (a Greek word meaning throat or gullet) or Adam's apple (so called because medieval legend had it that Adam choked on the apple and part of it stuck in his throat) you will notice that it has a notch on top. It is therefore shield-shaped, and was named the thyroid cartilage (the suffix -oid comes from the Greek *oeides* meaning form). Or, of course, it may simply be that the ancients thought of the Adam's apple as a door in the throat.

In any case, the name spread. A gland in the close neighbourhood of this cartilage was named the thyroid gland. In one sense the thyroid gland is indeed a shield, since it secretes a hormone (the most familiar form of which is called thyroxin or thyroxine) which regulates the rate at which the body burns its fuel and uses up oxygen. The gland thus shields us against improper rate of living. The discoverer (in 1915), Dr Edward C. Kendall of the Mayo Clinic, added the ox of oxygen and the -in suffix common for hormones to the thyr of thyroid and got the name.

An unfortunate child born without the ability to produce thyroxine turns into a dwarfed idiot called a cretin. This is a corruption of the French *chrétien,* meaning Christian. It is fairly common, you see, among people who do not understand mental ailments to consider a mentally deficient person to be under the special care of the gods. Our word silly, for instance, is a corruption of the Anglo-Saxon *saelig* (and the modern German *selig*) which means blessed. (Perhaps this is because mental deficients seem childishly happy at times and may seem unaware of the cares and troubles of the world. There may be blessing in that, after all.)

Tide

JUST AS the earth's gravity holds the moon in its orbit, so the moon's gravity has its effect on the earth. Our satellite pulls the water of the ocean towards itself slightly, so that the water piles up into a kind of hill in the direction of the moon. It pulls the earth itself away from the water on the side of the earth opposite to itself, so that another hill forms on that side.

If the earth were entirely covered with water these hills would seem to move around the earth as it rotated, keeping themselves always lined up with the moon, but lagging a little because of the earth's movement. If an island existed, then, twice a day (at progressively later times, since the moon was moving too, and not staying in one place), one of the water-hills would reach it. The water-level would rise and then recede again as the hill passed.

This actually does happen along shores, but because the land masses get in the way and break up what would otherwise be an orderly progression of the water-hills, the rise and fall of the water doesn't follow the moon quite as closely and as obviously as it would otherwise. Also, depending on the shape of the shore, the rise and fall can make a difference of 18 metres in a funnel-shaped bay like the Bay of Fundy or only a few centimetres in a nearly land-locked sea like the Mediterranean. So the connection of the water-hills and the moon went unnoticed until early modern times.

What was strongly noticed, however, was the regularity of the rise and fall. Those who sailed vessels had to know when the water was high and when low, and they found they could count on this in advance. So the rise and fall was named tide, from the Anglo-Saxon *tid,* meaning time. The tides were so regular they could be used as a measure of time. The old meaning of the word tide is still preserved in archaic words such as eventide and Yuletide. When we speak of time and tide we are using synonyms, as in spick and span, hale and hearty and nook and cranny.

TIROS

THE FIRST ARTIFICIAL SATELLITES, launched in 1957, scarcely did more than show that the feat was possible. Then came satellites carrying instruments designed to study the upper atmosphere and the space just beyond. It was not long, however, before men began to use satellites in an attempt to study earth itself.

We are right on the surface of earth and can only see a small portion of it even when we rise as high as we can in an aeroplane or balloon. In order to understand the system of earth's air circulation as a whole, it is necessary to take measurements in many places at many times and try to correlate the results. And yet it is only in certain parts of the world that such measurements are taken continually. Over vast stretches of sea and land few or no measurements are taken. This adds to the difficulties of long-term weather forecasting.

But what if a satellite, circling the earth at a height of some 300 kilometres above the surface, continually took pictures of the cloud cover? Information could be gathered in a few minutes that could not be gained in any other way. On April 1, 1960, a satellite was launched with two television-type cameras and with equipment for storing photographs and transmitting them to earth on command. It was called Television and Infra-red Observation Satellite 1 or, using the initials, TIROS 1.

TIROS 1 was completely successful, as were other satellites of the same type. Thousands of pictures of earth's cloud cover were taken and transmitted. The view of earth with its spiralling clouds became familiar to all. Hurricanes were seen at their very start, and were kept in view at all times. The overall changes of the cloud cover with the seasons were followed. It was clear that the satellite programme was no mere stunt; it could be, and was, of important and practical use. (See also NIMBUS.)

Tranquillizer

MEN HAVE ALWAYS looked for some way of inducing peace of mind and of making themselves feel better. In the physical sphere, they have used natural products for the purpose—smoked tobacco, chewed betel nuts and cacao leaves, drunk fermented fruit-juices and extracts of the poppy.

Modern chemistry invaded this area at the turn of the century. In 1882 a new organic chemical was discovered which was called barbituric acid. (The name was supposed to have arisen because the chemist who named it had a girl friend called Barbara.) In 1903 the German chemist Emil Fischer discovered that by attracting certain atom groupings to the barbituric acid molecule he obtained a substance that had a soothing effect when swallowed. This and other similar compounds are grouped together as barbiturates.

The barbiturates, when taken in the proper dosages, do not relieve pain (they are not analgesics, from Greek words meaning no pain), but they do seem to allay minor discomfort and anxiety and succeed in calming people. They are sedatives (from a Greek word meaning to calm).

371

Barbiturates induce sleep, however. The larger the dose the deeper the sleep, and the more difficult the arousal. Sleeping pills, if enough are taken, become a way of committing suicide.

In 1952 a drug was extracted from the dried roots of an Indian plant with the scientific name of *Rauwolfia serpentinum* (named after a sixteenth-century German botanist, Leonhard Rauwolf), which had been used as a calming influence by inhabitants of the regions in which it grew. The drug extracted was named reserpine (an abbreviation of the Latin name).

It was the first of the tranquillizers, drugs that reduced excitement and made a person tranquil without, however, inducing sleep or making him difficult to rouse.

Transduction

IN THE FIRST THIRD of the twentieth century, some biochemists were studying chromosomes in the cell nuclei, and others were studying viruses. Viruses are objects which are much smaller than cells—indeed, about the size of chromosomes. They don't seem to grow and multiply in the ordinary manner, but once they get inside a cell they multiply very efficiently.

Both chromosomes and viruses were found to contain nucleic acid as a key component, and biochemists began to speculate that a virus might be a kind of independent chromosome.

A chromosome is made up of a chain of genes which are capable of directing the formation of certain protein molecules, making use of the complex chemical machinery of the cell to do so. A virus when it gets into a cell somehow takes over, making use of the cell machinery to form proteins necessary for its own purpose (rather than the cell's). It is a kind of intracellular parasitism.

Viruses, it would seem, can be so intimately involved with the cell machinery as to become a more or less permanent part of it. One or more genes of a virus may permanently join the genes in the cell's chromosomes, dividing when the rest of the chromosome divides, and passing on into the daughter cells. This may explain the ability of some viruses to affect a host organism permanently after a single infection.

The phenomenon whereby a virus adds on to a chromosome and produces a permanent change in the characteristics of a cell—one that is carried on through generations—is transduction, from Latin words meaning to lead across. Possibly some day biochemists will use viruses to add a desirable gene to a cell nucleus that lacks it. This is one of the techniques which may be used to change and modify the machinery of inheritance in a human being—the genetic engineering of the future.

Transfer RNA

THE STRUCTURE of the various molecules of deoxyribonucleic acid (DNA) in the chromosomes determines the structure of the enzymes produced in a particular cell. The DNA molecules, made up of chains of nucleotides, are in the nucleus, however, while the enzymes, made up of chains of amino acids, are in the cytoplasm.

Molecules of ribonucleic acid (RNA) pass from nucleus to cytoplasm. A particular RNA molecule is formed with a structure modelled on that of a particular DNA molecule. This messenger RNA travels into the cytoplasm, where it controls the formation of an enzyme molecule.

For this every different combination of three nucleotides (a codon) along the chain of the RNA molecule must represent a particular amino acid—but how is that done?

In 1955 the American biochemist Mahlon B. Hoagland discovered small molecules of RNA (much smaller than messenger RNA) dissolved in the cell fluid of the cytoplasm. These came in a number of different varieties. Each variety had a particular combination of three nucleotides at one end, which would fit on one particular codon of the messenger RNA. At the other end of the small RNA molecule an amino acid could be fitted, but only a particular kind.

The three particular nucleotides at one end of the small RNA always went along with one particular amino acid. This meant that when a whole series of these small molecules fitted itself on to the messenger RNA chain, the nature and order of those molecules were determined by the nature and order of the codons on the messenger RNA chain; the nature and order of the amino acids at the other end of the small molecules were also determined; and a particular enzyme molecule was formed.

Because the small molecules transferred the information from the messenger RNA to the enzyme, they were named transfer RNA.

Transistor

FOR ELECTRONIC DEVICES to work well, electric currents must be delicately controlled. They must be started, stopped, amplified or altered, in a tiny fraction of a second. The first device that made this possible was an evacuated bulb called a valve in Great Britain and a tube in the United States. A heated filament within the valve released a flood of electrons that streamed across the vacuum within. It was this flow of light electrons that could be easily and quickly controlled. Valves, however, were bulky and fragile, and took time to heat.

The situation changed in the 1940s when an English-American physicist, William Shockley, and his co-workers studied semiconductors and their manner of working. The semiconductors existed in two types, some with a small surplus of electrons and some with a small deficit. In 1948 Shockley found that he could prepare combinations of these two types in such a way that the current passing through them could be controlled as conveniently and in the same manner as the electron flow through a vacuum.

Of course, the electron flow through a semiconductor did not progress with the almost total absence of resistance that characterizes electron flow through a vacuum. In the semiconductors, the electrons are transferred across a resistance. John R. Pierce, who worked for the same company as Shockley's group, suggested that the new device for controlling electron flow be given a name taken from the phrase *trans*-ferred across a re-*sist*-ance. It came to be called a transistor.

Where there had been tubes there could now be tiny transistors. As a result radios and other electronic equipment could be made much smaller than before, and more rugged. What's more, radios would start the moment they were turned on. Since there was no filament to heat, there was no warm-up period. Transistors also replaced the tubes in computers, and in consequence, these were drastically reduced in size.

Transuranium Elements

IN 1789 the German chemist Martin Klaproth discovered a new element which he named uranium after the planet Uranus, which had been discovered only eight years before.

In the first half of the ninteenth century the relative masses (or atomic weights) of the atoms of the various elements were determined. It finally appeared that uranium had the most massive atoms of any known element. Although many more elements were discovered in following decades, uranium retained this pride of place. Its atomic weight (238) was higher than that of any other element even as late as 1940.

When the English physicist Henry Moseley worked out a

method for determining the electric charge on an atomic nucleus (the atomic number) it turned out that uranium had the highest of all—92. By 1940 chemists were quite certain that uranium atoms were the most complex that occurred naturally on earth. If more complex atoms ever existed they were probably so radioactive that they broke down and were gone aeons before man appeared on earth. (Uranium itself is radioactive also, but breaks down very slowly—see HALF-LIFE).

Might not more complex atoms be made in the laboratory, though? In 1934 the Italian-born physicist Enrico Fermi bombarded uranium atoms with neutrons, hoping to form element 93. For a while he thought he had succeeded, and called the new element Uranium X. He was, however, causing the uranium atom to split (uranium fission), and this confused matters.

With Fermi switching to the study of uranium fission, it was left to two American-born physicists, Edwin M. McMillan and Philip H. Abelson, to isolate element 93 in 1940. They named it neptunium after Neptune, the planet beyond Uranus. Neptunium was the first of the transuranium elements (beyond uranium). Since 1940 no fewer than thirteen more, up to element 105, have been prepared in the laboratory.

Trypsin

AMONG THE EARLIEST enzymes discovered were those in the digestive juices which dissolved or liquefied meat. This ability was reflected in the naming of pepsin (see ENZYME) which is found in the stomach juices. Other digestive enzymes were later discovered, and the name pepsin inspired the later namings.

In 1874, for instance, the German physiologist Willy Kühne discovered an enzyme in the juice formed by the pancreas which resembled pepsin in action. Since he obtained this enzyme by rubbing and grinding the pancreas in glycerol, he named the enzyme trypsin, a combination of the Greek *tribein* (to rub) and *pepsin*. It was, in other words, a kind of pepsin obtained by rubbing. When a second and similar enzyme was found in the pancreatic juice it was called chymotrypsin, the prefix chymo-coming from the Greek *chymos* (juice).

The inner lining of the small intestine contains small glands that produce intestinal juice. In 1901 the German pathologist Julius Cohnheim isolated still another enzyme, related in action to the ones already mentioned. This he called erepsin, from the Greek *ereptesthai* (to feed upon) because its enzyme action was a kind of feeding upon substances from meat. (Still, I have a notion that the similarity of the name erepsin to pepsin also influenced him.)

All these enzymes work by splitting up the large protein molecules into smaller pieces. They are therefore called proteases. (The -ase suffix is now accepted as signifying an enzyme.) Eventually the individual amino acids making up the protein molecules are set free. One important amino acid was first isolated in 1900 by the British biochemists Frederick G. Hopkins and S. W. Cole from protein that had been digested by trypsin. They named the amino acid tryptophane, the suffix -phane coming from the Greek *phainein* (to appear). It had appeared, so to speak, as the result of trypsin's activity.

Tsunami

WHEN AN EARTHQUAKE takes place on one of the continents, it can do a great deal of damage. It might seem that a quake far out at sea, however, could be ignored. The water would shake a bit, but surely no one would be hurt. And yet an earthquake at sea can be more dreadful than one on land.

The seaquake will set up a wave that is not very high in mid-ocean but stretches across the surface for an enormous distance and therefore involves a large volume of water. Such a wave spreads outward in all directions from the point at which the quake took place. As it approaches land and as the ocean gets shallower, the stretch of water in the wave is compressed front and rear and piles up higher, then much higher. If the wave moves into a narrowing harbour its volume is forced still higher, sometimes fifteen to twenty metres high.

That tower of water, coming suddenly and without warning, can break over a city, drowning thousands. Before the wave comes in, the preceding trough arrives. The water sucks far out, like an enormous low tide, and then the wave comes in like a colossal high tide. Because of this out-and-in effect, the huge wave has been called a tidal wave. This is a poor name, though, for it has nothing to do with the tides.

In recent decades the name tsunami has been used more and more frequently. This is a Japanese word meaning harbour wave, which is an accurate description. The Japanese, living on islands near the edge of our largest ocean, have suffered a great deal from tsunamis.

The greatest tsunami in recent years was in 1883, when the volcanic island Krakatoa in the East Indies exploded and sent thirty-metre waves crashing into near-by shores. About 1400 B.C., an Aegean island exploded and a tsunami destroyed the civiliza-

376

tion on the near-by island of Crete. Yet a third famous earthquake
and tsunami destroyed the city of Lisbon in 1755.

UFO

ON JUNE 24, 1947, a Seattle businessman, Kenneth Arnold, saw
from a plane a series of shining disc-like objects moving through
the air in a skipping fashion. He reported this, and from his
description men began to speak of flying saucers.

Since Arnold's first sighting many thousands of reports of such
things have appeared. Some were easily explained, proving to be
anything from mirages to the sightings of bright planets. Others
have turned out to be hoaxes. And some, generally because of
insufficient information, remain unexplained.

Nevertheless a veritable hysteria seized some people, who were
sure that the objects were enemy planes or even extraterrestial
spaceships manned by other-world life-forms. The American
Government and sober scientists were accused of hiding the facts,
and people who tended to see conspiracies everywhere were
never-ending in their ridiculous theories and accusations.

Hounded onward by people ranging from well-meaning
unsophisticates to fiery-eyed crackpots, the Government insti-
tuted investigation after investigation, finding nothing and offering
only fresh fuel for cries of conspiracy. Since the term flying saucer,
though still popular, was clearly inadequate, investigators called
the things seen in the air unidentified flying objects, as objective a
term as possible. This was quickly abbreviated to UFO, sometimes pronounced as initials, sometimes as a word, and flying-
saucer enthusiasts or investigators are jocularly called ufologists.

Undoubtedly there remain phenomena in the atmosphere that
are as yet poorly understood by scientists, but the ufologist's
contention that the mysteries are to be explained by flights from
other worlds is unlikely in the extreme.

Ultracentrifuge

MOST PROTEIN MOLECULES have weights from thousands
to millions of times as great as that of the hydrogen atom. The
ordinary methods for determining molecular weight fail in the face
of such size.

Protein molecules are so large, in fact, that the endless motions
of the water molecules surrounding them when they are in solution
barely suffice to keep them distributed evenly. If they were some-

what larger they might actually settle out of solution under the pull of gravity. Naturally, the larger they are the more rapidly they settle out, and in that way one can determine their molecular weight—by the rapidity with which the settling takes place.

When anything settles out of solution it forms a sediment (from a Latin word meaning settling). The rate at which proteins settle out is therefore the sedimentation rate.

In order for actual proteins to have a measurable sedimentation rate, the force of gravity would have to be greater than it is. It is impossible to arrange this condition, but there is something which can be used to imitate gravity. If a container is whirled rapidly, its contents are forced away from the centre of rotation. This is the centrifugal effect (from Latin words meaning to flee from the centre), and the instrument is a centrifuge.

In 1923, the Swedish chemist Theodor Svendberg developed an ultracentrifuge (beyond the centrifuge) that whirled so rapidly it produced an effect large enough to force protein molecules to move outward through the water solution. A sedimentation rate was obtained which could be used to determine the molecular weight of proteins.

Nowadays ultracentrifuges are used which whirl at 75 000 times each minute and produce centrifugal effects that are 400 000 times as powerful as gravity.

Umbra

ANY DARK body in the neighbourhood of a sun (as the earth or moon, for instance) casts a shadow. To an object located anywhere within the shadow the sun is invisible.

When the moon is eclipsed it is because it passes through the earth's shadow. When the sun is eclipsed the earth is passing through the moon's shadow. The shadow of a world narrows as it extends outward because the world decreases in apparent size with respect to its sun as distance from it increases.

From the distance of the moon the earth still appears to be four times the diameter of the sun, so earth's shadow is thick enough to cover the entire moon. From the same distance, however, the moon, a smaller body, seems just the size of the sun, and its shadow is narrowed down to almost nothing. No sooner has it moved to cover up all the sun than, as it continues moving, the sun begins to emerge at the other side. For this reason any particular eclipse of the sun is confined to a small region of the earth, and only lasts for seven minutes at the most.

A world's shadow is its umbra (the Latin word for shadow),

while the region near it, where the sun is partly but not entirely covered, is the penumbra (from the Latin *paene*, meaning almost, hence the almost shadow). Similarly, the central part of a sunspot, which though glowing hot seems dark and shadowy against the still more glowing surface of the rest of the sun, is also called an umbra, and is also surrounded by a penumbra.

The word enters astronomy in still another way. Alexander Pope, in his poem *The Rape of the Lock,* invented a moody, sorrowing spirit, who lived in the shadow of grief, so to speak, and whom he called Umbriel in consequence as a sort of counterpart to the gay spirit Ariel (the name being chosen with airy in mind, probably) in *The Tempest.* In 1851, when William Lassell discovered the third and fourth satellites of Uranus, he decided to give them the names of spirits, since the first two had been named Oberon and Titania after the king and queen of fairies. Consequently, be gave to one the name Umbriel and to the other Ariel.

Uncertainty Principle

IN 1900 the German physicist Max Planck demonstrated that energy existed in tiny separate pieces. Each piece was called a quantum (a Latin word meaning how much), a term first introduced by Albert Einstein in 1905. The amount of energy in one quantum depended on the wave-length of the radiation associated with the energy. To calculate the energy per quantum from the wave-length of the radiation one had to use a very small number called Planck's constant, and it turned out that this number set important limits to certain phenomena.

For instance, it had always been assumed that scientists could make measurements that were as accurate as they wished. If they constructed tools that were perfect, then the measurements would be perfect too, and would have no uncertainty about them.

In 1927, however, a German physicist, Werner Heisenberg, challenged this assumption. Measurements always involved the use of energy in one fashion or another (even if only light to see by), and the energy had to come in certain sizes dictated by Planck's constant. If the measurement one tried to make was so small that the particle of energy used was large in comparison, then the measurement would be uncertain no matter how carefully it was made or how perfect the measuring instrument. (It would be like playing the piano with boxing gloves on. Even the most perfect pianist couldn't do a good job).

Heisenberg calculated in particular that if one tried to measure the position of an object with greater and greater accuracy, then the simultaneous measure of its momentum (its mass times its velocity) would be determined with less and less accuracy, and vice versa. The uncertainty of one measurement multiplied by the uncertainty of the other had to be no smaller than a certain fraction of Planck's constant. This was the uncertainty principle, and it is a limit set by the structure of the universe.

Ungulate

THE HORNY sheaths at the ends of a mammal's fingers and toes can take on several shapes. As almost every child learns by sad experience, a cat possesses claws. On the other hand, a horse possesses hoofs. This suggests a sensible way of making a division among mammals. After all, this is a considerable difference. Claws are offensive weapons and mark the aggressive meat-eater; while hoofs are devices for running away, and mark the timid plant-eater.

The English naturalist John Ray in 1693 did make just such a classification. The Latin word for fingernail is *unguis,* and the Romans made two diminutives of it; a small nail could be *unguiculus* or *ungula,* and Ray made use of both. A clawed mammal he called an unguiculate and a hoofed mammal he called an ungulate. The two terms still exist, embracing two of the sub-sections of the placental mammals (see MARSUPIAL). However, allowance is now made for mammals that don't fit well in either group.

For instance, there are mammals that have nails that are neither claws nor hoofs, but actual nails like ours. Such animals include ourselves, of course, and also the rather similar-appearing apes and monkeys. The name of these animals, however, has nothing to do with nail in either English or Latin. The group includes ourselves, and so, with simple and natural conceit, Karl von Linnaeus, the great classifier of plants and animals, called a nailed mammal a primate, from the Latin *primus* (first).

And then there are mammals who have so adapted themselves to the sea that they have left only two flipper-shaped forelimbs with no visible nails of any type. These include the whales, porpoises and dolphins. A non-nailed mammal like that is a cetacean, from the Greek *ketos* (whale).

Birds, by the way, also have claws. The unusual thing about birds, though, is that at least one toe on each foot points backward, so there is not less than one claw on a bird's foot where on a mammal the heel would be. A bird's claw is therefore a talon, from the Latin *talus* (heel).

University

A GROUP OF individuals acting in combination towards a single goal under a single direction behaves as though it were one person. In the Middle Ages such a group was called a *universitas,* from the Latin *unum* (one) and *vertere* (to turn); a group, in other words, had turned into one person. In the most general sense, this came to mean the group of everything in existence considered as a unit; that is, the universe.

It also came to be used in a more restricted sense. In the early Middle Ages, for instance, a school of higher learning was called a *studium,* from the Latin *studere,* meaning to be zealous or to strive after. From this come our words study and student. The Italian version of *studium* is studio, which has come into English as a place where the fine arts, particularly, are studied or practiced (a memorial to the fact that in Renaissance times Italy was the centre of the world of fine art).

A group of students at such a school would refer to themselves (with the usual good opinion of themselves that students always have) as *universitas magistrorum et scholarium*—a group of masters and scholars. They were a *universitas,* you see, because they were all pursuing the single goal of learning. And gradually the name of the group became, in shortened form, the name of the schools which, around 1300, began to be known as universities.

Meanwhile, within the university, groups of students following a particular speciality, as for instance law, would band together for mutual aid. They were a group of colleagues (Latin, *collegae*) from the Latin *com-* (together) and *ligare* (to bind). They were bound together for a common purpose. Such a group of colleagues formed a college (Latin, *collegium*), a word now used to denote a particular school within a university.

The older meaning as simply a group of colleagues persists today in the College of Cardinals of the Roman Catholic Church, and in America's electoral college which meets every four years to elect a president.

Uranium

IN 1781 the German-born British astronomer William Herschel discovered a new planet and caused a tremendous sensation in the scientific world. It was the first discovery of a planet in recorded history. To be sure, since the invention of the telescope, over 150 years earlier, four satellites had been discovered circling Jupiter, and four circling Saturn, but these were not independent worlds.

381

This new discovery was a true planet, circling the sun at twice the distance of the farthest anciently known planet, Saturn.

This new planet (the seventh) was named Uranus, after the Greek god of the sky, Ouranos, who, according to the Greek mythology, had been the father of Cronos (Saturn, according to the Romans, and the sixth planet) and the grandfather of Zeus (Jupiter, according to the Romans, and the fifth planet).

So much for that. Now, in 1789 the German chemist Martin Klaproth was working with a heavy black mineral called pitchblende. In it he found evidence of a new metal, hitherto unknown. It had been an old-fashioned habit of the medieval alchemists to refer to metals by the names of various heavenly bodies. Gold was called the Sun; silver, the Moon; copper, Venus; iron, Mars; and so on (see MERCURY). Now here was a new metal and a new planet, too. Klaproth therefore named the new metal uranium after the new planet.

A century and a half later there was an echo to this story. In 1940 American scientists at the University of California formed two new elements by means of nuclear reactions. Previously uranium (with an atomic number of 92) had been the most complicated known element. The two new elements, however, were more complicated still, with atomic numbers of 93 and 94. So they were named after the two planets beyond Uranus, planets that had been discovered since Herschel's day. Element 93 was named neptunium after Neptune (the eighth planet, and the Roman god of the sea) and element 94, plutonium, after Pluto (the ninth planet, and the Greek god of the underworld).

Vaccination

UNTIL 150 years ago smallpox was a common and dreaded disease. It often killed (as it did Louis XV of France) or, if it did not kill, it usually disfigured (as, for instance, George Washington), leaving the face pock-marked. (The word pock, from the Anglo-Saxon, is applied to any skin eruption resulting from disease or to the scar it leaves behind when it is gone.)

People who got smallpox and survived did not get smallpox a second time. They were resistant to it. Moreover, even in the 1700s, it was suspected that if a person caught cowpox (a much milder disease, caught from cows, and hence its name) and recover, that person was thereafter resistant not only to cowpox but also to the much more dangerous smallpox. (It is perhaps for this reason that milkmaids were always so beautiful in the eighteenth-century romantic comedies. They caught cowpox early in life, and were spared the disfigurements of smallpox.)

In any case, in 1796 a Scots physician, Edward Jenner, tested this theory by infecting a boy with material from a pock on the hand of a milkmaid who had cowpox. He continued such experiments, and it gradually became clear that people so infected were indeed resistant to smallpox.

The fancier medical term for cowpox is vaccinia, from the Latin *vacca,* meaning cow. A preparation containing material from a cowpock is a vaccine and the deliberate infection of a person with it is vaccination.

The term, however, has passed far beyond its original connection with cows. A generation later, the French chemist Louis Pasteur used the word vaccine to describe his own preparations of various germs with which he hoped to make human beings resistant to certain diseases, preparations that had nothing to do with cows. The best-known modern preparation was formerly one developed by Dr Jonas Salk containing dead or nearly dead poliomyelitis virus—the so-called Salk vaccine—but this has now been largely superseded by the Sabin-type vaccine—oral, unlike the Salk injection vaccine, and of live but weakened virus.

Van Allen Belts

ON OCTOBER 4, 1957, the age of space opened when the Soviet Union sent Sputnik 1, the first of the artificial satellites, into orbit. The United States was not far behind. Its first artificial satellite, Explorer 1, was sent into orbit on January 31, 1958.

Of course, satellites were not merely intended to go into orbit to prove the strength of their rocket engines. They carried instruments designed to make measurements of various kinds. One obvious type of measurement was that of counting the number of cosmic ray particles and other energetic particles in the regions through which the satellite passed. Explorer 1 was designed to do that.

As Explorer 1 rose higher, it recorded more and more energetic particles, then went dead and reported nothing. Explorer 3 was launched in March 1958 with a more rugged counter. That too went dead.

It occurred to the American physicist James A. Van Allen, who was in charge of this part of the work, that the trouble might be not that there were no particles above a certain level, but that there were too many. Perhaps the counters were flooded and couldn't work.

When Explorer 4 was launched on July 26, 1958, it contained a counter with a lead shield that would let through only a small portion of the particles. This time the results were conclusive. The number of energetic particles was indeed high, far higher than anyone had imagined.

Apparently energetic particles, originating for the most part from the sun, were trapped in earth's magnetic field and existed in a thick belt outside the atmosphere all around earth. On closer study there seemed two or even three belts surrounding earth, and these were first called the Van Allen belts. In later years, to adjust the name to those given to other portions of the atmosphere, the region came to be called the magnetosphere.

Vasopressin

IN THE EARLY 1940s it was found that the pituitary gland produced a substance that controlled reabsorption of water passing through the kidneys. By such control the body was kept from losing too much water through urination.

An extract from the pituitary (which see), called pituitrin, was found to contain the hormone. Any factor which increases urine volume is said to be diuretic, from a Greek word meaning to urinate. Pituitrin, by promoting reabsorption of water, lessened the need to urinate, and was therefore the opposite of diuretic. It was said to contain an anti-diuretic hormone (ADH).

Pituitrin possessed two more important abilities. It increased blood-pressure by bringing about the contraction of blood-vessels. This was called vasopressor activity, from Latin words meaning vessel-compressing. It also contracted the muscles of the pregnant uterus in preparation for birth. This was called the oxytocic effect, from Greek words meaning quick birth.

The American biochemist Vincent du Vigneaud obtained two pure substances from pituitrin in the early 1950s. Since one showed the vasopressor effect and the other the oxytocic effect, he called them vasopressin and oxytocin respectively. Vasopressin also produced the anti-diuretic effect, so it was also the anti-diuretic hormone.

Du Vigneaud analysed the two hormones and showed that each was made up of eight amino acids, which made them very simple proteins indeed. He identified the eight amino acids and worked out the order in which they appeared in the molecule. Then in 1954 he synthesized the hormones from the individual amino acids, which he put together in just the right order. He showed that the synthetic substances had all the properties of the natural hormones. It was the first time any natural protein (albeit a simple one) had ever been synthesized.

Vector

THERE ARE two broad classes of measurements in science. One is simply a how much measurement or a how many. You might say that there are two apples in a basket or that a line is 5·476 centimetres long or that an elephant weighs 2732 kilogrammes or that there are 60 minutes in an hour or that an angle is 45 degrees. These are scalar quantities.

The word scalar comes from the Latin *scala,* meaning a ladder or staircase; something, that is, with a succession of rungs or steps than can be counted off. That, after all, is all you are doing—counting off. It may be units or centimetres or kilogrammes or minutes or degrees, but it is only a matter of counting. A device for measuring one of the most common of the scalar quantities, weight, is called simply a scale.

Sometimes counting isn't enough. You have to ask not only how much but in what direction. You may exert a push of a kilogramme, for instance, but it is important to ask where you're pushing. In one direction you may be guiding a girl across a dance floor; in another direction you may fall out of a window. Forces, in other words, are not scalar quantities, but are vector quantities.

The word vector comes from the Latin *vehere,* meaning to carry (the past participle of the Latin word is *vectus*). The notion of carrying comes from the fact that in any vector quantity there is the implication of something being carried from here to there. Thus my kilogramme push carries an impulse from here (my body) to there (my dancing partner's body, or through the window glass, as the case may be).

Bacteriologists use the word in a sense that has a more direct connection with its derivation. Some diseases are carried from person to person (or from animal to person) by some intermediate creature. Some species of mosquito, for instance, may carry malaria from one person to another. Malaria is therefore said to have an insect vector.

Velocity

THE MOST common word for the rate at which a body moves—that is, the distance covered by it in a given time—is speed. This comes from an Anglo-Saxon word meaning success, and is still used in this sense when we speed the parting guest. We are not trying to get rid of him quickly, of course; we merely wish him success on his way. Presumably there was the general feeling that anyone who went about things quickly was more likely to succeed than another who was sluggish, so speed got the meaning of rate of movement, and began to imply quickness rather than greatness.

In ordinary speech, velocity (from the Latin *velox,* meaning swift and *velocitas,* swiftness) is used as a synonym for speed. If there is any difference it is that speed is usually applied to the motion of a living thing, like a racehorse, and velocity to an inanimate object, like a bullet.

In physics, however, there is an important difference. Speed is simply rate of movement. A car may have a speed of kilometres an hour. Velocity, however, implies not only rate of movement but direction of movement. Thus a car may have a velocity of forty kilometres due north. Two cars moving in opposite directions may have the same speed but must have different velocities.

A moving object may change its velocity as time goes by. It may slow down or speed up or change its direction of motion. The rate at which velocity is changing is acceleration, from the Latin *ad-* (to) and *celare* (to hasten). Acceleration, in other words, means to hasten towards.

Strictly speaking, acceleration refers to any change in rate of motion, but in popular speech the implication of the derivation holds and acceleration applies only to motion that is speeding up. A slowing of motion is referred to as deceleration. The Latin prefix *de-* can mean from, so that deceleration means taking quickness from motion.

Rocket experiments these days have introduced new phrases. Positive acceleration refers to either acceleration or deceleration that causes blood to move towards the head; negative acceleration to that which causes it to move towards the feet.

Venera

THE NEAREST SIZEABLE heavenly body to us is the moon, which is a mere 400000 kilometres away. Therefore man's ventures into space have naturally concentrated on the moon, more than on any other extraterrestrial body. In July 1969 men actually landed on the moon, and there have been additional trips there since then.

The next nearest bodies are the planets Venus and Mars. Venus can be as close to us as forty million kilometres and Mars fifty-six million. Of the two Mars is the less inhospitable, so that it is the next target for a manned vessel. Both, however, have been the target for unmanned probes.

The Soviet Union was not satisfied to have these probes merely fly near Venus, as did their first two, or crash on Venus, as did the third. Soviet scientists have worked to design rockets that would make a soft landing on the planet, and then radio back information from within the atmosphere. Those rockets intended for this purpose they called Venera, after the name of the planet.

The first attempts, Venera 4, 5, and 6, were sufficiently accurately aimed to strike Venus' atmosphere. The instrument package began descending through it by parachute. In each case, however, the atmospheric conditions were so extreme as to cause the transmitting devices to break down before the package had descended to within twenty kilometres of the surface.

Finally, on December 15, 1970, Venera 7 succeeded. It made a soft landing on Venus, the first time such a feat had been accomplished on any world in space other than the moon. The data sent back by Venera 7 showed the temperature on Venus's surface (at its point of landing) to be 474 °C, over a hundred degrees higher than is required to melt lead. The atmospheric pressure is about 9×10^6 newtons per square metre, or ninety times that of earth's atmosphere. Considering that Venus's atmosphere is almost entirely carbon dioxide, the planet seems inhospitable indeed.

Venera 8 landed on the planet's day side (all previous probes having landed on the night side) on July 22, 1972, and largely confirmed former findings.

Ventricle

THE LATIN word for belly is *venter,* and since the belly is a hollow place within the body, it came to be applied to other hollows as well. A small hollow would be *ventriculus,* or in English a ventricle.

The most important ventricles in the body are the two main chambers or hollows into which the heart is divided: a left and right ventricle. Above each ventricle is a smaller chamber into which the blood enters on the way to the ventricle. It is the antechamber of the ventricle, so to speak, and is often called the atrium, which is Latin for antechamber. However, the point where the blood enters each atrium bulges up like a small ear. The Latin word for ear is *auricula,* and that word has come to be applied to the entire atrium. The two small chambers are commonly called left and right auricle.

The auricle contracts and forces blood into the ventricle, which contracts in its turn and forces the blood out of the heart altogether into an artery (see ARTERY). When the ventricle contracts blood must not pass back into the auricle, and to prevent that there is a valve (from the Latin *valvae,* meaning folding doors) between each auricle and ventricle, which will allow blood-passage only one way, from auricle to ventricle, not vice-versa.

The valve between left auricle and ventricle consists of two triangular flaps which look when closed like a bishop's mitre, and is therefore called the mitral valve, or bicuspid valve, from the Latin *bi-* (two) and *cuspis,* genitive *cuspidis* (point). The valve between right auricle and ventricle is made up of three flaps and is the tricuspid valve. The Latin prefix *tri-* means three.

When the left ventricle contracts the blood is pushed into the aorta, the largest artery of the body. The aorta extends straight up at first, then loops and travels down the trunk. The first upward extension, however, makes the aorta look like a handle by which the heart is raised or held up, and, indeed, the word aorta comes from the Greek *aeirein,* meaning to raise.

Vertebra

AN AXLE is the shaft on which a wheel, or wheels, turns. The Latin word for axle is axis, and that is used specifically for the line (or imaginary shaft) about which a sphere turns—as, for instance, the axis of the rotating earth.

We have such lines of turning in our own body. The anatomist calls the armpit by its Latin name of axilla, which like *ala,* meaning wing, is related to the word axle. Raise your arm stiffly and you will see that it turns about a line (or axle) running through the shoulder above the armpit.

A more direct example exists in the backbone. This is also called the spine, from the Latin word *spina* (thorn), because the individual bones of the spine are irregular in outline and have projections like thorns. These individual bones are vertebrae (singular, vertebra), from the Latin word *vertere* (to turn), because joints between particular vertebrae allow the turning of the head.

In man there are 33 vertebrae altogether. The top 7 are the cervical vertebrae, from the Latin *cervix* (neck); then 12 thoracic vertebrae, from the Latin *thorax* (chest), also called dorsal vertebrae from the Latin *dorsum* (back); then 5 lumbar vertebrae, from the Latin *lumbus* (loin).

So far the names indicate the position of the bones. But then, we have 5 sacral vertebrae, from the Latin *sacrum* (sacred) because this portion of an animal was used in sacrifices, and finally 4 caudal vertebrae, from the Latin *cauda* (tail). These last 4 vertebrae do actually represent the last traces of a human tail.

We come back to axis because the second cervical vertebra has that as its special name. It is the axis, because it is upon that particular vertebra that the head (plus the first vertebra) turns. This first cervical vertebra is called the atlas, the Titan who in Greek mythology had to uphold the world. A man's brain is a sort of world, after all.

The four caudal vertebrae are sometimes lumped together under the single word coccyx, from the Greek *kokkyx* (cuckoo) because the early anatomists thought the bone combination resembled the beak of a cuckoo.

Virus

WHEN THE French chemist Louis Pasteur was studying the dreaded disease hydrophobia in the 1880s he noticed that the spinal tissues of affected animals could transmit the disease; yet he could detect no bacteria in the tissues. Pasteur had invented the germ theory of disease, and this observation did not shake him in his belief that diseases were carried by microscopic organisms. In this case, he speculated, the organisms were so small that they could not be seen by the microscope, that was all. And as it turned out, he was right.

In 1892 the Russian investigator D. Ivanovski made a mash of leaves from tobacco plants suffering from mosaic disease (in which their leaves were mottled and discoloured) and passed the liquid from the mash through filters so fine that even bacteria could not pass through. The bacteria-free liquid that emerged could, however, still pass the disease on to healthy tobacco plants.

As was true of many Russian discoveries, this one was ignored by the West until one of their own scientists had repeated it. In this case it was the Dutch investigator, M. W. Beijerinck.

At first this was puzzling. An infective liquid free of living agents of disease? The word *virus* in Latin means poison (as in virulent), and this name was often applied to something that seemed to carry disease. Now, the infective liquid was called a filtrable virus (poison that passed through a filter). In 1961, however, the American H. A. Allard used a finer filter than had any of his predecessors, and held back the virus. The liquid that came through was not infective. So the virus was still a living agent except that, as Pasteur had guessed, it was a particularly small one.

Finally, in 1935 the American biochemist Wendell M. Stanley separated out pure tobacco mosaic virus (Ivanovski's virus), crystallized it and showed it to be a complex protein molecule. Such molecules still retain the old mystery in name, at least. They are virus molecules.

Vitamin

BY 1900 it had become obvious that some diseases were connected with diet. For over a century scurvy had been prevented by the use of citrus juices (see ASCORBIC ACID). Similarly, after 1878, the Japanese Navy cut down on the incidence of the serious nerve disease beri-beri by forcing its sailors to eat brown rice rather than polished white rice. The Russian biochemist Nikolai Ivanovich Lunin showed in 1881 that rats died on a diet of purified carbohydrates, fats and proteins, but survived if a small quantity of milk were added.

In 1901 the Dutch biochemist Gerrit Grijns first suggested that diseases such as scurvy and beri-beri were caused by the lack of some particular chemical in the diet, needed in very small quantities, and supplied by foods such as milk, fruit juices, rice husks and so on. Because these diseases are caused by a dietary deficiency, they are called deficiency diseases.

The chemical in rice husks that prevented beri-beri proved to be an amine (see AMMONIA), and the Polish-born American biochemist Casimir Funk, supposing that the whole group of such chemicals might be amines, proposed in 1912 the name vitamine for the group, from the Latin *vita* meaning life. They were amines of life.

It was soon found out however that most of the chemicals in question were not amines. The e was therefore dropped to reduce the effectiveness of the allusion, and the word is now simply vitamin.

It early became obvious that there were at least two vitamins: one soluble in fat, not in water; the other soluble in water, not in fat. The American biochemist Elmer V. McCollum, in 1915, called these fat-soluble A and water-soluble B. By 1920, when the word vitamin was well established, the names vitamin A and vitamin B became common. In the years that followed vitamin B was shown to be a mixture of many substances which became known as vitamin B_1, vitamin B_2, and so on, up to vitamin B_{12} and even beyond.

Vitriol

IN ANCIENT times transparent objects (from the Latin *trans*, meaning across, and *parere*, meaning to appear; light can, in other words, travel across a transparent object and appear on the other side) were unusual, and names were derived from the fact (see CRYSTAL).

The Latin word for glass, for instance, is *vitrum*, so things that are glassy in appearance are vitreous. The jelly-like transparent fluid inside the eyeball is called vitreous humour, for instance. (Humour means fluid in this case; see HUMOUR.) The more watery liquid in front of the eye's lens is the aqueous humour, from the Latin *aqua* (water).

In the early Middle Ages vitreous or glass-like minerals came to be called vitriols. The first mineral to be named so (in about A.D. 600) was iron sulphate, which was called vitriol of Mars, Mars being the alchemical name for iron (see MERCURY).

In time a number of sulphates (minerals with molecules containing a sulphur atom and four oxygen atoms combined with any of various metal atoms) received the name, each being distinguished by colour. Copper sulphate, for instance, forms beautiful blue, semi-transparent crystals, and is called blue vitriol (or simply bluestone); while zinc sulphate forms colourless crystals and is white vitriol. The original vitriol, iron sulphate, forms green crystals and is green vitriol.

If vitriols are heated strongly vapours are given off. About A.D. 1200 it was discovered that if these vapours are trapped, cooled and dissolved in water an oily and very corrosive liquid is formed. It was called oil of vitriol, or sometimes simply vitriol, though its proper name today is sulphuric acid.

The term vitriolic, as applied nowadays to a caustic and cutting remark or to a sharp-tongued and bitter personality, refers to the strong and caustic sulphuric acid and not to the innocent glass from which all these names are derived.

Volcano

THE GOD OF FIRE in ancient times was Hephaistos to the Greeks and Vulcanus to the Romans. The ancients associated fire chiefly with metals, since it was necessary first in the smelting of ore to obtain metal and then in the melting or softening of the metal so that it could be shaped and moulded. It was natural, then, to think of the god of fire as being a divine and wonder-working smith. Hephaistos is pictured in just this way in Homer's *Iliad*.

At the same time the ancients could not help noticing Mt Etna in Sicily, a phenomenon among mountains. There were horrible noises within it, smoke issued from it, occasionally fire and molten rock belched out of it. Obviously it contained a gigantic forge, and later Latin poets made Etna the workshop of Vulcanus.

In Italian the name Vulcanus has become Vulcano or Volcano (it is simply Vulcan in English), and the word volcano came to be applied first to Mt Etna and then to any mountain that behaved like Etna.

The opening through which the fiery phenomena make their appearance is the crater. This is from the Greek *krater*, meaning a large bowl for mixing water and wine, which is actually what the volcano opening resembles during its quiet interludes. The hollowed mountains on the moon are called craters too, but this is a poor name because it implies volcanic action, whereas actually those craters were probably formed by meteor collisions.

There is another famous volcano, Mt Vesuvius, on the Italian mainland near Naples, and the Neopolitans have given us a well-known word in connection with volcanoes. The Italian word for to wash is *lavare*. A rainstorm so violent as to flood the streets and, so to speak, wash them under a current of water, was called a *lava* by the Neapolitans. The word was naturally applied to another much more horrible sort of stream supplied occasionally by the neighbouring Vesuvius, so that today lava means the stream of molten rock that pours out of an erupting volcano. In A.D. 79, when Titus was Emperor, there was a great eruption of Vesuvius which buried completely two cities, Pompeii and Herculaneum. These have been miraculously preserved, and have been excavated.

Volt

THE UNITS of electricity are a sort of Hall of Fame for various pioneers in the field (see POWER). In 1800, for instance, the Italian scientist Alessandro Volta produced electricity by piling up discs of silver and zinc alternately with brine-soaked paper between each silver-zinc pair (or cell). This was called a Voltaic pile. (Any group of similar objects is a battery of objects. The Voltaic pile was a battery of electric cells, and eventually simply a battery.)

The electromotive force (the Latin word *motivus* means causing to move—hence the force that causes electricity to move) increases with the number of cells in the battery. This force is measured in volts in honour of Volta.

The current strength (that is, the quantity of electricity moving through a conductor each second) is expressed in amperes. This is in honour of André Marie Ampère, a French physicist, who from 1820 on investigated the relationship of electricity and magnetism.

The total quantity of current moving through a conductor over a period of time is measured in coulombs. This commemorates the French physicist Charles Augustin Coulomb, who, from 1785 on, worked out the manner in which electric charges attract and repel each other.

In 1827 the German physicist Georg Simon Ohm published a pamphlet setting forth Ohm's Law. The law states that when electric current flows through a conductor the electromotive force divided by the current strength gives a figure which represents the resistance of the conductor. This pamphlet made no impression at the time, and the sensitive Ohm resigned his professorship at Cologne in mortification. However, the unit of electrical resistivity is today called the ohm in his honour. Moreover, electrical conductivity (1 divided by the resistivity) of any conductor is the mho, which is only ohm spelled backward. Thus a wire with a 5-ohm resistivity has a 1/5 mho conductivity.

Vulcanize

WHEN Christopher Columbus landed on the coast of Haiti he found the inhabitants bouncing balls of an elastic substance and playing games with them. Nothing like it was known in Europe. They obtained this material from a milky liquid that oozed out of a certain tree when the bark was cut, and they called the substance *cahuchu,* meaning weeping wood. (We have the more prosaic name of latex for the fluid, which is nothing more than the Latin word for fluid.)

The French call the bouncing substance *caoutchouc* and the Spaniards *cauch* from the original Indian word; but the English is far less romantic. The English chemist Joseph Priestley noted that the substance could be used to rub out pencil marks, so about 1770 he called it rubber and the name stuck. Occasionally it is called India rubber because it comes from the Indies, the old-fashioned name for the Americas, dating back to Columbus's belief that he had landed in India.

One of the first uses of rubber was as a waterproofing material. A Scot, Charles Macintosh, for instance, first prepared rubber-coated cloth, and raincoats are still sometimes called macintoshes for that reason.

However, there was this difficulty. Rubber got stiff and brittle in cold weather and soft and sticky in warm. The American Charles Goodyear was one of those who tried to correct this failing. He succeeded by accident. In 1839 he was mixing rubber and sulphur to see if that would help, and accidentally he tipped some over on to a hot stove. He got it off as quickly as he could (what a stink it must have made!) and when he did so he found he had a piece of rubber that stayed elastic in the cold and dry in the heat. Because Vulcan was the Latin god of fire (see VOLCANO), the rubber and sulphur, united successfully under the influence of fire, was said to be vulcanized rubber. The success of modern rubber depends on this one discovery, but Goodyear spent his life fighting for patent rights, and died loaded down with debts.

W-Boson

MOST SUB-ATOMIC PARTICLES act as though they are spinning. The spin gives the particle a property called angular momentum, and physicists can measure this. They use a system of measurement whereby some particles end with a spin expressed as a whole number. A photon, for instance, has a spin of 1. Others have spins expressed as a half-number. Electrons and protons have spins of ½.

Particles are distributed among different energy levels, and this distribution can be worked out by either of two kinds of statistical methods. One was worked out by the Italian-born physicist Enrico Fermi and the English physicist Paul Dirac. Such Fermi-Dirac statistics hold for particles with spins and half-values. Particles like the electron and proton are therefore called fermions.

The other statistical method was worked out by the Indian physicist Satyenda Bose and by the German-born physicist Albert Einstein. The Bose-Einstein statistics hold for particles with whole-number spins, and these are called bosons. The photon is a boson.

The photon is involved in electromagnetic interactions. Other kinds of interactions are also associated with special particles. The strong interaction associated with atomic nuclei involves the pion, which has a spin of 0 and is also a boson. Gravitational forces involve gravitons, which have spins of 0 and are bosons.

The only remaining interaction is the weak interaction associated with events such as radioactivity. Physicists feel this too must be associated with a particle, one that may be particularly massive, even more massive than a proton. Theory predicts it would have a spin of 1, so it would be a boson too. It is distinguished from other bosons by being called a W-boson, where the W stands for weak interaction. The W-boson has not yet been detected.

Weak Interaction

IN 1935 the Japanese physicist Hideki Yukawa worked out the manner in which the atomic nucleus, made up of protons and neutrons, held together. There was a nuclear interaction, setting up an attraction among the particles, and that proved to be the strongest known force, over a hundred times as strong as the electromagnetic interaction and billions of times as strong as the gravitational interaction.

The nuclear interaction is so strong that once two particles are close enough to allow the interaction, the result must follow in an unimaginably short period of time. Thus, once a pion is in the neighbourhood of an atomic nucleus a reaction follows in the space of ten quadrillionths of a second.

It would seem that any sub-atomic particle capable of being involved in the nuclear interaction would break up in that period of time. There are strong theoretical reasons for supposing so, and yet this does not happen. The pion, for instance, breaks down after a few thousand-millionths of a second. This seems short enough, but the pion is more than a hundred billion times as long-lived as it is supposed to be. A neutron can exist for ten minutes and more without breaking down.

Apparently, there must be another kind of interaction involving sub-atomic particles, much weaker than the one Yukawa had discovered. As early as 1934, the Italian-born physicist Enrico Fermi had worked out the theory behind such a weak interaction and Yukawa's came to be called the strong interaction in contrast. The weak interaction is much weaker even than the electromagnetic interaction, but is still billions of times stronger than the gravitational interaction.

There are some sub-atomic particles, like the electron and the neutrino, that are involved only in weak interactions, and that is why most ordinary radioactive breakdowns take place as slowly as they do.

Whiskers, Crystal

IN CRYSTALS the component parts—atoms, ions or molecules—make up an orderly array. The orderliness makes possible a strong cohesiveness that would not be possible if the various atoms, ions or molecules were arranged in a random heap. Still, when chemists calculated how strong a crystal ought to be on the basis of this orderly arrangement, it always turned out that the crystals in actual fact were far less strong than was to be expected.

When chemists learned to study the structure of crystals in detail, by X-ray diffraction, it turned out that the orderliness was never perfect. There were invariably small regions of disorder where there was a missing atom or a superfluous one, where lines of atoms didn't meet exactly or did so at a slight angle. When stress was placed on a crystal it gave at one of these points of weakness, long before the orderly portions were affected. Once a tiny crack appeared at these critical points it spread rapidly.

If a crystal could be formed free of all such defects it would be much stronger than ordinary crystals of that type. It had long been known that when crystals were slowly formed from solution tiny hair-like projections appeared on the crystals. These are called whiskers, because that is what they looked like. In the 1950s these whiskers were studied and found to be perfect crystals, and there-fore much stronger than ordinary crystals of the same size. Carbon whiskers have been found to have a great resistance to being pulled apart (tensile strength, from a Latin word meaning to be stretched). The tensile strength of carbon whiskers is from 15 to 70 times that of ordinary steel.

In 1968 Soviet scientists produced an ordinary, but very small, crystal of tungsten without imperfections. It could bear a load eight times as large as that which a piece of steel the same size could carry.

Wolf Trap

BY THE MID-TWENTIETH CENTURY, there seemed little hope, if any, for advanced life-forms anywhere in the solar system, except on earth itself.

But what about very simple forms of life, which are much hardier than complex animal forms? There are certain seasonal changes on the surface of Mars that might possibly be due to plant growth. Biologists have set up chambers containing the equivalent of a possible Martian atmosphere, maintained at a Martian temp-erature, and simple plants have managed to live and grow in secluded places on the moon where small amounts of air and water might have persisted under the immediate surface.

In recent years hopes have weakened. Men have actually landed on the moon and brough back samples of rocks—but these have shown no signs of organic matter. (To be sure, very little of the moon has yet been sampled.)

The Martian atmosphere has been found to be even thinner and dryer than had been thought. Still, hope has not vanished entirely with respect to Mars. Men won't set foot on Mars for years yet, but possibly some unmanned device could send back information. One possible gadget was devised by an American biologist, Wolf Vishniac, in 1960. An instrument would be soft-landed on Mars. It would suck a sample of neighbouring soil or dust into a clear solution containing chemicals on which life could live. If life-forms were included in the soil they might grow and increase in numbers, making the solution cloudy or increasing the acidity. In either case, information to that effect would be relayed back to earth. Since the device is designed to trap any primitive life present, it seemed appropriate to make use of the first name of the designer and to call it a Wolf trap.

X-rays

STREAMS OF ELECTRONS were first produced in evacuated tubes in the 1860s, and for thirty years physicists were fascinated by these cathode rays (so called because they emerged from the cathode; see ELECTROLYSIS) without really understanding them. In that period the discoveries that were made seemed all the more mysterious because of that lack of understanding.

In 1895, for instance, the German physicist Wilhelm Röntgen noticed that a certain chemically coated paper in his laboratory glowed whenever his cathode-ray tube was in operation, even when there was carboard between tube and paper. His cathode-ray tube would fog photographic plates too, even when the plates were protected by their wrappings.

He decided that some kind of radiation was formed in the cathode-ray tube that could pass right through glass, carboard and paper. He had no notion as to what this radiation might be, and he called it X-rays. This name suited the mystery surrounding the cathode rays, since X is the letter usually used by the mathematician to signify the unknown.

Nowadays we know that the cathode rays are really streams of electrons, and we know that X-rays are quite similar to ordinary light, except that they are much more energetic. Despite the greater understanding of today, Röntgen radiation is still the X-ray. Sometimes X-rays are called Röntgen rays after the discoverer, but the name is difficult to pronounce (except for Germans), and the habit has not caught on.

The same happened to another type of radiation similar to light, but one that is much weaker—the so-called radio-waves. The word radio is simply a general term that would fit any radiation. Here too there is an alternative name, Hertzian waves, after the discoverer, the German physicist Heinrich Hertz. And here too the meaningless name won out, and radio-waves is the term used by practically everybody.

X-ray Stars

UNTIL THE 1950s the only radiation we could detect from the sky was that of visible light, cosmic rays and certain radio-waves. Other kinds of radiation were mostly absorbed by the atmosphere.

In the 1950s, however, rockets carried instruments beyond the atmosphere and other radiations were detected. For instance, it was found that the sun radiated X-rays, which was very surprising since it was not thought hot enough to do so. It was discovered, though, that the sun's corona was at a temperature of one or two million degrees, and this was hot enough for X-ray emission.

The Italian-American physicist Bruno Rossi wondered if the solar X-rays might hit the moon and be reflected. In 1962 rockets were sent up with instruments designed to detect any X-rays coming from the night sky. They detected such radiation, but not from the moon. It was coming from the general direction of the centre of the Galaxy. Objects far beyond the solar system seemed to be radiating X-rays so strongly as to produce measurable quantities even across light-years of space.

Beginning in 1963, the American astronomer Herbert Friedman began a programme of rocket detection designed to cover the sky in order to locate regions particularly rich in X-ray emission. A number of such regions were located and these were called X-ray stars.

The nature of the X-ray stars proved puzzling, but one object which definitely radiates X-rays is the Crab Nebula. This is the remains of a gigantic stellar explosion, the effects of which reached earth a little over nine hundred years ago. Even after nine centuries, the energy of the explosion has not faded out and the nebula is a rich source of radio-waves as well as X-rays. It even emits cosmic rays, the most energetic radiation of all. In addition, the nebula contains a pulsar, and this may be responsible for at least part of the radiation.

Xerography

EVER SINCE WRITING was invented some five thousand years ago, people have been interested in copying what was written. For almost all of this time copying has been possible only painstakingly, by hand, and each copy was as hard to do as the original had been.

Once printing was invented any number of copies could be prepared, but to do that required a printing press, a lot of type, and considerable skill in setting up type. A mimeograph machine is much more of a small-scale operation, but this must use liquid ink and can be messy. Carbon paper is dry and prepares copies as one types, but only a few at a time.

What if one uses carbon powder instead of ink, and lets electrostatic attraction do the work? Suppose a sheet of white paper is electrically charged. The charge would attract any fine particles of carbon that might be present, and the entire sheet would be covered with a thin layer of carbon. Light, impinging on the paper, however, would cause it to lose the charge.

But suppose light shines through a paper with print on it and strikes the charged paper. Everywhere, except where the print casts a shadow, the charge is lost. Carbon powder clings only to the shadow. The paper is heated so that the powder (which contains a resin to make it stick) clings permanently. The second paper is then a copy of the first, and many copies can be made rapidly.

The process is called xerography, from the Greek words meaning dry writing, since nothing wet is used.

By 1960 the American inventor Chester Carlson had made the method practical enough for eventual use in almost every office, and this has revolutionized office procedure. The system most familiar to the general public has the trade name Xerox, from xerography.

Y-Chromosome

EVERY SPECIES of animal has a characteristic number of chromosome pairs in its cells. In the case of the human that number is 23.

In females each pair consists of two chromosomes that appear exactly alike. Not so in the male. After most of the chromosomes have been paired off two chromosomes are left that are distinctly different. One is rather longer than average, the other is a mere stub. The long chromosome is called the X-chromosome and the stub is the Y-chromosome. In the female there are two X-chromosomes. The X- and Y-chromosomes, which differ in arrangement in the two sexes, are called the sex chromosomes.

Egg cells and sperm cells when formed contain only 23 chromosomes, one of each pair. Since the female has two X-chromosomes (XX), all the egg cells have one X. Since the male has an

X-chromosome and a Y-chromosome (XY), half the sperm cells end up with an X and half with a Y. It is just random chance that determines whether an X-sperm or a Y-sperm fertilizes a particular egg cell. If the former the result is an XX fertilized egg, which develops into a female; if the latter an XY fertilized egg, which develops into a male.

Chromosome assignment is not always perfect in the formation of egg cells and sperm cells (which together may be called sex cells). Sex cells with abnormal numbers of chromosomes usually don't form a fertilized egg that can develop fully. In the late 1960s, however, it was discovered that sex cells with abnormalities in their sex chromosomes might form embryos that would develop and give rise to individuals with XXX in their cells, XXY, XYY, and so on. XYY individuals have been found to be males, generally tall and strong, and often given to irrational fits of violence. Society, in its treatment of offenders, must now take into account their chromosomal make-up in determining the extent to which they can be held responsible for their actions.

Yttrium

THE EARLY chemists used the name earth for any substance that did not dissolve in water and was not affected by heat. The five most common earths were silica, alumina, lime, magnesia and iron oxide. Together these make up 90 per cent of the earth's crust, so the word earth is fitting. Lime and magnesia can be brought into solution by treatment with chemicals and such solutions had alkaline (see POTASSIUM) properties. Lime and magnesia were therefore called the alkaline earths. When eventually calcium and magnesium were found to occur in those earths, they and certain related elements were termed the alkaline earth elements.

In 1794 a Finnish mineralogist, Johan Gadolin, investigated a new black mineral found seven years earlier in a quarry in Ytterby, a small village near Stockholm, Sweden. Gadolin decided that the mineral contained a new kind of earth, and reported this. The new earth was isolated, and named yttria after Ytterby. (It has since been renamed gadolinite.)

It did not take long to discover that there were a number of different earths in yttria and other similar minerals. To distinguish them from the very common earths that were already known, these new ones were called the rare earths, and when new elements were discovered in them these were called the rare earth elements.

By 1843 the Swedish mineralogist Carl Gustav Mosander had

divided yttria into three earths. For one he kept the name yttria, while the other two he called erbia and terbia, still after Ytterby. Then when in 1878 the Swiss chemist Jean Charles de Marignac discovered yet a fourth earth in what Mosander had called erbia, he named the fourth earth ytterbia, again after Ytterby. Eventually new metallic elements were discovered in each of these earths and they were named yttrium, erbium, terbium and ytterbium, so that the insignificant hamlet of Ytterby ended with four elements named in its honour.

Zero

THE ANCIENTS used letters of the alphabet to represent numbers. The most familiar example today is the Roman system, still sometimes used on inscriptions, clock faces, and the like. In the Roman system I is one, V is five, X is ten and so on.

Such a system did not involve positional values. In other words, it didn't matter where the X was in a number, it always stood for ten. Thus XXX was ten plus ten plus ten, or thirty.

In the Middle Ages a new system reached western Europe from India via the Arabs. In this system each number had a separate symbol (Arabic numbers) which were not letters of the alphabet and which had positional value. For instance, 555 is not five plus five plus five, as it would be in the Roman system. The 5 at the right is five indeed, but the 5 in the middle stands for fifty, while the 5 at the left stands for five hundred. The number is therefore five plus fifty plus five hundred, or five hundred and fifty-five.

This new system is so superior to the old that it is amazing the clever Greeks never thought of it. Apparently the catch was that they never thought of using a symbol for nothing at all.

For instance, how would you distinguish between fifty-five and five thousand and five? On the abacus (see CALCIUM), the two numbers look somewhat the same. For fifty-five, the two bottom rows of counters have 5 pebbles pushed to the right. For five thousand and five, also, two rows have 5 pebbles pushed to the right; but between the two rows are two untouched rows. The Hindus invented a symbol for such an empty or untouched row (a thing the Greeks never did), and the Arabs adopted it and called that symbol *sifr* (empty).

This has come down to us as cipher or, in more distorted form, zero. Now we write fifty-five as 55, while five thousand and five is 5005. The importance of zero to mathematics is shown by the fact that cipher has also come to mean to solve arithmetical problems.

Zinjanthropus

AFTER Charles Darwin had published his theory of evolution in 1859 there was naturally a search on for fossils of the ape-like precursors of mankind. In 1891 a Dutch palaeontologist, Eugene Dubois, discovered fossil remnants in Java of a man-like being with a skull capacity distinctly less than that of any normal modern man. Others were eventually found in China and Africa and the history of man and his immediate ancestors was moved back a million years.

Yet it seemed that the true origin of man must be looked for farther back in time. Two palaeontologists, the Kenya-born Englishman Louis Leakey and his wife Mary, were particularly interested in the Olduvai Gorge in Tanzania. This was a deep cut in the earth that exposed old layers which showed signs of being rich in fossils. Perhaps some might be pre-human. They searched meticulously, even using tweezers and small brushes.

Finally, on July 17, 1959, Mary Leakey crowned a search lasting over a quarter of a century by discovering fragments of a skull which when pieced together proved to encase the smallest brain of any man-like creature yet discovered. East Africa had been dominated by Arab traders before the coming of Europeans and the Arabic word for East Africa was Zinj. The Leakeys therefore called the new fossil find Zinjanthropus (East African man, the last part of the word being Greek).

Zinjanthropus was advanced enough to form and use primitive tools, though the rocks in which the fossil was found were nearly two million years old. Zinjanthropus does not seem to be a direct ancestor of modern man, but in 1961 Louis Leakey discovered another fossil, slightly older than Zinjanthropus, which may be in our direct line of ancestry. This new fossil he called *Homo habilis* (Latin words meaning skilful man), for it already seemed to have hands as nimble and skilful as those of modern man.

Zodiac

IT IS a natural thing in observing the stars to run them together into patterns that resemble familiar things. The best example to most of us is the Big Dipper. Ancient peoples did this quite elaborately. The Greeks, for instance, divided all the visible stars into groups which we now call constellations, from the Latin prefix *con-* (together) and *stella* (star). Constellations are stars grouped together, in other words.

In general the ancients were quite imaginative in what they saw in the constellations. They saw bears, dogs, winged horses, serpents, crows and assorted people.

Certain constellations were particularly important. The sun, the moon and the five visible planets, as they moved against the background of the stars, restricted their motion to a relatively narrow band in the sky, passing through only certain constellations. This band is called the ecliptic (Greek, *ekleipsis*) because in it eclipses of the sun and moon take place. The word eclipse (referring to times when the moon passes into the earth's shadow, or vice versa) comes from the Greek words *ek* (out) and *leipein* (leave). During an eclipse the sun or moon is left out of the sky, so to speak.

The stars of the ecliptic are divided into twelve constellations. The moon travels through all twelve in 27 days; the sun spends one month in each; the planets pass through the cycle in more complicated patterns. The constellations are Aries (the Ram), Taurus (the Bull), Gemini (the Twins), Cancer (the Crab), Leo (the Lion), Virgo (the Maiden), Libra (the Scales), Scorpio (the Scorpion), Sagittarius (the Archer), Capricornus (the He-Goat), Aquarius (the Water-Bearer) and Pisces (the Fishes).

Seven of the twelve constellations are animals. The Greek word for animal is *zoon*. A diminutive of *zoon* is *zodion* and the adjective from *zodion* is *zodiakos*. In talking of the constellations of the ecliptic you are talking mostly of animals, and they are known as the zodiac to this day.

Zone-refining

WITH ADVANCES in modern technology it has become more and more important to deal with extremely pure materials. In preparing transistors, for instance, even a very few atoms of impurities are liable to give off or absorb electrons in such a way as to disrupt the delicate electron movements on which transistor properties depend. Impurities must often be kept down to only a couple of atoms per thousand million.

The traditional way of freeing a material from impurities is to dissolve it and then subject it to some chemical treatment that will cause it, or the impurities, but not both, to crystallize out of solution. This is done over and over again, and at each step more impurity is removed. This fractional crystallization is, however, a tedious process.

A much simpler process was worked out when it was realized that an impurity might be more soluble in the liquid form of a material than in the solid form, or vice versa. Imagine a metal rod which can be heated at some particular section or zone. That zone is brought just to the melting point. The heating process is moved along the ingot so that the melted portion travels. In no zone does the metal stay molten long enough to drip.

As the melted zone travels along the ingot it collects impurities that dissolve more easily in the liquid than in the solid. By the time it has travelled from one end to the other the impurities are concentrated in that end. A second zone of melting follows, and another, each one flushing out more of the impurities and forcing them to the far end. If the impurities are more soluble in the solid the melted zone forces the impurities backward so that they collect in the near end. In either case the impure end is cut off, and what is left of the ingot is extremely pure.

This process of zone-refining adds no chemicals to the ingot, and so does not introduce new impurities.

ZPG

ONE MEASURE of man's success in the world is his numbers. There are now an estimated 3 600 000 000 people in the world, four times as many as there were less than two centuries ago.

Mankind's population has been steadily increasing since prehistoric times, as his restless brain and nimble hands steadily increased his control over the environment. The taming of fire, the herding of animals, the development of agriculture, metallurgy, writing—each contributed to the ability of mankind to multiply.

The steady population increase was accelerated in the nineteenth century because new lands were brought under large-scale agricultural production, because industrialization developed and spread, and because medical advances drastically reduced the death-rate.

As a result, the time it takes for earth's human population to double has steadily decreased. Now, with life expectancy at the seventy-year mark even in many regions which while not industrialized have taken advantage of modern notions of hygiene and medicine, man's population is expected to double by the year 2000.

Increasing population and developing technology, however, have put an even greater strain on earth's resources and on its capacity to absorb industrial wastes. We seem to be corroding the quality of the environment faster than we can repair the damage.

Many feel that we must reorder our technology to make its first priority preservation of the environment and that we must call a halt to the continuing wild increase in human population. It is thought that there must be zero population growth (usually abbreviated to ZPG) if we are to survive. If this is so, then scientists face a vital problem. How are the physical, psychological and sociological sciences to be used in such a way as to preserve the environment and halt man's population increase in an efficient and humane way?

Index

Claude, Georges, 294
Clausius, Rudolf, 69, 142
Clavicle, 345
Clay, 80
Clostridium botulinum, 60–1
Cloud, 90–1, 112–13, 346;
  chamber, 64, 252
Clusters, galactic, 226
Coal, 60, 123
Cobalt, 93, 155
Coccus, 354
Coccyx, 389
Cocoon, 199
Codon, 373
Cohen, Paul J., 17
Coherent light, 215
Cohesion, 69
Cohnheim, Julius, 375
Colchicine, 253
Cold light, 136
Cole, S. W., 376
Coleoptera, 217
Collagen, 171
Columbium, 263
Combustion, 346
Communication, 292
Communications satellite,
  196
Competitive inhibition, 98
Complex number, 193
Composite number, 313
Computers, 55–6, 288, 297;
  analogue, 22; digital, 22,
  123–4
Conductivity, 356
Conductors, electrical, 341
Cones, 25, 333
Conic section, 282
Conifer, 365
Coniine, 261
Constellation, 402–3
Continental drift, 244
Continuous maser, 234
Cook, James, 232
Cook, Newell C., 241
Cooper, L. Gordon, 237
Copp, D. Harold, 66
Copper, 241, 357
Core, nickel-iron, 246
Coronagraph, 101
Coronary artery, 74
Corpuscle, 178
Cortex, 101
Cortical steroids, 102
Corticoids, 102
Corticosterone, 102
Cosmic rays, 300, 328, 383
Cosmochemistry, 166
Cosmological principle, 100
Cosmology, 100
Cosmos, 103, 104
Cosmotron, 117
Cotton gin, 141
Cotyledon, 25
Coulomb, 393
Coulomb, Charles Augustin,
  393
Counting, 16
Courtois, Bernard, 201–2
Cowan, Clyde L., 258
Cozymase, 95

Crab Nebula, 267, 321, 339,
  398
Cranium, 345
Crater, 392
Crescent, 68
Cretin, 369
Cronstedt, Axel Fredrik,
  93
Cross-section, nuclear, 46
Crossing over, 88, 89
Crossopterygii, 94
Crust, earth's, 247
Cryonics, 108
Cryosurgery, 108
Crystallization, fractional,
  403
Crystalloid, 96
Crystals, 298; electricity and,
  298
Cube, 352; root, 326, 352
Cuneiform, 183
Curie, Marie, 18, 161, 298,
  328
Curie, Paul, 297, 347
Curie, Pierre, 18, 155, 161,
  297, 328, 347
Curie temperature, 155
Cuvier, Georges, 149, 157
Cyanidin, 114
Cyanogen, 113
Cyanometer, 114
Cyclotron, 358
Cyg XR-1, 340
Cyg XR-2, 340
Cysteine, 117
Cytoplasm, 239, 275

DALE, HENRY, 9
Dalton, John, 40, 225
Darwin, Charles Robert,
  149, 401
Davy, Sir Humphry, 311
Dead room, 23
De Bakey, Michael E., 25
De Bary, Heinrich Anton,
  221
De Berthollet, Count Claude
  Louis, 82
Debierne, André, 328
De Boisbaudran, Lecoq, 161,
  360
Deceleration, 386
Deciduous, 25
Decimetre, 242
Deficient number, 288
Degree, 76
Deimos, 293
Dekametre, 242
De la Salle, Poulletier, 85
Delta, 120
Delta Cephei, 77
Delta particles, 274
De Marignac, Jean Charles,
  401
Dement, W., 332
Dementia praecox, 337
Demokritos of Abdera, 40
Demon Star, 16
De Morveau, Guyton, 327
Demotic, 183
Dendrites, 257

Deoxyribonucleic acid
  (DNA), 84, 88, 239–40,
  268, 275, 373
Deoxyribose, 268
Dephlogisticated air, 278
Deryagin, B. V., 309
Descartes, René, 299
Desoxyribose, 268
Destriau, Georges, 137
Deuterium, 29
Deuteron, 29, 122
Devonian, 306
De Vries, Hugo, 253, 254
Dewar, James, 108
D'Hérelle, F., 45
Diabetes mellitus, 200
Diagonal, 122
Dialysis, 96
Diameter, 122
Diamond, 11, 60
Diatom, 130
Diatomaceous earth, 130
Diborane, 59
Dichlorodiphenyltrichloro-
  ethane (DDT), 119
Dicotyledon, 26
Diethyl ether, 147
Digestion, 143
Digit, 123, 124
Digital computer, 22
Digitigrade, 304
Dipeptide, 251
Diplococcus, 355
Diplogen, 121
Diplopoda, 77
Diptera, 217
Dirac, Paul, 29, 250, 394
Dirigible, 128
Disaccharide, 251
Distillation, 15
Diterpene, 256
DNA—*see* Deoxyribonucleic
  acid
Döbereiner, J. W., 74
Dodecaborane, 60
Dole, Stephen, 134
Dollfus, Audouin, 208
Dolphins, 57, 133
Domagk, Gerhard, 355
Domain, magnetic, 155
Doppler, Christian, 330
Doppler effect, 330
Dorn, F. W., 264, 328
Dorsal vertebrae, 388
Double refraction, 307
Douglass, Andrew E., 120,
  121
Down, John, 213
Down's syndrome, 213
Dreams, 332
Dry ice, 91
Dubois, Eugène, 401
Duck-billed platypus, 230,
  325
Duct, 54
Ductless gland, 200
Duodenum, 129
Dust, atmospheric, 346
Du Vigneaud, Vincent, 384
Dynamo, 131
Dyne, 140

408

Pterosaur, 320
Pulmonary artery, 73
Pulsar, 59, 260, 398
Pulse, 24
Pupa, 199
Purkinje, Johannes, 319
Pyrolusite, 228
Pyroxylin, 329

QUADRANGLE, 205
Quadrilateral, 205, 283
Quadrillion, 245
Quadrivium, 220
Quadruped, 76
Quantum, 379; theory, 274
Quartz, 172, 298
Quasar, 142, 173, 324, 330; nature of 342–3
Quicksilver, 236
Quinine, 261

RABIES, 188
Radar, 244
Radiation, 327; electromagnetic, 250
Radio, 327; astronomy, 207, 244, 342; telescopes, 207, 321; waves, 54, 244, 398
Radioactivity, 180, 332, 366
Radium, 181, 328
Radius, 327
Radon, 264
Ramsay, Sir William, 182, 264
Ranger: VII, 231; VIII, 231; IX, 231
Rare earths, 11, 400
Ratio, 158, 203
Rational, 203; number, 159
Rauwolf, Leonhard, 372
Rauwolfia serpentinum, 372
Ray: alpha, 18; beta, 18; cathode, 397; cosmic, 103, gamma, 18; heat, 197; Röntgen, 61, 397
Ray, John, 82, 233
Rayleigh, Lord, 264
Real number, 193
Rectangle, 283
Red blood corpuscles, 179, 343
Red dwarfs, 198
Red giants, 198
Red shift, 324
Reflex angle, 26
Refract, 350
Refraction, 307
Refrigerators, 107, 108
Reich, Ferdinand, 194
Reines, Frederick, 258
Relativity, theory of, 89, 104, 173, 224, 252, 280, 358
Renal artery, 73
Reptile, 126
Reserpine, 372
Resin, 365
Resistance, electrical, 356
Resonance, 333; particles, 354
Retina, 57, 333
Retort, 107

Rhesus, 334
Rhine, Joseph B., 285
Rhodopsin, 333
Rhombus, 205
Rh positive, 334
Ribonucleic acid (RNA), 239, 240, 268, 335, 373
Ribose, 116, 239, 275
Ribosome, 275
Rice, 167
Richter, Hieronymus T., 194
Rickets, 65
Ricketts, H. T., 45
Rickettsia, 45
Right: angle, 190; triangle, 190
RNA—*see* Ribonucleic acid
Robot, 57
Rochester, George D., 189
Rocket, 39, 203, 314-15; fuel, 60, 315
Rodent, 193
Rods, 333
Roemer, Olaus, 270
Röntgen, Wilhelm, 61, 397
Röntgen ray, 397
Rose, Heinrich, 263
Rosin, 365
Rossi, Bruno B., 398
Rubber, 135, 393
Ruben, S., 328
Runge, F., 65
Rust, 34
Rutherford, Daniel, 264
Rutherford, Lord, 49, 121, 216, 259, 268, 269, 274, 318, 339
Rutherfordium, 216

SACCHARINE, 316
Sachs, Julius von, 84
Sacral vertebrae, 388
Sacroiliac, 345
Sal ammoniac, 345
Salk, Jonas, 383
Sarcoma, 72
Satellites, 37; artificial, 56, 105, 169, 231, 233, 237, 262, 313-14, 350, 370-1, 383; Galilean, 363; natural, 208
Saturn, 208
Scalar, 385
Scale, 385
Scandium, 11
Scapula, 345
Schaefer, Vincent, 91
Scheele, Karl Wilhelm, 82, 113, 264
Schiaparelli, Giovanni, 151
Schleiden, Matthias, 318
Schmidt, Bernard, 337
Schmidt, Maarten, 324
Schmidt telescope, 337
Schönbein, Christian, 62, 165
Schrödinger, Erwin, 275
Schultze, Max, 305
Schwann, Theodor, 143, 318
Schwarzschild, Karl, 59
Schwinger, Julian, 250

Seaborg, Glenn T., 11, 161
Seamounts, 177
Sea urchin, 132
Section, 282
Sedatives, 371
Sediment, 192
Sedimentary rock, 192
Sedimentation rate, 378
Seminal fluid, 315; vesicle, 315
Serine, 118
Serpent, 126
Servomechanism, 114
Sesquiterpene, 256
Sets, infinite, 17
Sex cells, 399-400
Seyfert, Carl, 343
Shapley, Harlow, 78
Sharp, Abraham, 297
Sharpey-Schafer, Sir Edward, 184, 200
Shell numbers, 227
Shidarshan, E. C. George, 359
Shockley, William, 341, 374
Sigma particle, 189, 274
Silanes, 59
Silica, 67
Silicates, 80, 344
Silicon, 59, 67, 341, 344
Silurian, 306
Silver, 356; iodide, 91
Sinanthropus pekinensis, 301
Skull, 345
Sleep, 331-2, 372
Sleeping pills, 372
Slide rule, 22
Small intestine, 129
Smith, J. L. B., 94
Smith, William, 157
Smoke, 112, 346
Socrates, 261, 338
Soda, caustic, 50
Soddy, Frederick, 206
Solar flares, 346; system, 134; life in, 151
Solenoid, 351
Solids, 125, 305
Sorption, 13
Sound, 347; electricity and, 298; motion and, 330; reflection of, 23, 133; speed of, 225, 349
South latitude, 144
Space, 39; biology, 56; flight, 7, 30-1, 39, 56, 105, 163, 203, 350, 386-7; medicine, 56; molecules in, 38; travel, 39; walk, 163
Space-time, 126
Spallanzani, Lazzaro, 132
Spectral tarsier, 217
Spectroscope, 351
Speech, 292, 306
Speed, 385-6
Spermatophyte, 25
Sphere, 39
Spheroid: oblate, 166; prolate, 166
Spine, 388
Spirilla, 355

414